# CISM COURSES AND LECTURES



The series presents lecture notes, monographs, edited works and proceedings in the field of Mechanics, Engineering, Computer Science and Applied Mathematics.
Purpose of the series is to make known in the international scientific and technical community results obtained in some of the activities organized by CISM, the International Centre for Mechanical Sciences.

INTERNATIONAL CENTRE FOR MECHANICAL SCIENCES

COURSES AND LECTURES - No. 419

# SEMI-RIGID JOINTS IN STRUCTURAL STEELWORK

EDITED BY

MIKLOS IVANYI
BUDAPEST UNIVERSITY OF TECHNOLOGY AND ECONOMICS

CHARALAMBOS C. BANIOTOPOULOS
ARISTOTLE UNIVERSITY OF THESSALONIKI

Springer-Verlag Wien GmbH

This volume contains 238 illustrations


Originally published by Springer-Verlag Wien New York in 2000
SPIN 10776865

In order to make this volume available as economically and as rapidly as possible the authors' typescripts have been reproduced in their original forms. This method unfortunately has its typographical limitations but it is hoped that they in no way distract the reader.

ISBN 978-3-211-83331-5 ISBN 978-3-7091-2478-9 (eBook)
DOI 10.1007/978-3-7091-2478-9

*To the memory of Professor P. D. Panagiotopoulos (1950-1998)*

## *PREFACE*

*The notion of semi-rigid connections in steel structures, although known since the fifties, has been recently re-introduced into the structural steelwork praxis by Eurocode 3: "Design of Steel Structures" where, these connections, as exhibiting a behaviour intermediate between that of classical pinned and rigid, are characterized as semi-rigid. Having as scope to discuss in depth all the aspects of the notion of semi-rigidity in steel structural connections - from the fundamental notions till the applications of this theory -, an Advanced Professional Training Seminar coordinated by the editors of the present hook was organized at the International Center for Mechanical Sciences (CISM) in Udine the period September 20-24, 1999. The present book is the fruit of this Seminar and includes the texts of the lectures presented there.*

*The Book presents a comprehensive survey of all the relevant topics: the definition of semi-rigid steel structural connections, their classification and their influence to the structural response of sway and non-sway steel frames. The sources of connection compliance, ductility and the application of the component method for characterization of the joint properties are some of the topics presented. The verification procedures for the available and the required capacity of joints and the design of semi-rigid steel structural connections are also discussed. In addition, the simulation of the structural response of semi-rigid connections by means of appropriate numerical methods that take into account all prominent phenomena (cf. e.g. contact, friction and plasticity) is also presented. Analysis techniques and design procedures for beam-to-column, beam-to-beam, column-base-plate and other special types of conventional or hollow section connections are topics included among others to the chapters of the present Book.*

*Part I, by Miklos Ivanyi from the Technical University of Budapest (with the collaboration of L. Hegedüs, M. Ivanyi Ir. and G. Varga in chapter 4, and with the collaboration of L. Hegedüs in chapter 5) presents the interaction of the steel frames and their joints, and describes an acceptable method of joint design.*

*Part II, by Jean -Pierre Jaspart from the University of Liège, describes the sources of deformability and gives information on how to integrate the actual joint hehaviour into the frame design and analysis process. In particular, four main aspects are dealt with: the so-called joint characterization, classification, modelling and idealization.*

*Part III, by Ulrike Kuhlmann and Frank Kühnemund from the University of Stuttgart, presents a comparison between the definition and the verification of rotation capacity of steel joints and members and suggests a coherent definition. In particular, the dominant influence of the component "column web in compression" is given special consideration in the case that the rotational capacity of a joint is derived.*

*Part IV, by Frantisek Wald from the Technical University of Prague, gives an overview of the recent progress made on the investigation of column bases with end plates and embedded column bases. In particular, the column base detailing, tolerances and modelling by application of the component method to column bases are introduced and the particular components including their influence on the connection behaviour are also described. In addition, the basic principles in background of the column base classification and the strength design are presented.*

*And finally, Part V, by Charalambos Baniotopoulos from the Aristotle University of Thessaloniki, gives certain basic principles, numerical techniques and algorithmic models which are applied to the simulation of the structural response of steel semi-rigid connections. Nonlinearities introduced by unilateral contact and friction over the connecting interfaces, as well as material nonlinearities including yielding are considered. The direct application of the Finite Element Method, as well as mesh size, choice of finite elements and other parameters affecting the accuracy and the numerical behaviour of the proposed models are discussed.*

*The editors wish to thank the above three colleagues and Professor P. Mäkelainen (Helsinki Technical University) for the excellence of the work they carried out during the preparation of the present Advanced Professional Training Course, which is well reflected in the monograph at hand.*

*The editors wish also to extend to the CISM General Secretary Professor G. Bianchi, the CISM Rector Professor S. Kaliszky, the Editor of the Series Professor C. Tasso, as well as to all the CISM staff in Udine their sincerest thanks.*

*The present book is dedicated to the memory of one of the founders of Nonsmooth Mechanics, Professor P.D. Panagiotopoulos (1950-1998) who unexpectedly passed away on August 12, 1998. Miklos Ivanyi, having for the last fifteen years close scientific contacts with Professor Panagiotopoulos, applied some of Panagiotopoulos' ideas to the solution of steel structures problems where "softening" phenomena arise. Charalambos Baniotopoulos, being student and collaborator of Professor Panagiotopoulos, has been inspired and guided by him along the lines of the herein presented research for more than twenty years.*

*Miklos Iványi*
*Charalambos C. Baniotopoulos*

CONTENTS

# PART I

# SEMI-RIGID CONNECTIONS IN STEEL FRAMES

**M. Ivanyi**
**Budapest University of Technology and Economics, Budapest, Hungary**

**Abstract:** The purpose of this chapter is to present the interaction of the steel frames and their joints and to describe an acceptable method of joint design. A key problem in dealing with joints is their classification, the basis of which is described in the Eurocodes and other available design codes in a variety of different ways. Eurocodes take into account whether the joint is applied within a frame with fixed nodes or within one with sway nodes. The method is based on the results of experiments, therefore it is of great importance to perform large scale experimental tests under both monotonic and cyclic loading. Engineering methods help us with establishing the load–displacement diagrams of frames by using simple techniques, in a way that local "softening" effects occurring in the vicinity of joints can also be taken into account. Engineering design is an activity of fair complexity, thus it is important to establish direct design methods which, while simple, take into consideration certain complex phenomena such as the stiffness and strength properties of the joints (including beam-to-column joints as well as column bases). The preparation of this chapter has been supported financially by the National Scientefic Research Fund of the Republic of Hungary (OTKA) under grant No. T020358.

## 1 Introduction: A Reasonable Principle for Connection Design

### 1.1 Introduction to Connection Design

The design of connections can be approached from a number of directions: the type of structure, the type of fastener, the type of loading, and the designer's special interest. The dominant concerns in the design of connections in buildings, bridges, and towers. Bolts, welds, and devices such as cable sockets transmit forces in different ways. Static loads, dynamic loads, and the expected number of repetitions of either, pose different problems. Structural engineers and fabricators have shared interests and responsibilities, but the focus of the former may be on obtaining a desired type of behaviour and that of the latter on ensuring practicable fabrication and erection (McGuire, 1992).

### 1.2 Basic Criteria for Structural Behaviour

Relevant properties of steel are its strength, its stiffness and its ductility or deformation capacity. These properties can be demonstrated in compression and tensile as well as connection test, see Figure 1 (ESDEP, 1994) (Colson, 1996). A well designed steel structure should possess the same good properties.

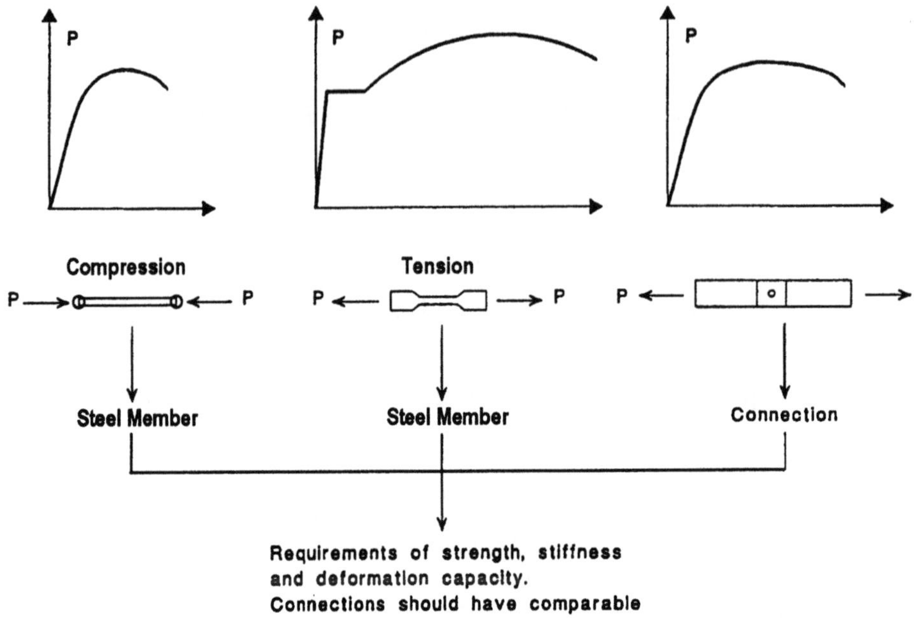


**Figure 1.** Strength, stiffness and deformation capacity of steel and connections.

The basic requirements for structural elements/connections concern the strength, the stiffness and the deformation capacity.

Figure 2 defines the strength, stiffness and deformation capacity of a beam-to-column connection designed to transfer a moment from the beam into the column.

The rotation capacity is a measure of the deformation that can be obtained before failure somewhere in the connection causes a drop in the moment resistance.

The strength, stiffness and deformation capacity of connections are discussed in greater detail below.

**Strength.** For the determination of the forces on the connection, a static analysis must be carried out. Such analysis includes the determination of the design loads and the modelling of the structure, see Figure 3. In the schematization of the structure, the stiffness of the connections is an important element. Connections can be assumed rigid, as hinges or as having a stiffness between these two. The deformation capacity of the members (beams and columns) and of the connections plays an important role in the ultimate distribution of forces in the structure.

All parts of the structure must be designed so that they can resist the calculated forces and have a deformation capacity that is consistent with the assumptions made in the global analysis.

**Stiffness.** The stiffness of the connection affects the level of loading for which it should be designed. A connection of low rotational stiffness does not attract major bending moments and therefore may be assumed as a pinned connection in the schematization of the structure.

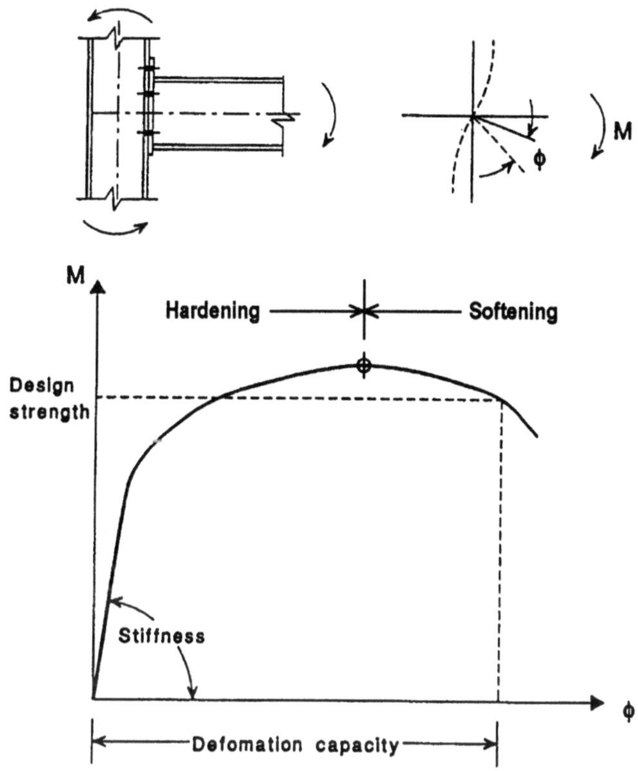


**Figure 2.** Moment–rotation diagram of a beam-to-column connection ($M$–$\phi$ curve).

Of course, the stiffness of the connections affects the deflections of the beams. Especially in non-braced frames the stiffness of the connections may have a major effect on the deflections of the structure as a whole and on its stability.

If the connections are assumed rigid in the modelling for the static analysis of the structure, then consequently the form of the connections should be such that their deformations have a negligible influence on the load distribution and the deformations of the structure.

On the other hand, if pinned connections are assumed, they should have sufficient flexibility to accommodate rotations without causing significant bending moments that may lead to premature failure of (parts of) the connection or connected members.

It is obvious that a structure with pinned connections will deflect more than one consisting of the same members but rigid connections. The connection stiffness also affects the distribution of forces in the structure and thus the relative effectiveness of its members.

An elementary example of the influence of connection deformation on system resistance is illustrated in Figure 4. If a slender column is connected to a stiff beam by a fully rigid connection as in Figure 4a, the elastic critical load for in-plane buckling is

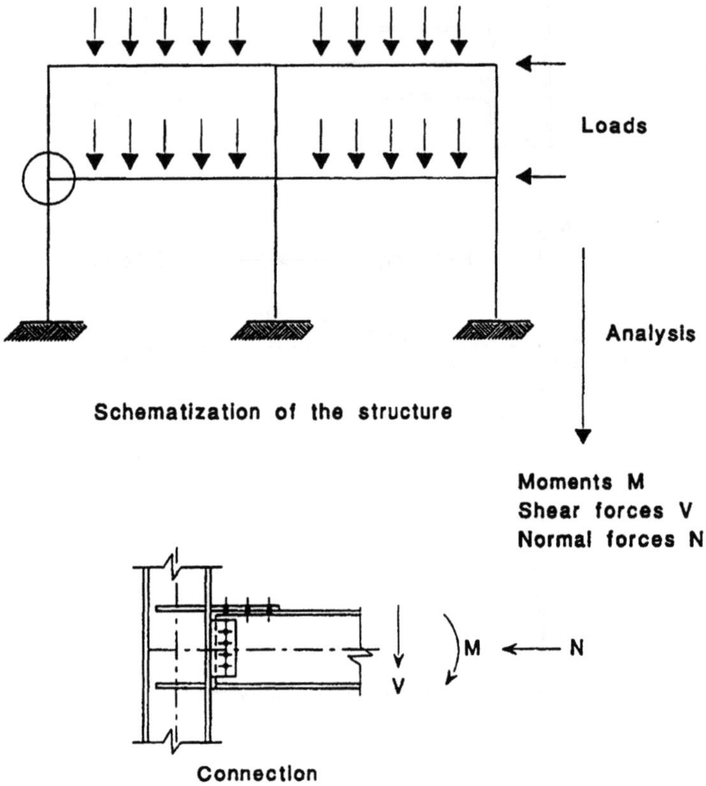


**Figure 3.** Analysis of the forces on the connection.

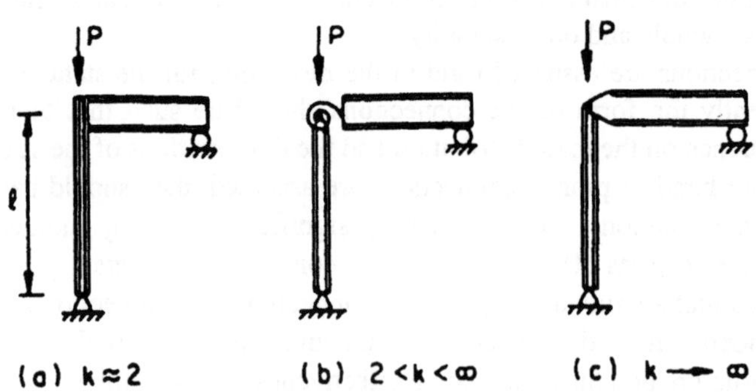


**Figure 4.** Frame stability.

$$P_E = \frac{\pi^2 EI}{(2l)^2}. \tag{1}$$

If, on the other hand, the connection is a true pin as in Figure 4c, the system has no stability at all. For a connection that offers partial restraint, the critical load is between these two extremes. Stated in another way, the effective length factor, $k$, may be anywhere between the limits of 2 and infinity, depending on the connection stiffness (McGuire, 1992).

**Deformation Capacity.** The requirements for strength and stiffness are clear. They result from the static calculation.

The requirement for deformation capacity is more qualitative. In practice it is sometimes difficult to check this requirement.

Ductile connections that have a great deformation capacity contribute to the overall safety of the structure in the event that the connection becomes overloaded. Such connections may also be a design requirement in certain instances, e.g. when plastic design is employed with plastic hinges forming in the connections.

### 1.3 Classification as a Basis for Design

Connections should be designed according to their required behaviour. Many design approaches are possible for stiffness and strength. In Figure 5 three designs for a beam-to-column connection are given with their moment–rotation characteristics. These characteristics may be described as follows:

1. Both the rotational stiffness and the moment resistance are small and may reasonably be neglected, leading to the concept of a pinned or hinged connection. Only vertical shear is required to be transferred, with the value being equal to the beam end reaction necessary to resist the beam loading.
2. Behaviour intermediate between characteristics 1 and 3, in which the connections possess a finite moment resistance less than the full moment resistance of the beam and a rotational stiffness that permits some relative rotation. (Such connections are usually both semi-rigid and partial strength. However it is possible to have connections that are full-strength and semi-rigid or fully rigid but only partial strength.)
3. The rotational stiffness is very high and the connection's moment resistance is at least that of the beam. Continuity is thus preserved with no rotation of the beam relative to the column; both the beam end reaction and its end moment are transferred to the column.

All three possibilities can be applied in multi-storey frames. Type 3 can be applied in both braced and unbraced frames; Type 1 is only suitable for braced frames. Type 2 may be used for both braced and unbraced frames, but in the latter case the influence of connection flexibility on frame behaviour needs to be considered. Figure 6 gives the schematisation of rotational stiffness for the frame analysis.

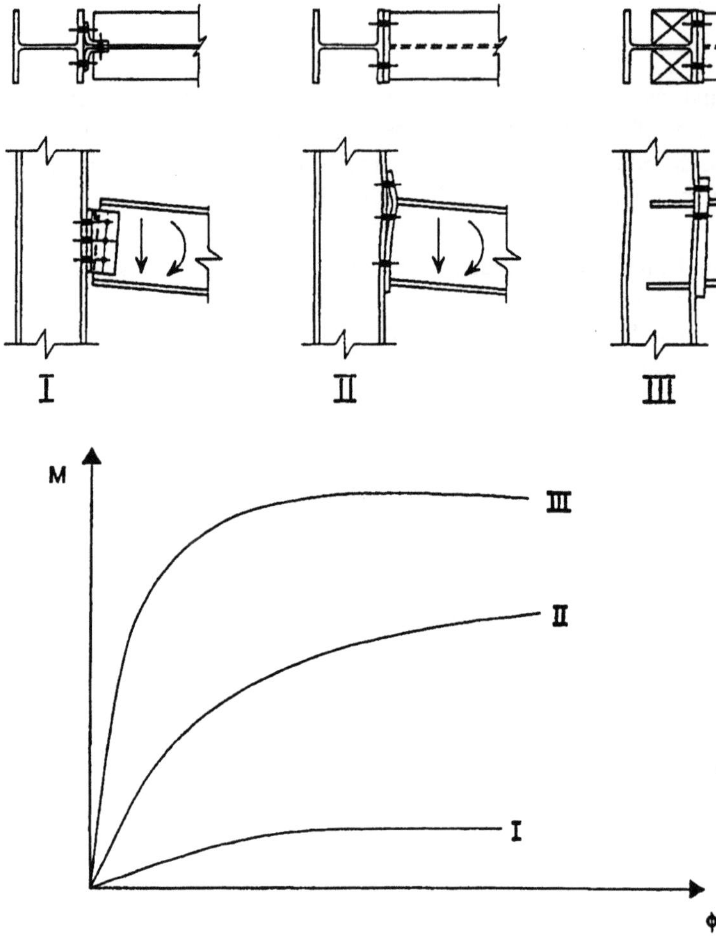


**Figure 5.** Moment–rotation diagrams ($M$–$\phi$ curves).

Eurocode 3 (EC3, 1993) provides design rules for both the rotational stiffness and the moment resistance.

$M$–$\phi$ curves as measured in tests are in general non-linear. Another phenomenon is that stiff connections have sometimes a low moment resistance and vice versa. Some examples are shown in Figure 7. $M$–$\phi$ curve $E$ is characteristic for some types of bolted connections in clearance holes. At small moments, slip causes a rotation before a higher moment is obtained.

For practical application, it is necessary to idealise the $M$–$\phi$ curves. For design a non-linear idealisation or linear idealisations like the bi-linear or tri-linear ones in Figure 8 may be used.

Eurocode 3 (EC3, 1993) gives a classification on the basis of strength and stiffness.

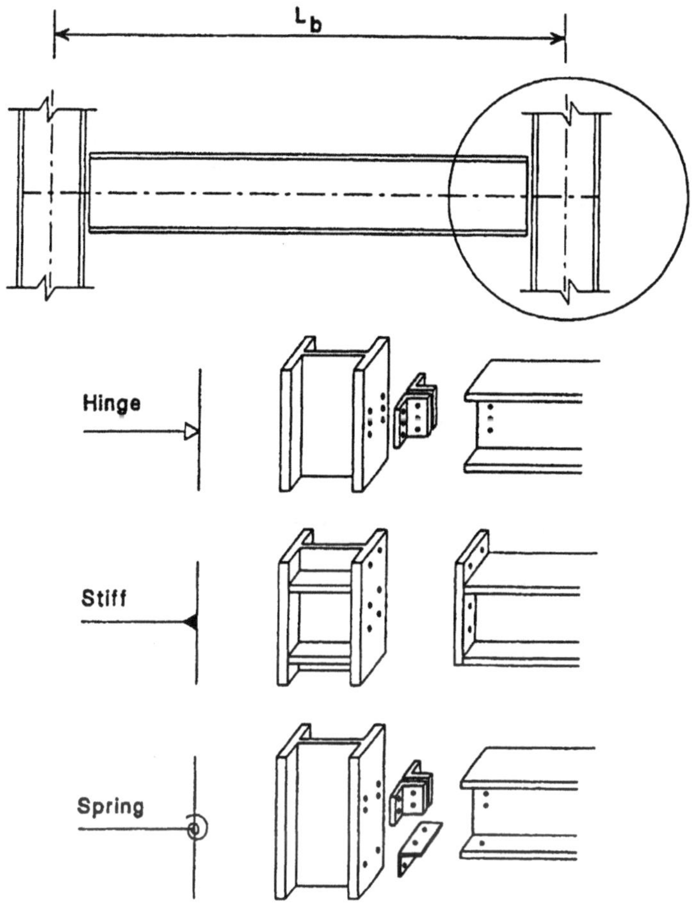


**Figure 6.** Schematization of rotational stiffness.

## 1.4 An Adapted Design Philosophy for Connections

Any rational design philosophy has to take account of both the complexity and variability of practical connection behaviour. It is the uncertainty that presents the greater difficulty. If connection behaviour was merely complex it would be possible for designers to make a suitable choice for a particular situation. Detailed analysis could be used that took account of the complexity, thus permitting economic design with small load factors. Alternatively, simple analysis could be used in conjunction with higher load factors to account for possible variations between analysis and behaviour. Because of the variability of behaviour the first alternative cannot lead to economic and safe design. However refined the analysis, it would not be appropriate to reduce the load factors significantly (Owens, Cheal, 1989).

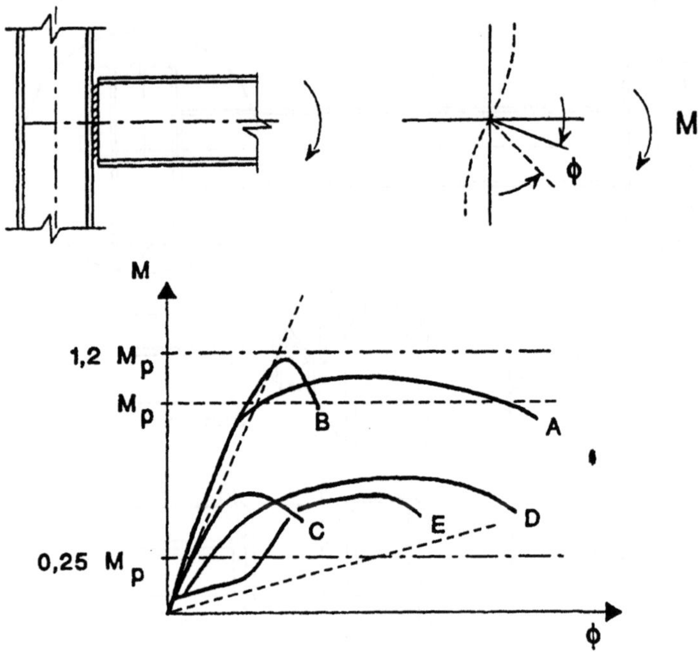


**Figure 7.** Various forms of $M$–$\phi$ curves.

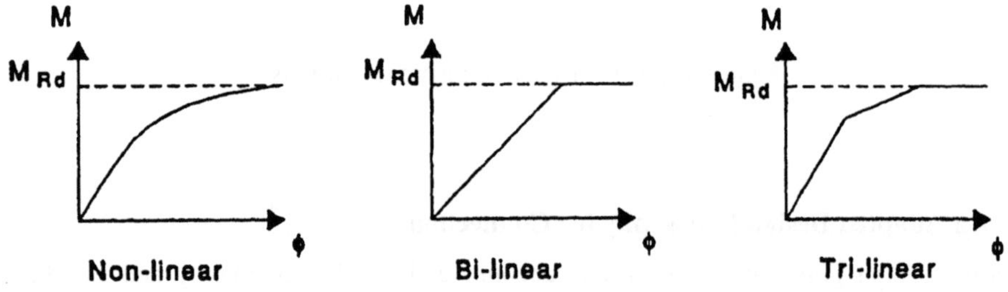


**Figure 8.** Possible idealisations for $M$–$\phi$ curves.

The method is presented below and its application in outline is demonstrated in the following section.

*1. Taking account of overall connection behaviour, carry out an appropriate simple analysis to determine a realistic distribution of forces within the connection.*

For many connections this analysis should be based on the concept of "force paths". Here the overall loads acting on the connection are replaced by equivalent systems of forces which can then be assigned specific paths through the connection.

It is most important to ensure that the analysis is consistent throughout the connection. In general, this is achieved by carrying out a single analysis of the most critical part of the connection and using that to determine the distribution of forces in other parts of the connection. Surprisingly, it is not uncommon to see designs where serious inconsistencies in analysis have occurred. These most commonly arise when more than one analysis has been used to determine the distribution of forces. For example, in the end-plate bracket connection shown in Figure 9 it would not be correct to use separate conventional analyses to determine both the distribution of forces in the bolts and in the weld attaching the end plate to the beam. Such separate analyses would assign different proportions of the tensile and compressive forces to different levels in the beam and imply instantaneous redistributions of these forces at the plane of contact between the weld and the end plate, as shown in Figure 9c.

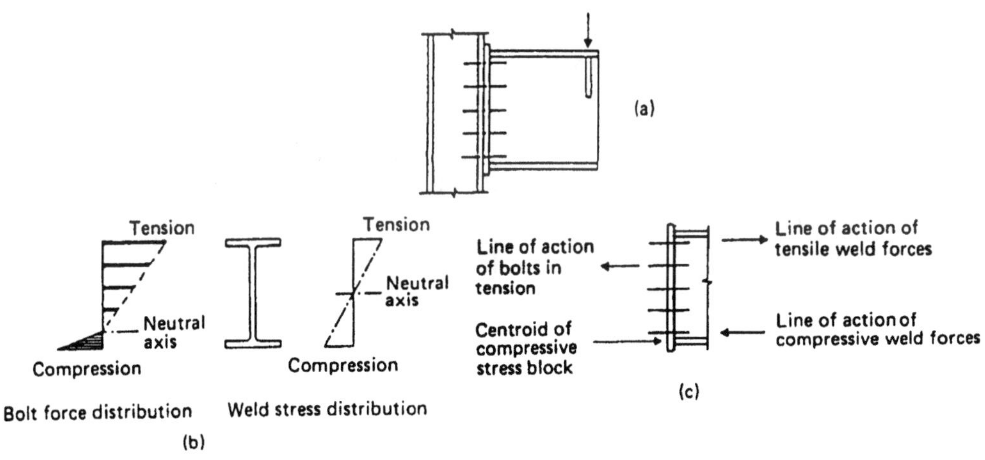


**Figure 9.** Inconsistency in connection analysis.
(a) Bracket connection; (b) conventional elastic analyses; (c) stress resultants

*2. Ensure that each component of each force path has sufficient strength to transmit the required force.*

This is self-evident, and yet it is surprising how frequently designers leave a weak link somewhere in a connection. A major disadvantage of traditional methods of analysis is that they concentrate on the distribution of forces in the connectors. Many codes of practice only give guidance on connector strength.

3. *Recognising that the above procedure can only give a connection where equilibrium is capable of being achieved but where compatibility is unlikely to be satisfied, ensure that the components are capable of ductile behaviour.*

4. *Recognising that the preceding steps only relate to static ultimate capacity, ensure that the connection will achieve satisfactory serviceability, fatigue resistance, etc.*

## 1.5 Application of the Design Philosophy

Most connection design is very straightforward and satisfies the preceding criteria by implication rather than by specific calculation (Owens, Cheal, 1989). For example, conventional design of the simple beam-to-column connection shown in Figure 10a is carried out in a few lines of calculation (or by means of design tables) as a designer/detailer determines the number of bolts necessary to resist the eccentric load shown on the basis of their shear and bearing capacity. Here the simple analysis is based on the assumption that the eccentricities between the lines of bolts *A* and *A'* may be discounted because of symmetry in a connection of normal proportions with two web cleats. Experience indicates that the only critical checks are the bolts in shear and the cleats and beam web in bearing. Other "weak links" are likely be designed out by minimum pitch criteria and practical edge distance requirements. Experience has also indicated that such connections have sufficient ductility to accommodate both lack of fit within the connection and the beam end rotation as it takes up its deflected profile.

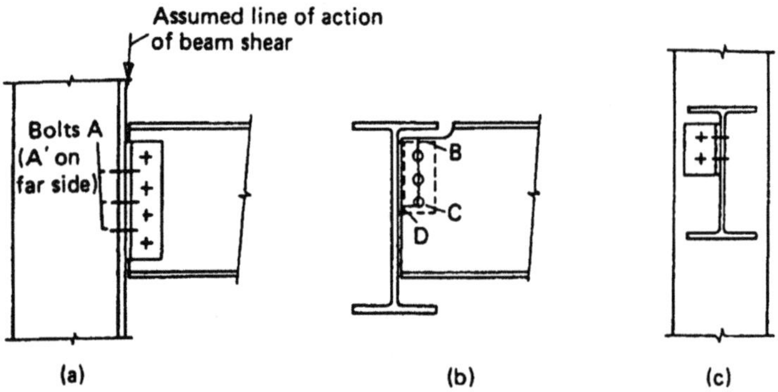


**Figure 10.** "Simple" beam-to-column and beam-to-beam connections.
(a) Conventional beam-to-column connection with double-web cleat;
(b) beam-to-beam grillage connection with double-web cleats; (c) single-web cleat connection

If all connection design were as straightforward there would be little point in the explicit procedure presented in the preceding section. However, it is only necessary to vary some details in this simple example to illustrate the importance of a sound appreciation of connection design. Figure 10b shows the beam end detail that results if the top flange has to be notched to accommodate a flange in the supporting beam. Line *BCD* becomes a very important critical section to be checked in shear and tension. If a single short web cleat were to be used, as shown in

Figure 10c, the same local moments that were reasonably ignored in the double-cleat connection may at least cause unserviceability as the beam twists and could lead to an unacceptable reduction in strength. Similarly, the same simple approach to analysis could lead to an unacceptable reduction in strength if a fillet welded beam end connection were used, because of lack of ductility in the weld.

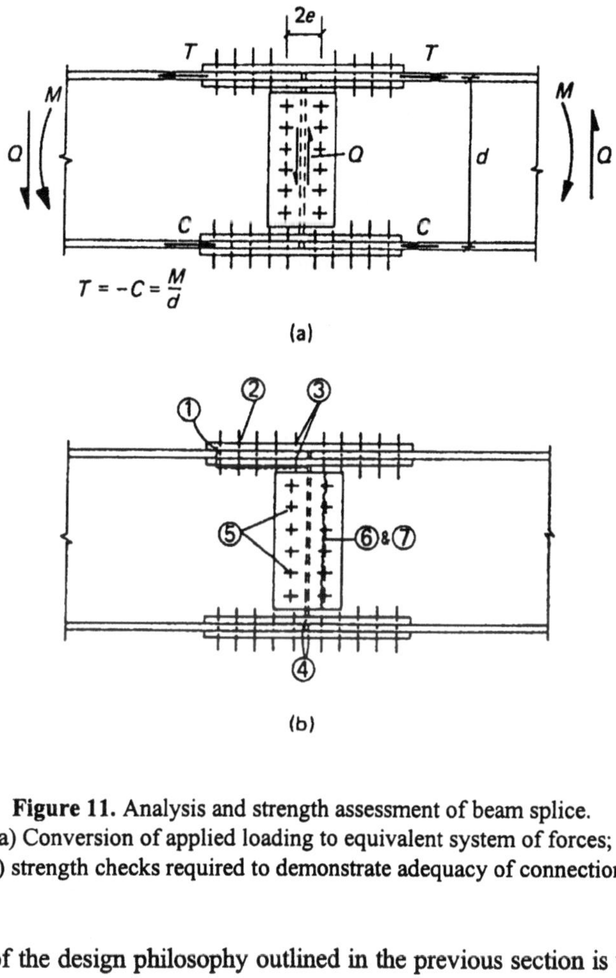


**Figure 11.** Analysis and strength assessment of beam splice.
(a) Conversion of applied loading to equivalent system of forces;
(b) strength checks required to demonstrate adequacy of connection

The direct use of the design philosophy outlined in the previous section is well demonstrated on a beam splice with HSFG bolts (Owens, Cheal, 1989). Figure 11a shows the straightforward set of forces that can replace the applied moment and shear. (This simplification with its implied redistribution is only applicable for building structures. In bridges such redistribution is not usually permitted, primarily because of general concern over fatigue.) Figure 11b enumerates the checks that have to be carried out to ensure that there are no weak links within the connections, and these checks are as follows:

1. The capacity of the flange to resist the tensile force. The critical section is the vertical net section through the first line of holes, in tension, together with the horizontal web/flange intersection in shear. In addition, the effective section of the flange through the first line of

holes should be checked under the flange stress resultant, without any redistribution of bending moment to the flanges. The effective section is defined as the net area times a coefficient greater than unity which recognises that nominal stresses slightly above yield may occur on a net section without detriment;

2. The frictional capacity of the HSFG bolts;
3. The net sections of the splice plates, assuming that the force $T$ is equally divided between the pair of cover plates.

Note that because of the symmetry of the connection these checks also demonstrate the adequacy of most of the compressive flange splice. The only additional checks are:

4. The compressive capacity of the splice plates, free to buckle vertically (this is satisfied by maximum pitch criteria);
5. The frictional capacity of each web bolt group under a load $Q$ at an eccentricity $e$;
6. & 7. The capacity of the net section of the web and splice plates. (These are most unlikely to be critical unless the shear is a very high proportion of the beam capacity.)

Figure 12 shows the application of the design philosophy to an exterior beam-to-column connection subject to moment and shear, one of the most complex connections that commonly occurs in practice (Owens, Cheal, 1989).

Figure 12a shows the analytical procedure that is appropriate for this connection for values of the factored applied moment up to approximately 70% of the plastic moment capacity of the beam. Above that value it will be necessary to mobilise some of the bending capacity of the web in order not to overstress the flanges.

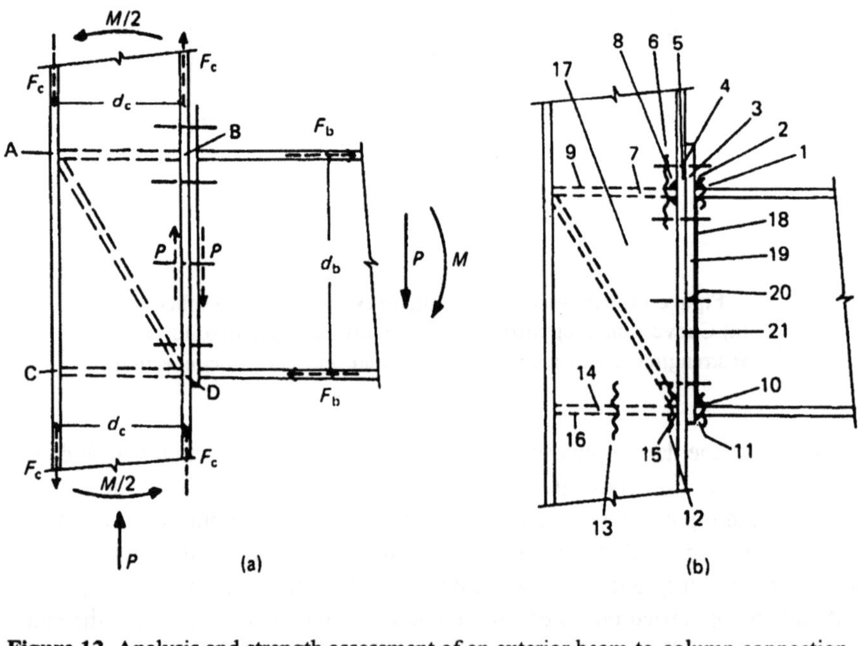


**Figure 12.** Analysis and strength assessment of an exterior beam-to-column connection.

Remote from the connection, the beam and column flanges will be making the greatest contribution to the moment resistance of the respective members and the beam shear will be concentrated primarily in the web. These distributions are used as the basis of the simple analysis. Thus the forces within the connection transmitting the bending moments are given by:

$$F_b = \frac{M}{d_b} \quad \text{and} \quad F_c = \frac{M}{2d_c}, \tag{2}$$

where $d_b$ and $d_c$ are the distance between flange centroids of the beam and column, respectively. The shear $P$ is assumed to remain in the web. An alternative analysis could be based on some notional centre of rotation at the end-plate/column face contact surface and an assumed linear response of the bolts, in accordance with traditional elastic analysis.

The next step is to trace the paths of these forces through the connection. The bending forces $F_b$ and $F_c$ may be considered to continue on their original lines of action until they reach the panel *ABCD*, whose corners are defined by the intersections of these lines of action, and thus has dimensions $d_b \bullet d_c$. Overall equilibrium of these forces is achieved by shear on this panel. The shear flow on *AB* and *CD* is thus

$$\frac{F_b}{d_c} = \frac{M}{d_b \cdot d_c}. \tag{3}$$

The vertical (complementary) shear flow is

$$\frac{2F_c}{d_b} = \frac{M}{d_c \cdot d_b}. \tag{4}$$

The beam shear is simply transferred straight through to the near column flange and hence down the column. It is customary to ignore the eccentricity between the column face and the column centreline. This leads to an apparent lack of equilibrium, which arises because, in the earlier calculation of bending forces, the applied moment $M$ is taken as the value at the intersection of the column and beam centrelines rather the value at the column face.

It is now possible to check the elements on the various force paths. Thus, starting with the path of the tensile force $F_b$, in the top flange of the beam and tracing it through to the panel *ABCD*, it is necessary to check:

1. The tension capacity of the flange adjacent to the connection. The effective width of the flange may be limited by the bolt layout;
2. The welds attaching the beam flange to the end plate;
3. The end plate in flexure as it disperses $F_b$ into the bolts;
4. The bolts in tension. Account should be taken of prying action as appropriate;
5. The column flange in flexure as it transfers the bolt loads into the column web. If this is inadequate, extra capacity may be obtained by providing a stiffener along *AB*. In extreme cases the column flange may still be inadequate with a stiffener;
6. The tensile capacity of the effective length of the column web. If this is inadequate a stiffener has to be provided;
7. The tensile capacity of stiffener *AB* if this has been provided to satisfy either 5 or 6;
8. The welds connecting the column flange to the stiffener, in tension;
9. The welds connecting the column web to the stiffener, in shear.

The detailed check is then continued by examining the path of the compressive force $F_b$ from the beam bottom flange into the bottom of panel *ABCD*:

10. The compressive capacity of the beam flange. This is affected, in the absence of stiffener *CD*, by the stress concentration caused by the column web. If the flange is overstressed, a stiffener should be provided;
11. The flange/end-plate welds;
12. Local crushing of the column web. If there is an overstress a stiffener should be provided;
13. Buckling of the column web. If there is an overstress a stiffener should be provided;
14. The compressive capacity of stiffener *CD*, if this has been provided to satisfy any of 10, 12 or 13;
15. The stiffener/column flange welds, in compression;
16. The stiffener/column web welds, in shear.

Finally, to complete the force paths associated with the moments:

17. Panel *ABCD* in shear. If this is inadequate a diagonal stiffener should be provided along *AD*. Both this stiffener and its welds should be checked in accordance with 14-16 above.

The shear force path requires less attention for this particular connection. It is only necessary to check:

18. The beam web/end-plate welds in shear;
19. The end plate in shear and bearing;
20. The bolts in shear;
21. The column flange in bearing.

The design procedure is presented as a straightforward sequence. In practice, there may be an interaction between sections; for example, if a stiffener has to be provided along *AD* one will not generally be necessary along *CD*. This change in geometry will affect the force paths and should be acknowledged by a modification to the overall analysis.

When carrying out these detailed checks it is important to keep in mind the third part of this design procedure, namely that the components of the connection should be ductile. For this connection this means complying with certain criteria for stiffener proportions, minimum weld sizes and bolt/end-plate thickness ratios.

This example looks very daunting; apparently twenty-one separate calculations have to be carried out! However, it should be appreciated that many of the checks are very straightforward and may either be satisfied by inspection or by very simple calculation. Experienced designers will recognise that for a connection of conventional proportions, only checks 3, 4, 5, 13, 17 and 20 require significant computation. Checks 1, 6, 7, 10, 11, 12 and 14 can be dealt with by a single line of calculation; the remainder are either governed by ductility criteria or can be satisfied by inspection. However, the principle of considering each link in the chain explicitly would seem to be a very sound discipline for all connections. It is certainly essential for connections of unusual proportions (Owens, Cheal, 1989).

# 2 Classification of Connections by Structural Eurocode

## 2.1 Classification of Framed Systems

**Introduction.** In common practice, (continuous) framed systems are considered to be assemblies of beams and columns in which all the joints are completely continuous. In contrast truss or simple framed systems are those in which all the joints can be considered hinged (Fig.13) (ESDEP, 1994).

A purely rigid jointed frame system is characterised by joints between frame members and by no additional bracing system. The frame itself has to resist all the actions, vertical as well as horizontal. At the same time, it has to provide the required stiffness to the structure in order to limit deformations within the allowable values. Even though the detailing of all the connections results in a less economic structure, framed systems have some benefits:

1. the connections are more ductile and therefore the structure performs better in earthquakes;
2. from the architectural and functional points of view, it can be advantageous not to have any triangulated truss in the structure.

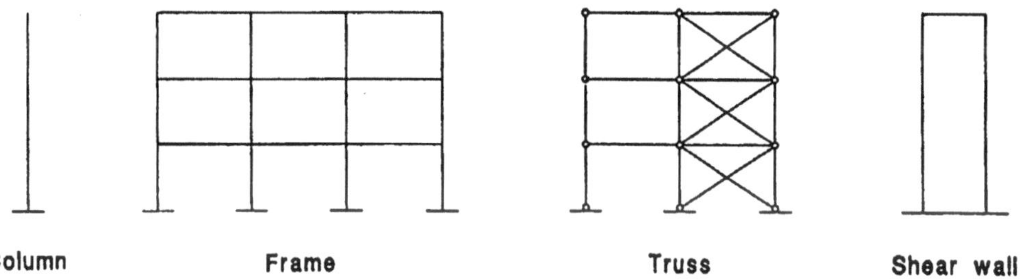


**Figure 13.** Continuous and hinged systems.

Actual structures do not always fall into the two categories defined above. Connections themselves actually behave semi-rigidly and therefore the hinged and framed conditions are only idealisations. An engineering definition is needed to define when a semi-rigid connection can be assumed to be hinged and when it can be assumed to be rigid.

The practice in different countries in the design of connections and framed systems varies. The approaches used are different in different parts of the world. In some countries, e.g. USA, the concept of semi-rigid connections dates back to the 1930's when the first studies on semi-rigid riveted connections were carried out by Johnston (Johnston, Mount, 1941). In these countries also the code allows the use of semi-rigid connections (type 3 connections) and introduces the concept of wind design (type 2 connections). In wind design the connection is assumed to be capable of transmitting only part of the bending moments (those due to the wind and not those due to vertical loads). [In the USA it was adopted as three "Types of Construction": Type 1 (rigid frame), Type 2 (simple framing), and Type 3 (semi-rigid framing).]

The approach of semi-rigid connections, in use for several years, is also adopted, for example, in UK, Australia, Canada, and Netherlands. In other countries, e.g. Italy, France, Spain, Greece, these concepts have not been introduced and therefore semi-rigid connections are not widely adopted although they are included in Eurocode 3 (EC3, 1993).

*Engineering Definition* (ESDEP, 1994). To determine whether a system can be considered continuously framed or not, the effects of the connection on the frame behaviour has to be considered, taking into account that the ideal condition of hinged or rigid connection does not correspond to reality. These effects are not for immediate evaluation since the behaviour of the frame can be different depending on what is being considered:

- vertical load-carrying resistance;
- horizontal load-carrying resistance,
- stability;
- performance in seismic situations.

In some cases the strength of the connection is the important consideration whilst in other cases the flexibility of the connection plays a major role. Sometimes the elastic behaviour of the connection is sufficient to determine its effects on the behaviour of the frame whilst in other cases the complete inelastic behaviour of the connection is needed for evaluating frame effects.

In recent years this subject has been given much attention from research devoted either to the analytical and to the experimental aspects. The effects on frame behaviour are described and some indicative values for connection characteristics are suggested as a basis for assuming a system as a rigidly jointed frame.

A connection between two or more different members has to transmit all the internal actions through the members, i.e. axial load, shear load and bending moment in the case of a plane frame.

However, the term "semi-rigid connection" is only used to refer to the bending flexibility in the discussion below.

In Figure 14, taken from (Cosenza, De Luca, Faella, 1989), connection behaviour is shown qualitatively by definition of the regions in which the connection can be assumed as hinged, flexible or semi-rigid. This qualitative representation suggests that fully welded connections, extended end plate, and top and bottom flange splices could be considered as rigid restraints.

For partial strength semi-rigid connections, some results (Cosenza, De Luca, Faella, 1989a) indicate that, for rigid jointed frames, within the ranges indicated before, some degree of flexibility can be allowed, but no partial strength is advised. Partial strength connections are considered to be within the semi-rigid or flexible range (Fig.14).

*Eurocode Definition.* In Eurocode 3 (EC3, 1993) it is first stated, within the Design Assumptions (5.2.2), that:

"the assumptions made in the global analysis of the structure shall be consistent with the anticipated type of behaviour of the connections".

Then, in 5.2.2.2, simple framing is defined for frames where the connections between the members may be assumed not to develop moments. In the global analysis, members may be assumed to be effectively pin connected.

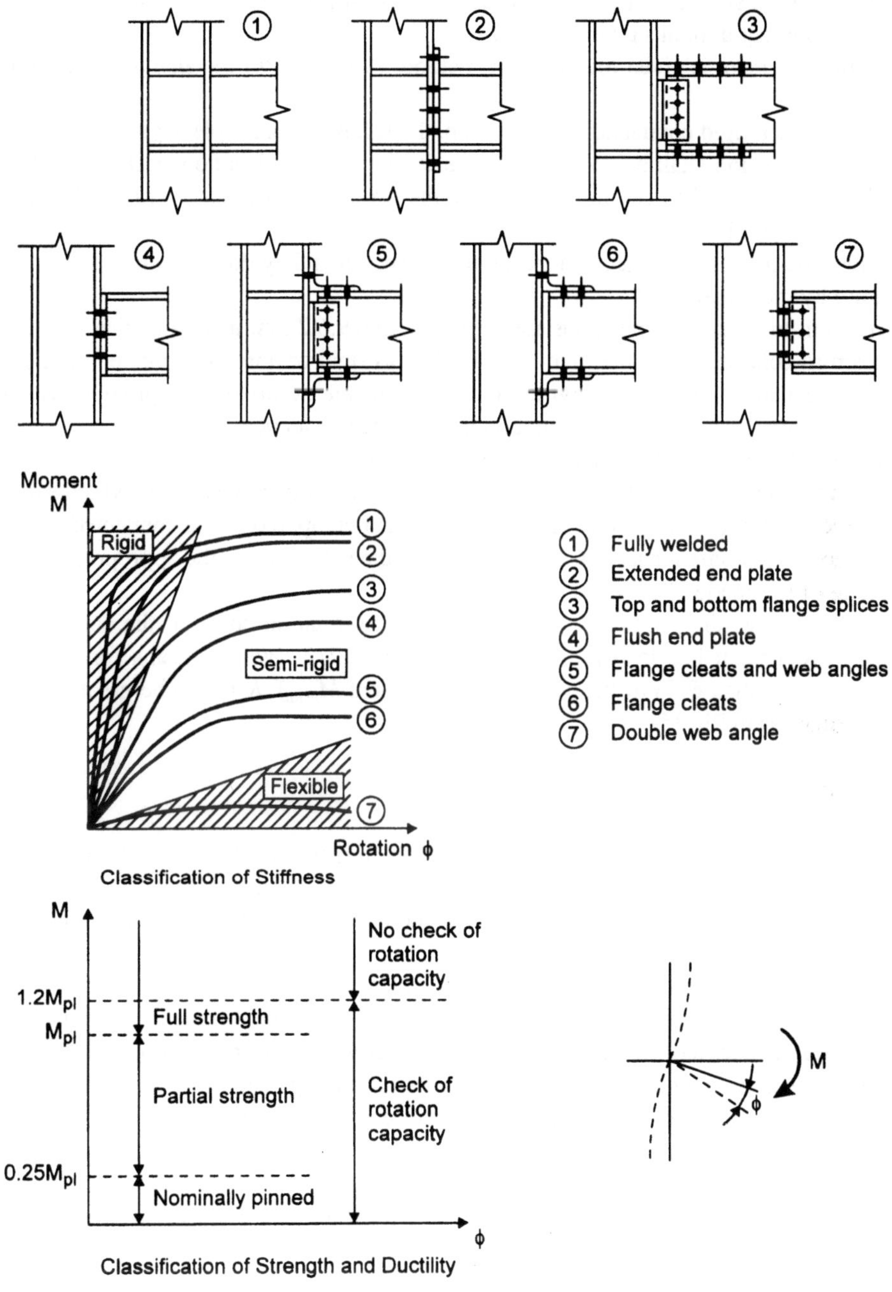


**Figure 14.** Experimental $M$–$\phi$ relations of connections.

Further, in 5.2.2.3, when continuous framing is defined, it is stated that:

"Elastic analysis should be based on the assumption of full continuity, with rigid connections which satisfy the requirements given in 6.4.2.2".

The same statement is given for the other methods suggested for the analysis, i.e. rigid plastic and elasto-plastic.

In 6.4.2.2, when rigid connections are defined, the following principle is given:

"A rigid connection shall be so designed that its deformation has no significant influence on the distribution of internal forces and moments in the structure, nor on its overall deformation".

The application rules provided state:

"The deformation of rigid connections should be such that they do not reduce the resistance of the structure by more than 5%".

When beam-to-column connections are classified, in 6.9.6.2 first it is stated that:

"A beam-to-column connection may be classified as rigid or nominally pinned on the basis of particular or general experimental evidence, or significant experience of previous satisfactory performance in similar cases or by calculation based on test evidence".

Then, the following is suggested as an application rule:

"A beam-to-column connection in a braced frame, or in an unbraced frame which satisfies the condition specified in (5), may be considered to be rigid compared to the connected beam, if the rising portion of its moment characteristics lies above the solid line on the appropriate diagram of Figure 6.9.8. of Eurocode 3" (Fig.15).

This rule suggests that, in order to have the moment–rotation curve of the connection lying above the solid line, it has to result, for values of $\overline{m} < 2/3$ (which means for values of the moment $M$ smaller than 67% of the plastic beam moment $M_{pl,Rd}$), $K$ has to be greater than 25. In fact (see definitions in Fig.15):

$$K = \frac{\overline{m}}{\phi} = \frac{M}{M_{pl,Rd}} L_b \frac{M_{pl,Rd}}{EI_b \phi} = \frac{M}{\phi} \frac{L_b}{EI_b} > 25 \tag{5}$$

This value of 25 guarantees the stiffness characteristic of the connection. Some other requirements are also given in order to ensure sufficient strength. The fact that the entire moment–rotation curve has to be above the solid line indicates that the entire non-linear curve of the connection is to be checked.

The further requirement for unbraced frames, given in Clause (5) of 6.9.6.2 of Eurocode 3, discussed below, is somewhat cumbersome. The meaning of this requirement is that, for unbraced frames, the beams must have an adequate stiffness otherwise the frame will be too flexible, even with rigid connections. This consideration is obvious for an engineer who knows which typology to choose for each case. However, since all the serviceability and ultimate limit states have to be satisfied, this requirement might be considered automatically fulfilled.

Clause (5) states:

"The line given in Figure 6.9.8(a) (Fig.15 here) for unbraced frames may be used only for frames in which every storey satisfies:

$$\frac{K_b}{K_c} > 0.1; \tag{6}$$

where $K_b$ is the mean value of $I_b / L_b$ for all the beams at the top of that storey, $K_c$ is the mean value of $I_c / L_c$ for all the columns in that storey, $I_b$ is the second moment of area of a beam, $I_c$ is the second moment of area of a column, $L_b$ is the span of beam and $L_c$ is the storey height of a column.

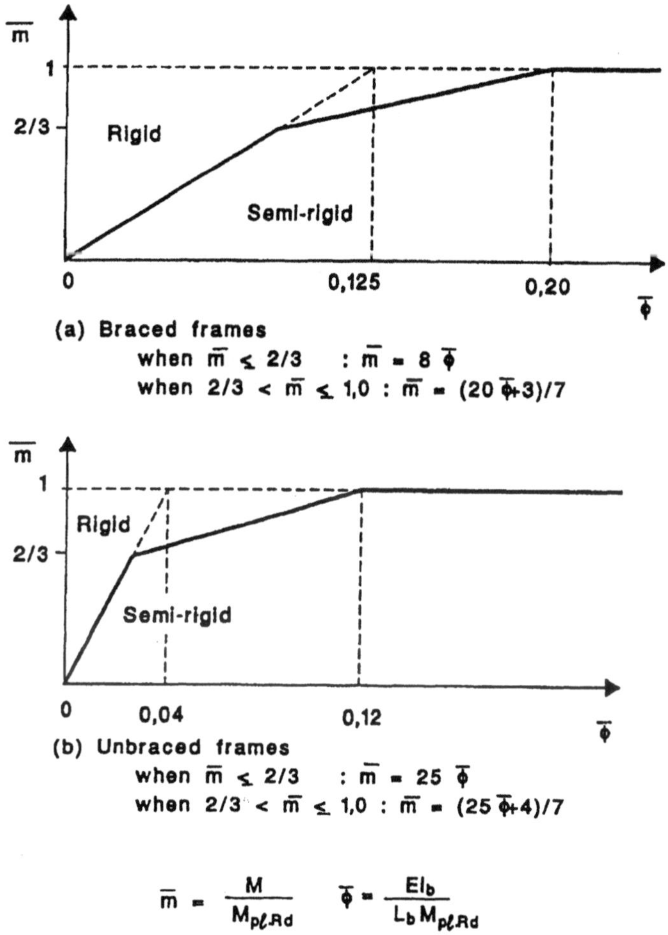


**Figure 15.** Eurocode classification boundaries for rigid beam-to-column connections in unbraced and braced frames.

**Braced and Unbraced Frames.** Sometimes the term "bracing system" is inappropriately identified with the term "braced frame". It is clear that the definitions of the two terms are different. The word "braced" in the second case is used as an adjective to the word "frame" and therefore at least two structures have to be identified: a bracing and a frame shown in Figure 16 (ESDEP, 1994).

A braced frame is commonly intended as a frame to which a triangulated truss is attached. The fact that in reality there is no clear cut distinction between hinged structures with bracing systems and purely rigid jointed framed structures calls for a more exact definition which allows distinctions to be made between:

- purely hinged bracing systems;
- rigid jointed framed systems;
- semi-rigid frames;
- braced frames.

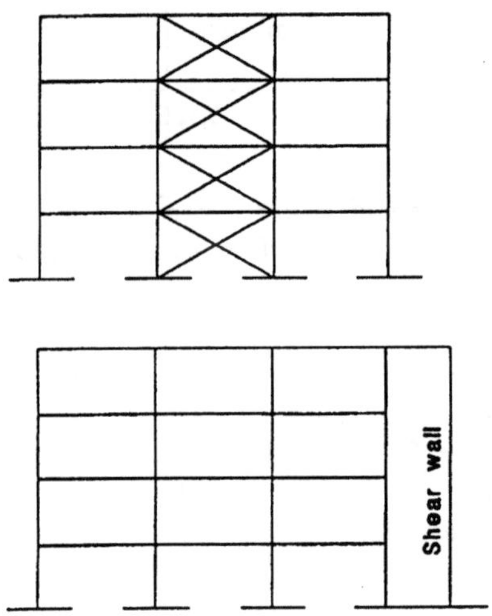


**Figure 16.** Common bracing systems.

*Engineering Definition* (ESDEP, 1994). The main function of a bracing system is: to resist horizontal actions, and is derived from the separation of the resisting systems: vertical and horizontal. In some cases the vertical system also has some capability to resist horizontal actions. It is necessary therefore, from an engineering point of view, to identify the two sources of resistance and to compare their behaviour with respect to the horizontal actions. Sometimes this identification is not obvious since the bracing is integral within the frame and therefore there is only one structure. However, even in this case, it is possible to make some assumptions in order to define the two structures to be compared. The examples given below clarify these concepts.

Figures 17 and 18 represent structures in which it is easy to define, within one system, two sub-assemblies which identify the bracing system and the system to be braced. In particular, a structure is shown in Figure 17 where there is a clear separation of functions: the horizontal loads are carried by the first hinged sub-assembly (A) and the vertical loads are carried out by the second one (B). In Figure 18, in contrast, since the second sub-assembly (B) is able to resist horizontal actions as well as vertical actions, it is necessary to assume that practically all the horizontal actions are carried by the first sub-assembly (A) in order to define this system as braced. In this case the first sub-assembly is defined as a bracing system if its lateral stiffness expressed by the spring constant $K_a$ is considerably higher than the one of the second sub-assembly $K_b$ (in this case a braced frame or system):

$$K_a >> K_b \tag{7}$$

This relation can be easily applied to the system of Figure 17 since the constant $K_b$ is equal to zero and therefore the relation is certainly satisfied. For the system in Figure 18, the stiffnesses of both sub-assemblies have to be calculated and compared.

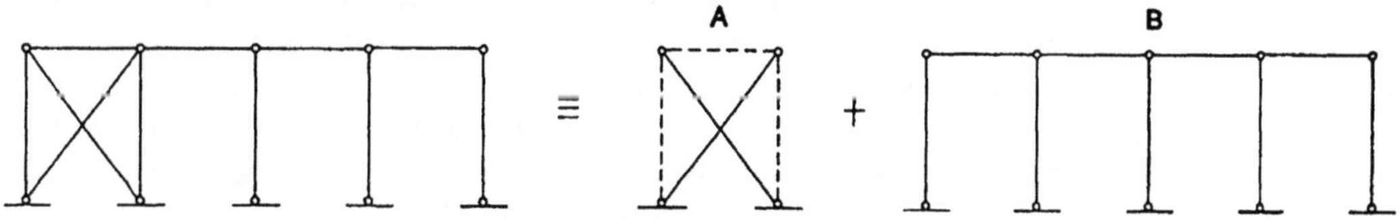


**Figure 17.** Pinned connection structure split into two sub-assemblies.

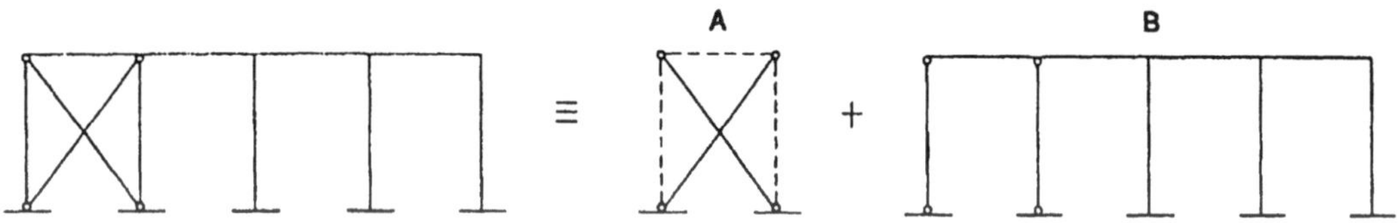


**Figure 18.** Partly framed structure split into two sub-assemblies.

*Eurocode Definition.* The following definition is provided in 5.2.5.3 of Eurocode 3 (EC3, 1993):

"the frame can be classified as braced if the bracing system reduces its horizontal displacement by at least 80%"

which means that the stiffness of the two systems have to be compared and the following relationship satisfied:

$$K_a > 0.8(K_a + K_b) \tag{8}$$

or

$$K_a > 4K_b. \tag{9}$$

**Sway and Non-sway Frames.** Before defining sway frames and non-sway frames, it is useful to note the common design practice for evaluating safety of structures against stability. It is often convenient to isolate the columns from the frame and treat the stability of columns and the stability of frames as independent problems. For this purpose it is assumed the columns are restricted at their ends from horizontal displacements and therefore are only subjected to end moments and axial loads as transferred from the frame. It is then assumed that the frame, possibly by means of a bracing system, satisfies global stability checks and that the global stability of the frame does not affect the column behaviour. This gives the commonly assumed non-sway frame. This approach has led to years of research spent in the field of behaviour of columns and beam-columns.

Design books, guidance documents, and even codes and recommendations, when speaking of stability of columns or stability of frames, commonly use the terms: "sway frames", "non-sway frames", "sway restricted columns" and "sway columns". To explain the concept of sway, as opposed to non-sway, figures such as Figures 19 and 20 are used. The frame of Figure 19 is considered to be the non-sway type and the one of Figure 20 is considered to be the sway type. This form of representation, which is based on common practice and common engineering sense, leads to the erroneous assumption that non-sway frames and braced frames are perfectly equivalent and therefore that one definition can be used instead of the other without causing any misunderstanding.

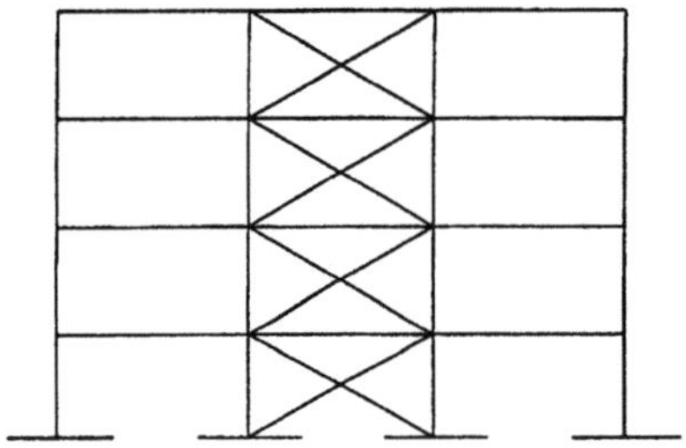

**Figure 19.** Braced frame (but may be a sway frame if bracing is very flexible).

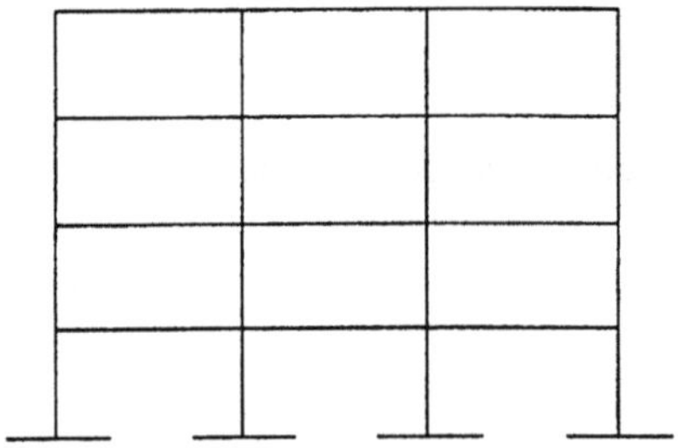

**Figure 20.** Unbraced frame (but may be a non-sway frame if it is sufficiently rigid i.e. insensitive to horizontal loading).

*Engineering Definition* (ESDEP, 1994). The equivalence between "braced" and "non-sway" frames cannot be established in general since the two terms refer to different aspects of the behaviour of the structure. The fact that the definitions of "sway" and "non-sway" appears when the problem of stability of columns and frames is evaluated suggests that these definitions are part of a simpler treatment of this problem.

The concept of braced and unbraced frames can be defined in engineering terms by means of comparison of the stiffness of the systems, and has no straightforward implications for stability. The concept of sway frames is not intrinsic to the structure: it is based only on its mechanical properties.

Another reason for defining "sway" and "non-sway frames" is the need to adopt conventional analysis in which all the internal actions are computed on the basis of the undeformed shape of the structure. To make this assumption it is necessary that second order effects are negligible, i.e. no significant moments arise due to the action of vertical loads on the deformed shape of the structure. This definition can be shown to be equivalent to the previous one since the vertical design loads cause no significant moments if their value is not close to the elastic critical load of the structure.

When there is interaction between global and column behaviour, it is not possible to isolate the column. The column or the frame then has to be assumed to be the "sway" type. Unfortunately, research has been limited in this field and therefore extrapolation of the same procedures already used for non-sway frames to sway frames has been used. As a result inaccuracies occur also due to the fact that the actual behaviour is inelastic and is therefore affected by all types of imperfections, i.e. cross-section, column and frame imperfections. In addition, the inelasticity in the columns prevents the use of the familiar concept of "effective length". The design of sway frames has to consider the structure as a whole.

On the basis of those considerations, the following definitions can be established for sway and non-sway frames:

A non-sway frame is a structure which, from the points of view of stability and the definition of the internal action, can be considered to have small interstorey displacements. Therefore column buckling is independent by frame buckling, i.e. the problems can be uncoupled. This definition will be true if the safety factor against overall buckling is sufficiently large that global buckling can be neglected when carrying out the check against column buckling. On the basis of this definition, it is clearly that to be a non-sway frame is not a characteristic intrinsic of the frame since the safety factor against critical load depends on the magnitude of the design vertical loads acting on the structure.

Whilst it is possible to define whether a frame is braced or not by evaluating the stiffness of its members, in order to evaluate whether a frame is the non-sway type, i.e. second order effects can be neglected, the design vertical loads have to be known. This is understandable since even a very flexible structure has no second order effects if the vertical loads are practically equal to zero.

*Eurocode Definition.* The definition provided by 5.2.5.3 of Eurocode 3 (EC3, 1993) is:

"A frame can be classified as non-sway if its response to in-plane horizontal forces is sufficiently stiff for it to be acceptably accurate to neglect any additional internal forces or moments arising from horizontal displacements of its nodes".

Examination of this definition does not immediately reveal the relation between sway and instability. However, the Eurocode also provides the following application rule:

"A frame may be classified as non-sway for a given load case if the elastic critical load ratio $V_{Sd}/V_{cr}$ for that load case satisfies the criterion:

$$\frac{V_{Sd}}{V_{cr}} \leq 0.1, \tag{10}$$

where $V_{Sd}$ is the design value of the total vertical load and $V_{cr}$ is its elastic critical value for failure in a sway mode."

This application rule confirms that the definition of a frame as non-sway depends on the vertical loads. Furthermore it establishes that a safety factor against overall buckling equal to 10 is enough for considering the problem uncoupled from column buckling.

## 2.2 Influence of Connections on the Behaviour of Frames

**Introduction.** The main structural elements of steel framed multi-storey structures are the columns, the beams and their connections. Conventionally the beam-to-column connections are considered to be either pinned or rigid.

In the case of pinned or "simple" connections, the frames have to be stabilised by appropriate bracing systems. Such frames are named braced frames by Eurocode 3.

The term "rigid" in this context implies that the connection is capable of resisting moments with a high stiffness, i.e., the connection flexibility has a negligible influence on the distribution of movements in the frame connections. When the connections are rigid, the overall stability may be provided by the frame itself without the inclusion of specific bracing systems. These rigid jointed or moment resisting frames are treated in this part.

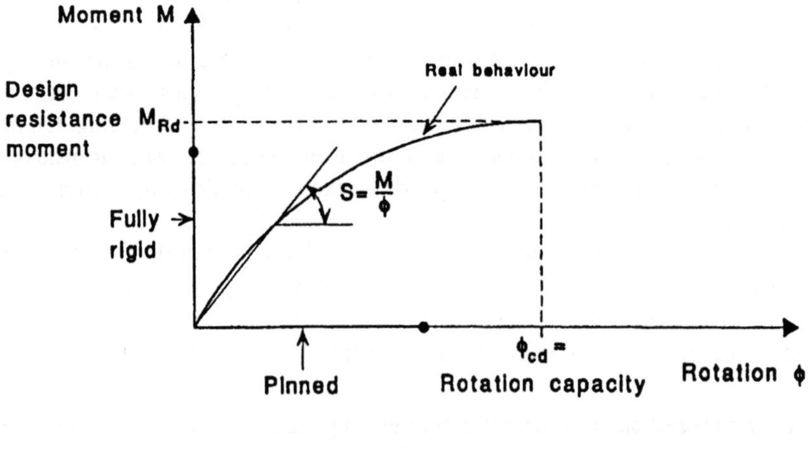


**Figure 21.** Characteristics of beam-to-column connections.

Although the idealisation of connection stiffness as pinned or rigid has been applied exclusively in the past it is generally recognised that the real behaviour of the connections is never as ideal as assumed in the analysis (Fig.21). The two cases, pinned and fully rigid, actually represent extremes of connection behaviour. In reality, the connections behave somewhere between those limits, that is they behave as semi-rigid (Chen, 1987) (Bjorhovde, Brozzetti, Colson, 1988).

A further classification of moment resisting connections relates to their strength. A "full-strength" connection is a connection that can at least develop the bending strength of the elements it connects. A "partial-strength" connection has a lower design strength than that of the elements it connects.

The rotation capacity of a moment-resisting connection can also be important. For example a beam with partial-strength end connections can be designed plastically if the connection rotation capacity is sufficient to ensure the development of an effective hinge at midspan.

Figure 22 shows the moment–rotation diagram of a beam-to-column connection. For design purposes the real connection behaviour can be represented by a bi-linear or tri-linear diagram in which the following properties can be distinguished:

- The design resistance of the connection;
- The stiffness of the connection when subjected to small moments;
- The stiffness of the connection when subject to ultimate moments;
- The rotation capacity.

The influence of connections on frame behaviour is treated separately for partial-strength and semi-rigid connections.

Braced frames are in general designed based on strength conditions and unbraced frames are generally designed based on stability and deformation conditions. Therefore partial-strength connections are mainly applicable for braced frames and semi-rigid connects for unbraced frames.

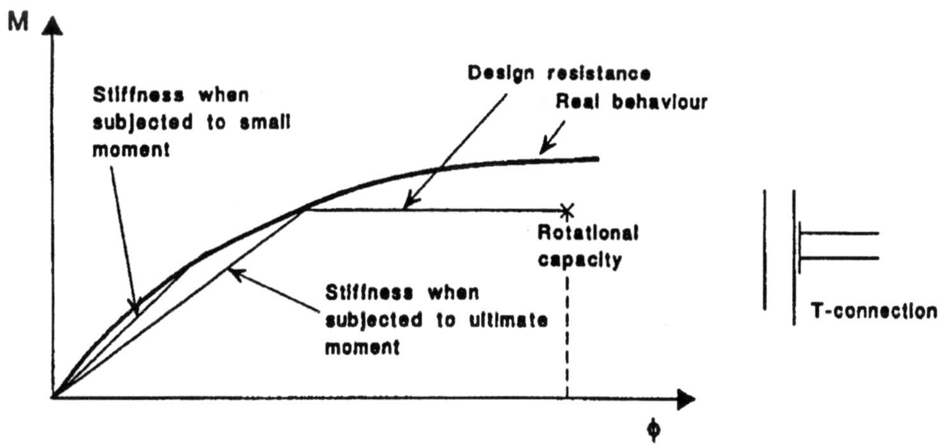


**Figure 22.** Design parameters for moment-resisting connections.

**Classification of Connections.** The main objective of a classification system is to define appropriate boundaries for behavioural classes of joints as a function of the mechanical properties of the connected members. The aim is to establish useful criteria allowing for a safe use of continuous or simple modelling of joints (ESDEP, 1994).

*Influence of Connection Flexibility on Elastic Frame Stability.* The end rotation of a simply supported beam under a uniform vertical load and external negative moments $M$ (Fig.23) is given by

$$\phi = \frac{q\ell_b^3}{24EI_b} - \frac{M\ell_b}{2EI_b}. \tag{11}$$

The $M$–$\phi$ response represented by Equation 11 is a straight line which is called the beam line as show in Figure 23b. In real frames the end moment restraint is provided by the rigidity of the connection $s = M/\phi$ as shown in Figure 23c and therefore the actual end moment and the end rotation of the beam is given by the intersection of the beam line with the connection characteristic as shown in Figure 23b.

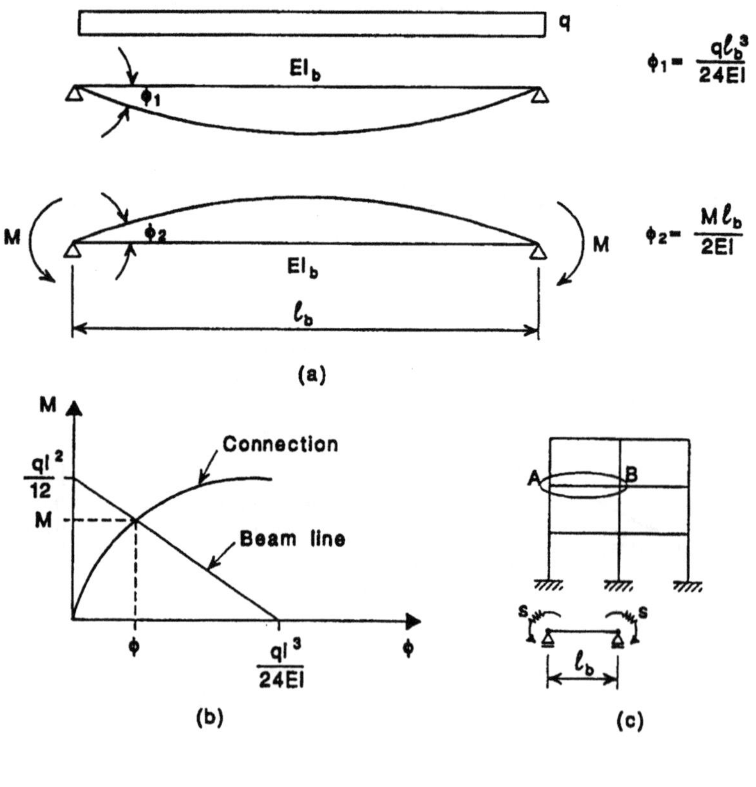


**Figure 23.** Beam-line and connection behaviour.

For practical design situations the actual non-linear connection behaviour has to be approximated. Various approximate connection characteristics are in Figure 24. The connection behaviour is characterised by its moment resistance $M_{Rd}$, its rotational capacity $\phi_{cd}$ and its rigidity $S = M / \phi$.

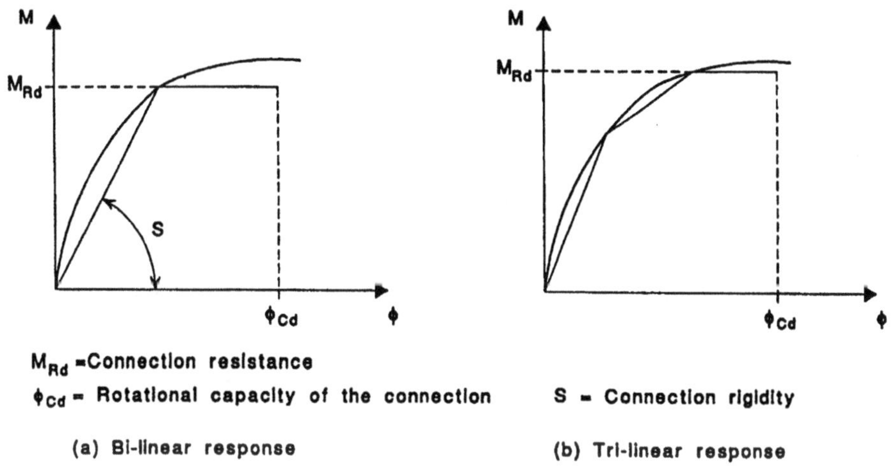


**Figure 24.** Derivation of approximate moment–rotation characteristics.

In order to determine if the connection flexibility $S^{-1}$ needs to be included in the overall frame analysis it is important to examine its influence on the behaviour of the frame (Bjorhovde, Colson, 1989). This is studied subsequently for single storey braced and unbraced frames as shown in Figure 25a (ESDEP, 1994).

Figure 25b presents the relationship between the relative connection-to-beam rigidity $\overline{S}$ and the relative beam-to-column rigidity ρ in order that the flexibility of the connection reduces the Euler buckling load of the rigid frame by 5%.

The structural behaviour will be analysed by showing the relationship between the parameters $\overline{S}$ and ρ. The parameter $\overline{S}$ is the relative rotation stiffness:

$$\overline{S} = \frac{S\ell_b}{EI_b}, \qquad (12)$$

in which $S$ is the rotation stiffness of the beam-to-column connection and $EI_b / \ell_b$ is the flexural stiffness of the beam. The parameter ρ is the ratio between the flexural stiffness of the beam and the column, i.e.:

$$\rho = \frac{EI_b \cdot \ell_c}{EI_c \cdot \ell_b}, \qquad (13)$$

in which $EI_b / \ell_b$ is the flexural stiffness of the beam and $EI_c / \ell_c$ is the flexural stiffness of the column.

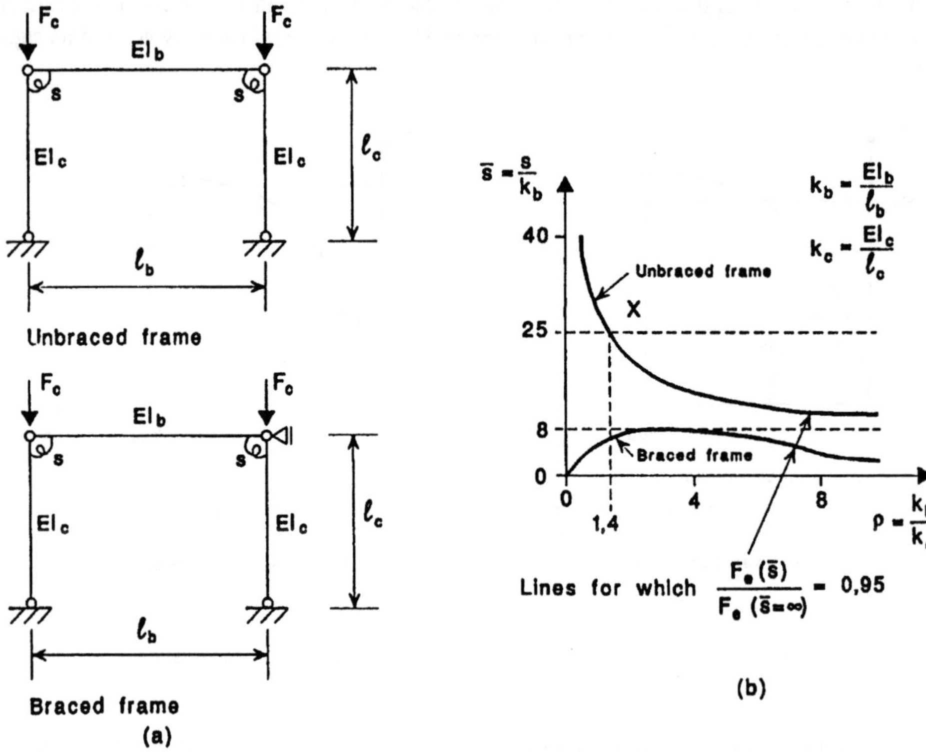


**Figure 25.** Influence of connection rigidity on frame behaviour.

An example of the evaluation of the curve for an unbraced frame is given by considering point $x$, where $\rho = 1.4$ and $\overline{S} = 25$ (Fig.25b).

1 First examine case with *fully rigid connection* ( $s = \infty$ ).

$k_1 = k_b = 10$ .

From EC3 Annex E:

$$\eta_1 = 1.0; \ \eta_2 = \frac{k_c}{k_c + k_1} = \frac{7.14}{7.14 + 10} = 0.417; \ \frac{\ell}{L} = 2.391.$$

2 Examine case with *semi-rigid connection.*

For symmetrical double curvature in beam, end rotational stiffness is equal to $4EI_b / \ell_b$, therefore effective stiffness of connection in calculation of $k_1$ is $S/4$ based on ratios of $EI_b / \ell_b$ to $EI_c / \ell_c$.

Hence,

$$k_1 = \frac{\frac{S}{4}}{\frac{S}{4}+k_b}\cdot k_b = \frac{\frac{250}{4}}{\frac{250}{4}+10}\cdot 10 = 8.62\,.$$

From EC3 Annex E:

$$\eta_1 = 1.0;\ \ \eta_2 = \frac{k_c}{k_c+k_1} = \frac{7.14}{7.14+8.62} = 0.453\,;\ \ \frac{\ell}{L} = 2.447\,.$$

Hence reduction of elastic critical load in the presence of such connection flexibility is given by:

$$\frac{\lambda_{crit,\bar{s}=25}}{\lambda_{crit,\bar{s}=\infty}} = \frac{\left(\frac{\ell}{L}\right)^2_{\bar{s}=\infty}}{\left(\frac{\ell}{L}\right)^2_{\bar{s}=25}} = \left(\frac{2.391}{2.447}\right)^2 = 0.95\,,$$

i.e. for a frame with $\rho = 1.4$, a 5% reduction in elastic critical load factor is caused by $\bar{S}$ of 25.

*Influence of Connection Flexibility on Frame Strength.* In real frames the most important quantity is the ultimate limit load $F_u$ of the frame rather than the Euler buckling load $F_E$. This can be found by the Merchant-Rankine formula for the unbraced frame. If $F_{pl}$ is the plastic failure load of the frame then according to the above formula the ultimate load is given by:

$$\frac{1}{F_u} = \frac{1}{F_{pl}} + \frac{1}{F_E}\,. \tag{14}$$

From Equation 14 it is obvious that a 5% drop of the Euler buckling load $F_E$ due to connection flexibility leads to a drop in load carrying capacity of the frame $F_u$ by not more than 5% (ESDEP, 1994).

The previous discussion leads to the observation that for a given frame configuration, depending on the parameter ρ, the minimum required connection stiffness $\bar{S}$ may be determined from Figure 25b that does not lead to a decrease of the frame ultimate capacity by more than 5%. If the actual stiffness of the connection is smaller than that one determined from Figure 25b, the connection flexibility shall be taken into account in the frame analysis.

In Eurocode 3, a further simplification is made, see Figure 26. Constant boundary factors $\bar{S}$ are chosen for the braced and unbraced frames rather than values dependent on ρ.

For braced frames the boundary value of $\bar{S}$ is equal to 8 as shown in Figure 25b. That means that if $S \geq 8k_b$ a connection may be treated as rigid, otherwise as semi-rigid. This is illustrated in Figure 26.

For unbraced frames the relevant value of $\bar{S}$ is 25, which means that if $S \geq 25k_b$ a connection may be treated as rigid, otherwise as semi-rigid. Figure 26 illustrates this case as well.

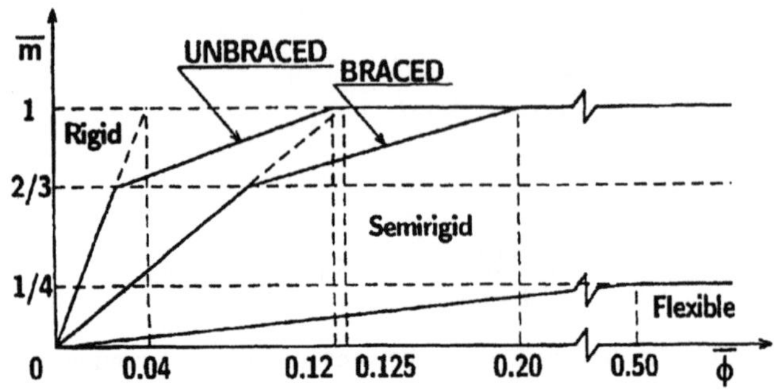


**Figure 26.** Classification boundaries for connections in respect to their rigidity.

According to Figure 25b the boundary value $\overline{S} = 25$ for unbraced frames covers only the cases when ρ≥1.4. For ρ<1.4 this value is unsafe. This situation is investigated below.

Frames for which ρ<0.1 are not realistic, so the value ρ=0.1 can be used as a boundary.

Figure 27 shows the relationship between ρ and the quantity of

$$(F_{E(\bar{s})} \,/\, F_{E(\bar{s}=\infty)}) \cdot 100\% \,.$$

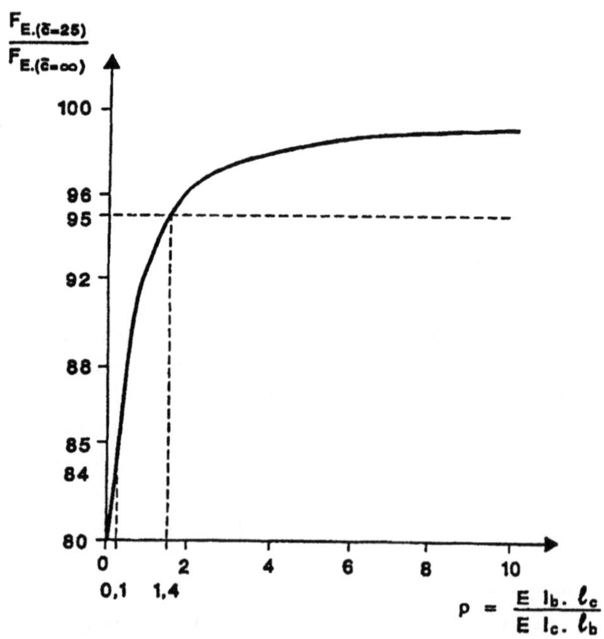


**Figure 27.** Relationship between ρ and $\dfrac{F_{E(\bar{s}=25)}}{F_{E(\bar{s}=\infty)}}$.

When ρ=0.1, the Euler buckling load based on $\overline{S} = 25$ is 85% of the Euler buckling load for $\overline{S} = \infty$.

The carrying resistance of the frame based on the Merchant-Rankine formula has reduced as follows:

$$\frac{1}{F_{u(\bar{s}=\infty)}} = \frac{1}{F_{pl}} + \frac{1}{F_{E(\bar{s}=\infty)}}; \tag{15}$$

where

$F_{u(\bar{s}=\infty)}$

is the carrying resistance of the frame with rigid connections;

$$\frac{1}{F_{u(\bar{s}=\infty)}} = \frac{1}{F_{pl}} + \frac{1}{XF_{pl}} \tag{16}$$

with:

$$X = \frac{F_{E(\bar{s}=\infty)}}{F_{pl}}; \tag{17}$$

$$\frac{1}{F_{u(\bar{s}=\infty)}} = \left(1 + \frac{1}{X}\right) \cdot \frac{1}{F_{pl}}; \tag{18}$$

$$F_{u(\bar{s}=\infty)} = \frac{X}{X+1} \cdot F_{pl}. \tag{19}$$

The following relationship holds for connections where $\overline{S} = 25$:

$$\frac{1}{F_{u(\bar{s}=25)}} = \frac{1}{F_{pl}} + \frac{1}{0.85XF_{pl}}; \tag{20}$$

$$\frac{1}{F_{u(\bar{s}=25)}} = \left(1 + \frac{1}{0.85X}\right) \cdot \frac{1}{F_{pl}}; \tag{21}$$

$$F_{u(\bar{s}=25)} = \frac{0.85X}{0.85X+1} \cdot F_{pl}. \tag{22}$$

The reduction of the carrying resistance is:

$$\Delta = \frac{F_{u(\bar{s}=\infty)} - F_{u(\bar{s}=25)}}{F_{u(\bar{s}=\infty)}} \cdot 100\% = \left(1 - \frac{X+1}{X+1/0.85}\right) \cdot 100\%. \tag{23}$$

Even for a very slender frame, namely when

$F_{u(\bar{s}=\infty)} = F_{pl}$,

so $X = 1$, the reduction $\Delta = 8\%$ and for $X = 2$ the reduction $\Delta = 5.6\%$.

It can be concluded that $\overline{S} = 25$ is a sufficiently safe boundary value for practical frames for the rotation stiffness of beam-to-column connection in unbraced frames in order to consider them as rigid.

In Figure 26 the boundaries for the distinction between rigid and semi-rigid behaviour of the connection follows a tri-linear rather than a bi-linear characteristic. The reasons are:

1. Experimental evidence shows that a beam-to-column end plate connection behaves elastically for up to at least 2/3 of its moment resistance.
2. The beam characteristic is also not linear up to the plastic moment. Plastification in the beam theoretically starts at about $W_{el}/W_{pl} \approx 0.9$ for I sections.

Due to residual stresses a more practical figure is 70% of the beam plastic moment. It is therefore considered reasonable to alleviate the bi-linear characteristic of the connection by a third branch.

*Influence of Connection Strength on Frame Behaviour.* As far as the moment resistance of beam-to-column connections is concerned, classification is simpler. If the moment resistance of the beam-to-column connection is equal to the plastic resistance of the connected beam, the connection is considered as full-strength. If not, the connection is considered as partial-strength. On the basis of the boundary for the rotation stiffness and the boundary for the moment resistance a bi-linear moment–rotation characteristic is achieved.

If the moment–rotation characteristic of a beam-to-column connection lies on the left hand side and above the boundary lines, this connection can be classified as rigid and full-strength.

**Modelling of the Connection.** Connections may be designed either by elastic or by plastic theory. Since they are modelled mostly by rotational springs as shown in Figure 24a, it is important to determine the spring characteristic representing the behaviour of the connection. From the lay-out of the connections, its moment–rotation characteristic may be determined on the basis of Annex J of Eurocode 3. However, the initial estimate of the characteristic has to be modified with the type of global analysis (elastic or plastic) and the method of connection verification (elastic or plastic) as shown in Figure 28 reproduced from Annex J.

If the global analysis is elastic the spring behaviour is considered to be linear since the same applies also to the behaviour of the other structural elements (beams and columns). For this type of analysis the joint may be verified either on the basis of elastic or on the basis of plastic theory. In the former case, only the linear part of the connections characteristic is taken into account i.e., the design moment of the connection is only 2/3 of its ultimate moment and its stiffness is equal to the initial value (Figure 28a). In the latter case the design moment of the connection is equal to its plastic moment but its stiffness is reduced by the factor $\chi$ as shown in Figure 28b to take into account the non-linearities.

In the case of plastic global analysis, with partial-strength joints, the spring behaviour of the connection is considered as bi-linear, since the same applies also to the behaviour of the beams and the columns of the frame. In that case the non-linear part of the connections is taken into account as shown in Figure 28c.

Obviously such connections should possess sufficient rotation capacity in order to be able to undergo the resulting plastic relations.

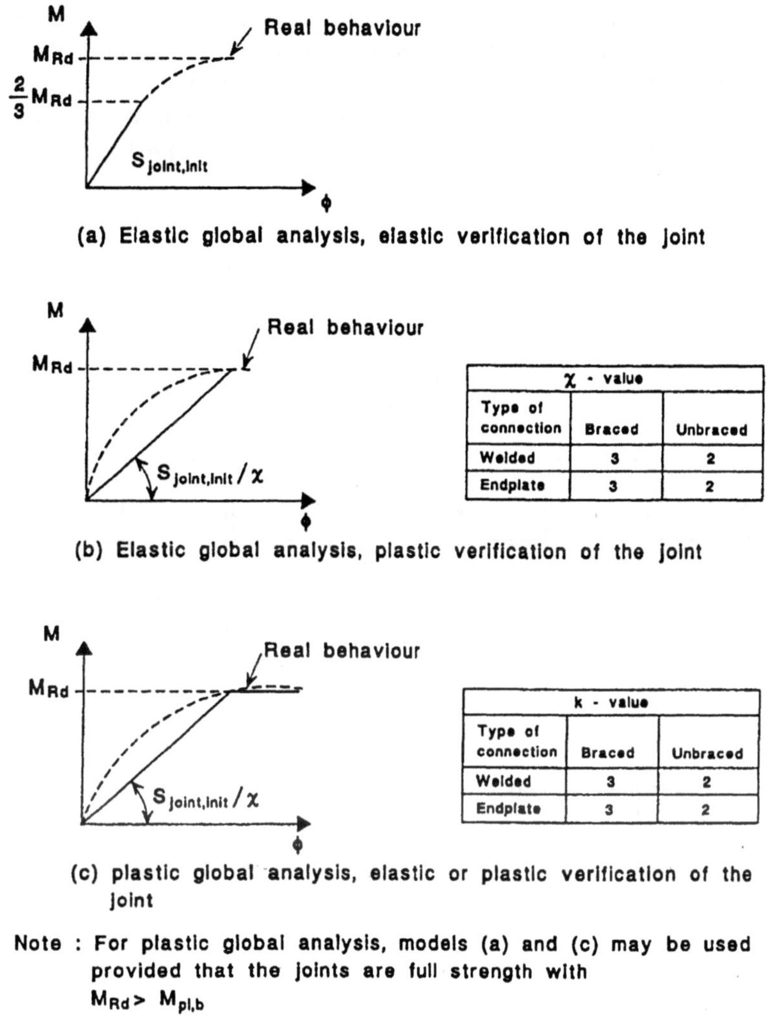


(a) Elastic global analysis, elastic verification of the joint

| $\chi$ - value | | |
|---|---|---|
| Type of connection | Braced | Unbraced |
| Welded | 3 | 2 |
| Endplate | 3 | 2 |

(b) Elastic global analysis, plastic verification of the joint

| k - value | | |
|---|---|---|
| Type of connection | Braced | Unbraced |
| Welded | 3 | 2 |
| Endplate | 3 | 2 |

(c) plastic global analysis, elastic or plastic verification of the joint

Note : For plastic global analysis, models (a) and (c) may be used provided that the joints are full strength with $M_{Rd} > M_{pl,b}$

**Figure 28.** Modified connection characteristics.

A last case is when plastic global analysis is applied with full-strength joints. In this case no plastic hinge is formed in the joint and the joint model is linear according to Figure 28a or 28b. In order to be sure that no plastic hinge will form in the joint and bearing in mind the possible overstrength of the beam, the following conditions shall apply:

- For Figure 28a: $\frac{2}{3}M_{Rd} \geq 1.2M_{pl,b}$;
- For Figure 28b: $M_{Rd} \geq 1.2M_{pl,b}$.

The greatest problem in the practical design of frames with semi-rigid connection lies in the determination of the connection flexibility since this is only possible once the connection is designed. This means that there must be an iterative process between frame analysis and joint dimensions until a connection is found with the characteristics assumed in the analysis i.e., one that is able to transmit the forces and moment and undergo the required rotation determined by the analysis.

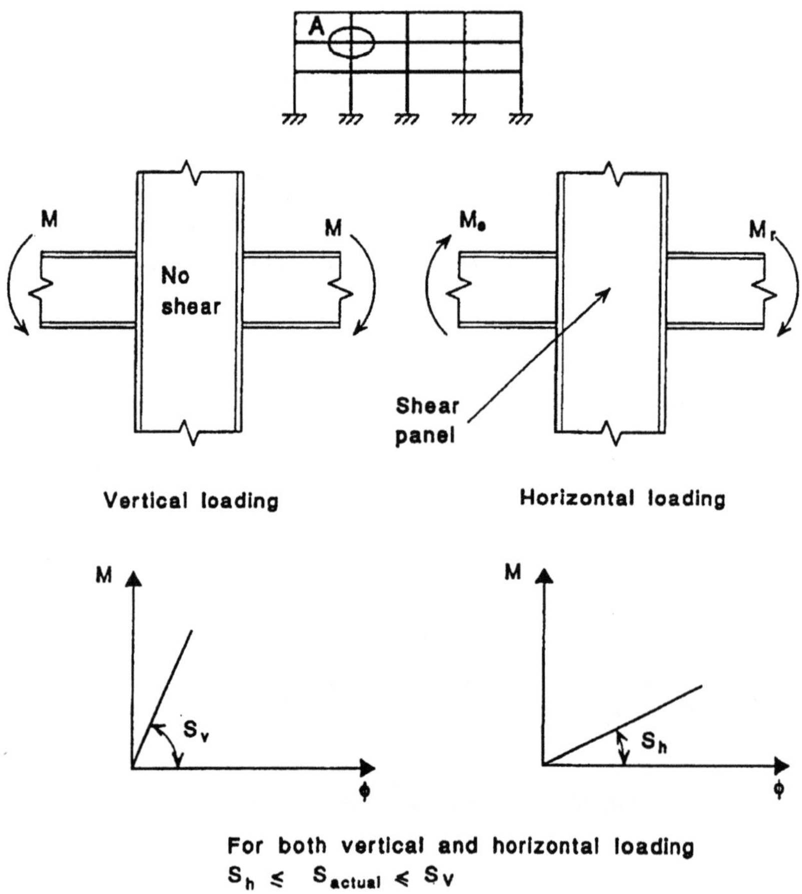


**Figure 29.** Connection rigidity for unbraced frames.

Specifically in the case of unbraced frames an additional parameter shall be taken into account in the modelling of connections. As illustrated in Figure 29, the deformation characteristic of the joint is different for the various loading conditions (vertical loading, horizontal loading) due to the large flexibility of the column web when submitted to shear. Since this case appears only for horizontal loading the joint behaviour is much softer in that case than in the case of vertical loading. In the normal case of both vertical and horizontal loading, the actual spring stiffness lies between the two values. This makes the practical design of unbraced frames with semi-rigid connections almost impossible.

**Relation between Frame and Connection Behaviour.** Frames and connections can be designed either by elastic theory or by means of plastic theory. It is even possible to perform the calculations for frames on the basis of elastic theory and for connections on the basis of plastic theory, or the other way around (ESDEP, 1994) (Bijlaard, Zoetemeijer, 1986).

There is, however, one limitation. A connection which remains elastic up to failure can only be used in frames designed by plastic theory provided that the calculated moment resistance is higher than that of the connected beam, and the beam sections have to be of the Class 1 type. Therefore, a full-strength connection is required in order to achieve sufficient plastic deformation adjacent to the connection.

When the force distribution in a connection is based on elastic theory, the connection will in general be stiffer than a connection designed in accordance with plastic theory. Clearly, this statement only holds when the connection in reality fulfils the assumptions used in the elastic or plastic theory calculations. It can be illustrated as follows. In using elastic theory, the hypothesis of Bernoulli holds (cross-sections remain plane). Such a hypothesis is employed to calculate the force distribution in the bolts connecting the end plate to the column flange. In reality, the end plate and the column flange have to remain plane. If this is realised, the connection is stiff because the only deformation is caused due to the elongation of the bolts.

Using plastic theory, it is necessary for the components in the connection to deform sufficiently in order to obtain a redistribution of forces and the formation of a failure mechanism inside the connection. The consequence is that such a connection is in general less stiff than a connection designed according to elastic theory.

*Plastically Designed Connections in Elastically Designed Frames.* In calculating the force distribution in a frame using elastic theory, the ultimate carrying resistance is not calculated. Instead the forces and moments due to the design load are computed. If the ratio $n$ between the Euler buckling load and the design load is small, say smaller than 10, second-order effects have to be taken into account. This can be done by multiplying the components of the moments due to side-sway (in the case of unbraced frames) by the amplification factor $n/(n-1)$. If the moments, calculated in this way, are smaller than the ultimate moment resistances of the various connections, then the connections meet the design requirements.

In the connections, designed in accordance with plastic theory, plastic deformations will occur when the ultimate moment resistance is reached. This effect is taken into account in calculating the Euler buckling load using the bi-linear approximation of the connection behaviour, which leads to a safe result (Bijlaard, Zoetemeijer, 1986). In the case of braced frames, the moments in the connections due to loading will perhaps be underestimated if a low rotational stiffness is used.

Therefore, one should take an upper bound for the connection stiffness when calculating safe values for the moments in the connections. This approach is in contradiction to the advice given earlier, namely the use of the secant stiffness (bi-linear approximation).

It is shown in the literature that using the bi-linear approximation for the rotational stiffness of the connection in calculating the moments in the connection is still safe provided that the connection possesses sufficient deformation capacity. In the case of unbraced frames the use of a lower bound for the rotational stiffness of the connection will lead to higher values for the moments in the connections due to the increased second-order effects. However, the first-order elastic moment in the connection decreases when a lower rotational stiffness is used. A lower bound for the rotational stiffness of connections in an elastic design of an unbraced frame does not necessarily lead to a safe elastic calculation of the moment in the connection. It is therefore necessary to use connections with sufficient rotational capacity even in an elastically designed frame.

*Elastically Designed Connections in Plastically Designed Frames.* In frames where more than one plastic hinge is necessary in order to reach the plastic failure mechanism, the first, second and subsequent plastic hinges have to rotate until the last plastic hinge is formed. This requirement holds for braced as well as for unbraced frames. Partial-strength connections which remain elastic up to failure are not to be used, because they possess insufficient deformation capacity. A method is given above for the calculation of the required rotational capacity of connections in braced frames. The required rotational capacity in unbraced frames is larger, and should in fact be calculated for the actual geometry of the unbraced frame. In most cases, however, an upper bound for the required rotational capacity is about 0.04 radians (ESDEP, 1994).

# 3 Recent Developments on Classification of Connections

## 3.1 Current Proposals for the Classification of Joints

**Introduction.** The behaviour of a steel structure with semi-rigid and/or partial strength joints may be substantially different from the corresponding to a structure with rigid and full-strength joints (Fig.30).

Ultimate load of these structures may be found performing a second order elasto-plastic analysis taking into account the actual behaviour of the joints. However, to evaluate the ultimate load in a simplified way, some codes accept the use of the Merchant-Rankine formula (here presented with the modification introduced by Wood):

$$\frac{1}{\alpha_u} = \frac{1}{\alpha_{cr}} + \frac{0.9}{\alpha_{pl}}, \tag{24}$$

where $\alpha_u$, $\alpha_{cr}$ and $\alpha_{pl}$ are the ultimate, the critical and the plastic load multiplier of the frame, respectively.

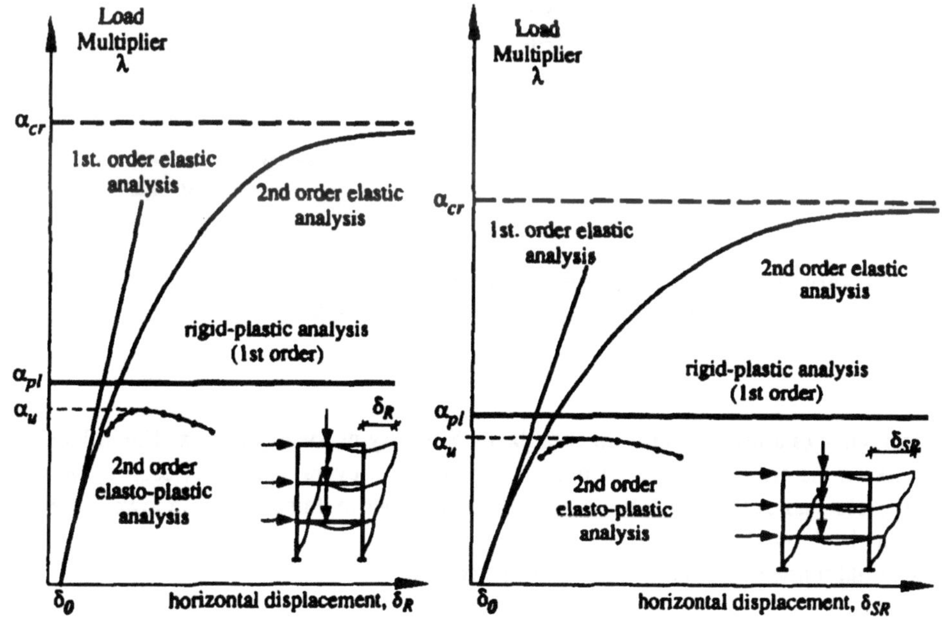


**Figure 30.** Load multiplier versus horizontal displacement of a frame.

Joint stiffness affects the critical multiplier, and thus the structural collapse load. It is then important to assess this influence, for which a lower bound will be derived.

Another implication of joint flexibility is the increase of sway deflections, that frequently governs structural design.

Recognising these implications of joint characteristics on the frame behaviour, Eurocode 3 proposes, on its revised Annex J, a classification of beam-to-column joints by strength and by stiffness. For the stiffness classification, the boundaries between nominally rigid and semi-rigid are defined as straight lines in the moment–rotation diagram, Figure 31, where the non-dimensional joint stiffness $S$ is obtained dividing joint stiffness $S_j$ by the beam stiffness $EI_b / \ell_b$:

$$\bar{S} = \frac{S_j}{EI_b / \ell_b}. \tag{25}$$

In the following, a new proposal (Gomes, Neves, 1996) for these boundaries is presented, based on the influence of joint flexibility on frame resistance.

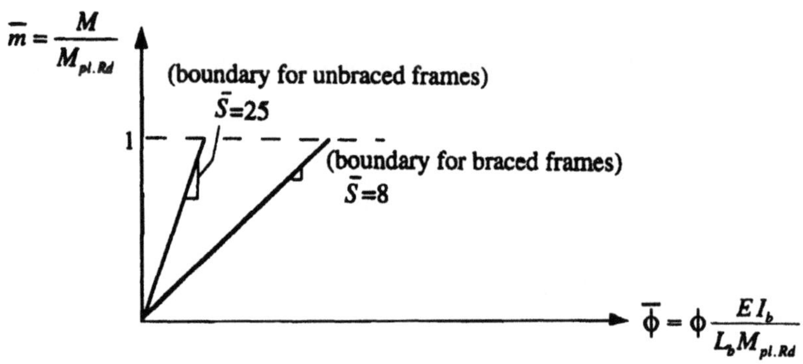


**Figure 31.** Classification by stiffness in the revised Annex J of Eurocode 3: boundaries between rigid and semi-rigid.

**Influence of Joint Stiffness on the Elastic Stability of Sway Frames.** The influence of joint stiffness on the critical load was studied previously by several authors (Cosenza, De Luca, Faella, 1989). Here an example of a portal is analysed in order to illustrate how a lower bound for this effect may be derived.

For the portal frame in Figure 32, we define the parameter:

$$T = \frac{I_b / L_b}{I_c / L_c}. \tag{26}$$

In the case of rigid joints, the horizontal displacement is given by:

$$\delta_R = \frac{HL_c^3}{12EI_c} \cdot \frac{1+2T}{T}. \tag{27}$$

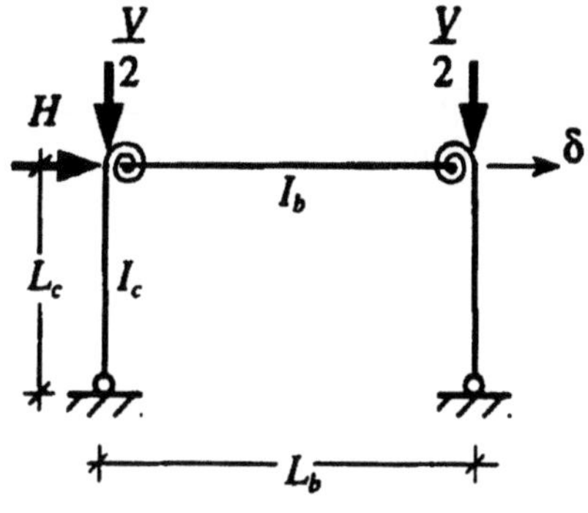


**Figure 32.** Portal frame.

Introducing the notation

$$T_{SR} = T\frac{\bar{S}}{\bar{S}+6}, \tag{28}$$

the horizontal displacement in the case of semi-rigid joints with non-dimensional stiffness $S$, is given by:

$$\delta_{SR} = \frac{HL_c^3}{12EI_c}\cdot\frac{1+2T_{SR}}{T_{SR}} = \frac{HL_c^3}{12EI_c}\cdot\frac{6+\bar{S}(1+2T)}{\bar{S}T}. \tag{29}$$

The critical load may be evaluated, for the rigid jointed frame, by Horne's method:

$$V_{cr,R} = 0.9H\frac{L_c}{\delta_R} = \frac{10.8EI_c}{L_c^2}\cdot\frac{T}{1+2T}, \tag{30}$$

while for semi-rigid joints

$$V_{cr,SR} = 0.9H\frac{L_c}{\delta_{SR}} = \frac{10.8EI_c}{h^2}\cdot\frac{\bar{S}T}{6+\bar{S}(1+2T)}, \tag{31}$$

In order to evaluate the effect of joint stiffness on structural behaviour, it is interesting to evaluate the amount of decrease of the critical load of the semi-rigid structure $V_{cr,SR}$ with respect to the critical load of the rigid structure $V_{cr,R}$ in the sway mode. The ratio between these two critical loads is then

$$\frac{V_{cr,SR}}{V_{cr,R}} = \frac{\bar{S}(1+2T)}{6+\bar{S}(1+2T)}. \tag{32}$$

When $T \to 0$ we find a lower bound:

$$\left(\frac{V_{cr,SR}}{V_{cr,R}}\right)_{\text{lower bound}} = \frac{\bar{S}}{\bar{S}+6}. \tag{33}$$

This lower bound may also be found from the exact solution for the critical load (Chen, 1987).

**Proposal for a New Classification of Joints.** (Gomes, Neves, 1996) For the case of rigid joints ( $S = \infty$ ) the Merchant-Rankine formula gives

$$\frac{1}{\alpha_{u,R}} = \frac{1}{\alpha_{cr,R}} + \frac{0.9}{\alpha_{pl,R}}, \tag{34}$$

and for the case of semi-rigid joints:

$$\frac{1}{\alpha_{u,SR}} = \frac{1}{\alpha_{cr,SR}} + \frac{0.9}{\alpha_{pl,SR}}. \tag{35}$$

In order to derive the range of joint stiffness that may be classified as *nominally rigid* we assume the following rule: "The deformation of rigid joints should be such that they do not reduce the resistance of the frame by more than 5%" (Eurocode 3 §6.4.2.2(2)). Thus the boundary stiffness will be found for

$$\alpha_{u,SR} = 0.95\alpha_{u,R}. \tag{36}$$

Rewriting Equation 35 for the limit case of Equation 36, knowing that $\alpha_{pl}$ is not influenced by the joint stiffness:

$$\alpha_{pl,SR} = \alpha_{pl,R}, \tag{37}$$

we obtain

$$\frac{1}{0.95\alpha_{u,R}} = \frac{1}{\alpha_{cr,SR}} + \frac{0.9}{\alpha_{pl,R}}. \tag{38}$$

Eliminating the quantity of

$$\frac{0.9}{\alpha_{pl,R}}$$

from Equations 35 and 38 we obtain

$$\frac{\alpha_{cr,SR}}{\alpha_{cr,R}} = \frac{1}{1 + \frac{0.05}{0.95}\frac{\alpha_{cr,R}}{\alpha_{u,R}}}. \tag{39}$$

As

$$\frac{\alpha_{cr,SR}}{\alpha_{cr,R}} = \frac{V_{cr,SR}}{V_{cr,R}} \tag{40}$$

from Equations 39 and 33 we obtain the particular value of the stiffness:

$$\bar{S}_b = \text{[illegible]} \frac{\alpha_{u,R}}{\alpha_{cr,R}} \tag{41}$$

defining a boundary between rigid and semi-rigid that do not reduce the resistance of the structure by more than 5%.

This simple formula may be rewritten as a function of the elastic critical parameter $V_{Sd}/V_{cr}$ (see Eurocode 3, § 5.2.6), assuming that the ultimate load multiplier is equal to the design value of the applied load multiplier. Then

$$\bar{S}_b = 114\frac{V_{Sd}}{V_{cr,R}}, \quad \text{but } \bar{S}_b \geq 8 \tag{42}$$

where $V_{Sd}$ is the design value of the total vertical load and $V_{cr,R}$ is its elastic critical value for failure of the rigid frame in a sway mode.

The limit value of

$$\bar{S}_b \geq 8$$

refers to the case of braced frames (Fig.31).

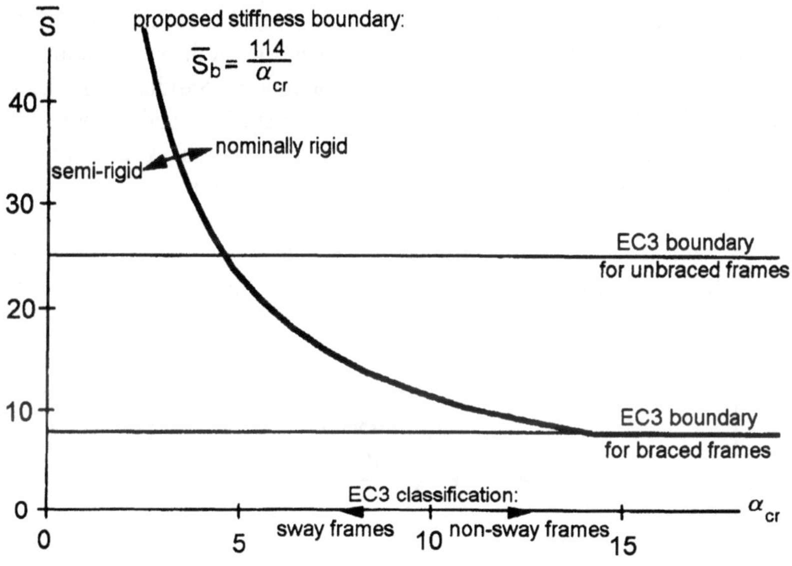


**Figure 33.** Comparison between proposed and Eurocode stiffness boundaries.

From the proposed formula of Equation 43 it may be concluded that a beam-to-column connection in an unbraced frame may be classified as rigid if its rotational stiffness satisfies the condition represented in Figure 33:

$$S_j = 114\frac{V_{Sd}}{V_{cr}} \cdot \frac{EI_b}{L_b}, \quad \text{but } S_j \geq \frac{8EI_b}{L_b} \tag{43}$$

### 3.2 Isolated Classification System for Connections

The classification system of (Bjorhovde, Colson, 1989) and (Bjorhovde, Colson, Brozzetti, 1990) is intended for the case when prior knowledge of the frame layout is not available. The connections are placed in one of three categories: rigid, semi-rigid and pinned, based on both the strength and the stiffness of the connections. The connection stiffness is represented as a ratio of the connected beam rigidity (*EI*) divided by a reference length ($L_r$). The reference length is taken as a multiple of the connected beam depth (*d*). If the reference length $L_r$ for a connection is smaller than $2d$, the connection is classified as rigid. If the equivalent reference length $L_r$ is greater than $10d$, the connection is classified as pinned. If the reference length is between $2d$ and $10d$, the connection is considered as semi-rigid. The moment capacity requirements for rigid, semi-rigid and pinned connections in this system are greater than 70%$M_{pl}$, between 20% and 70%$M_{pl}$ and smaller than 20%$M_{pl}$, respectively, where $M_{pl}$ is the plastic moment capacity of the connected beam. It has been demonstrated in (Bjorhovde, Colson, Brozzetti, 1990) that it is possible for a connection to be classified into different categories at the serviceability and ultimate limit states when both connection stiffness and strength are considered in determining the connection category. In addition to the limits for connection stiffness and strength, a requirement for connection rotation capacity has also been defined in this system. A diagram illustrating the system is given as Figure 34. Since it is not necessary to know the configuration of the whole frame in this classification system, it is relatively easy to use. However, because the overall behaviour of the frame has not been considered, the connection may be misclassified if the structural layout differs significantly from that assumed in the creation of this classification system.

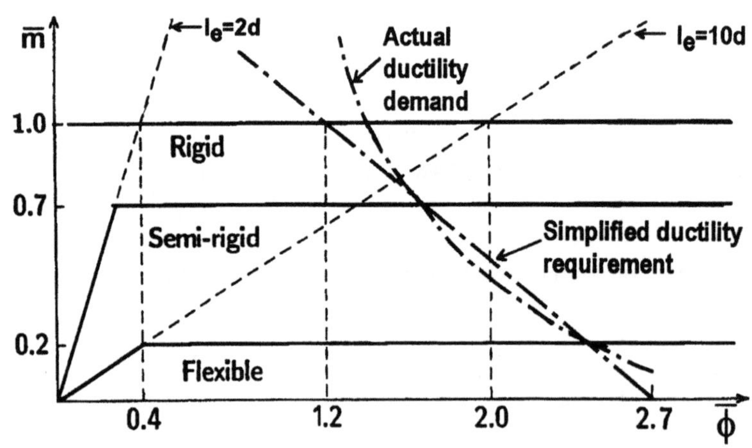


**Figure 34.** Bjorhovde's classification system.

### 3.3 Unified Classification System for Beam-to-Column Connections

**Introduction.** Proposals for the classification of beam-to-column connections have been presented independently by Bjorhovde et al. (Bjorhovde, Colson, Brozzetti, 1990) and in (EC3, 1993). However, the proposed methods are not entirely consistent with traditional ideas of rigidly-jointed and pinned-jointed frames. For example, if the connections of a frame are classified as rigid, the internal force distribution within the frame should be very close to the theoretical result obtained from traditional rigid frame analysis in both the elastic and the plastic states. On the other hand, if the connections are classified as pinned, the internal force distribution should be very close to that given by traditional pinned frame analysis. The two existing classification methods cannot always meet these expectations and may therefore cause confusion to design engineers. The purpose of the unified system is to develop a classification method for beam-to-column connections which will ensure results consistent with the expectation of traditional rigid and pinned analyses (Nethercot, Li, Ahmed, 1998).

The purpose of connection classification is to choose a suitable basis on which to conduct the overall frame analysis. Therefore, it is necessary that a classification system be able to reflect the connection assumptions used in the frame analysis. A connection may respond differently at the serviceability and the ultimate limit states, therefore it should be possible for the connection assumptions used in the frame analysis to recognise this. Accordingly, a classification system should consider both the serviceability limit state and the ultimate limit state. This has been mentioned in Bjorhovde et al.'s classification system, but is not addressed in the EC3 system. In addition, a classification system should also be able to place a connection into a single category for the same limit states and not allow it to fall into different categories and to give ambiguous results. The separate use of the stiffness and strength criteria in the EC3 system may give such ambiguous results. For example, a connection can be rigid and partial strength according to the EC3 classification system and this may be hard to understand because a partial strength connection cannot be rigid at the ultimate limit state. On the other hand, a semi-rigid connection cannot be full-strength since large connection moments may not be possible because of the low connection stiffness. In addition, the use of semi-rigid connections may be controlled by their rotational capacities, therefore it is necessary to specify a rotational requirement for all connections. Although the classification system proposed by Bjorhovde et al. has included the rotation requirement, it is not so accurately addressed and is not necessarily always on the conservative side as discussed in (Goto, Miyashita, 1995). For these reasons, a more comprehensive classification system is required and the development of such a system is needed.

The classification system proposed herein is termed a "unified classification system for beam-to-column connections". In this system, connection stiffness and strength are considered simultaneously in determining the category and thus terms such as "rigid" and "full-strength" will not be appropriate for naming the connection category because they each represent only one aspect of the connection behaviour. The connections are therefore placed into one of four categories (Nethercot, Li, Ahmed, 1998):

- fully-connected connections;
- partially-connected connections;
- pin-connected connections;
- non-structural connections.

Following the practice with other classification systems, the strength and stiffness of a connection under shear and axial loading are assumed to be sufficiently large that they will not affect the load carrying capacity and deformation of the frames. Therefore, they are not to be considered in the criteria for connection classification.

Since it is impossible for a connection to have an infinite stiffness as assumed in traditional rigid frame analysis, it has to be accepted that the moment transferred by a fully-connected connection will be slightly lower than the moment obtained from traditional rigid frame analysis. If this cannot be accepted, then all connections must be treated as partially-connected and will require semi-rigid frame analysis. Therefore, some level of approximation must be accepted. An assumption of this type is implicit in the two other classification systems (Bjorhovde, Colson, Brozzetti, 1990) (EC3, 1993). A 5% variation of connection moment between that in the fully-connected frame and that obtained by treating the frame as rigid jointed is adopted in determining the connection categories. Since the assumption of pin-connected connections would err significantly on the conservative side of reality, a larger tolerance from behaviour assuming actual pins will be used when defining pinned-connected connections.

**The Need to Consider Connection Stiffness and Strength Simultaneously.** It is known that the internal force distribution within a framed structure (except for a statically determinate structure) is dependent on the relative stiffness of each of the members. This indicates that the stiffer components will attract large internal forces and the flexible components will undergo large deformations and develop small internal forces. Accordingly, the moment developed at a connection will be dependent on its relative stiffness. As demonstrated in (Li, Choo, Nethercot, 1995), if a connection has a large moment capacity but a relatively low stiffness, it will not be possible for the connection to fully develop its full moment capacity. This suggests that if a connection does not have sufficient stiffness, even though it is a full-strength connection, the strength may not be fully utilised in the design. On the other hand, if the moment capacity of a connection is relatively small, no matter how high its stiffness, it will not be possible for the connection to act as rigid because the connection could fail before the moment level determined from rigid frame analysis is achieved. These features can be better appreciated from the following example of a single span subframe illustrated in Figure 35 with flexible beam-to-column connections. According to elastic analysis, the moment at the connection can be obtained from Equation 41:

$$M_j = \frac{qL^2}{12}\left(\frac{\alpha}{2+\alpha+2\alpha\dfrac{EI}{KL}}\right) \tag{44}$$

in which $M_j$ is the connection moment; $q$ is the uniformly distributed load; $L$ is beam span length; $EI$ is beam section rigidity; $K$ is connection stiffness; $\alpha$ is column-to-beam rotational stiffness ratio, expressed as

$$\alpha = \frac{K_c}{EI/L};$$

$K_c$ is the sum of the rotational stiffnesses of all the members connected to the connection except the beam considered. If the connection is an internal column connection $K_c$ should normally be taken as infinite. For this example, $K_c$ is equal to $K_c$=$8EI_c/L_c$. Here $EI_c$ is the column section rigidity and $L_c$ is column height as shown in Figure 35.

It is known that if the beam-to-column connection is ideally rigid, the moment at the connection will be

$$M_{c,r} = \frac{qL^2}{12}\frac{\alpha}{2+\alpha}. \tag{45}$$

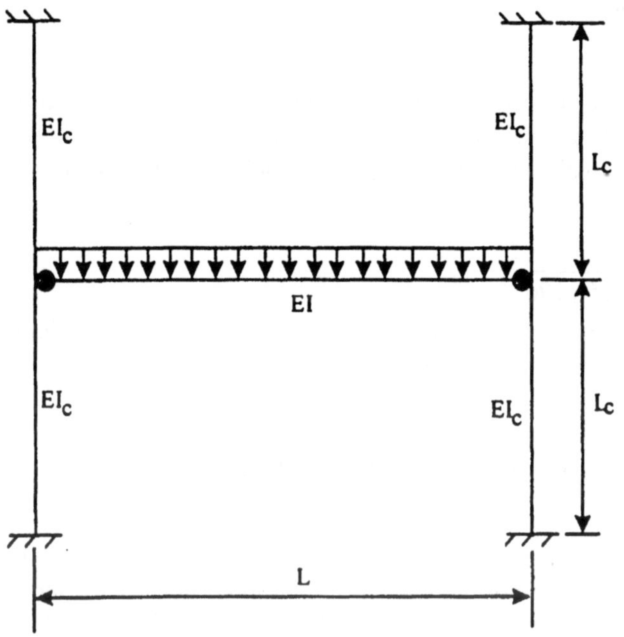


**Figure 35.** Single span subframe.

According to EC3, if the connection moment capacity is not less than 95% of the negative moment capacity of the beam (e.g. $M_{c,r}$ when the beam is fully designed), it may be classified as a full-strength connection. The concept of a full-strength connection implies that:

1. it is possible to utilise the full connection moment capacity in the design;
2. the beam and frame deformations will not be affected by the connection properties.

However, this may not be true if the connection stiffness is relatively low compared with its strength. For example, if the connection stiffness in Figure 35 is relatively small (say $K = 8EI / L$), then the maximum moment that can be developed at the connection is only

$$M_{c,s} = \frac{qL^2}{12}\frac{4\alpha}{8+5\alpha}. \tag{46}$$

If the column/beam stiffness ratio α is greater than 0.54, which is very common in practice, the connection moment will never be able to exceed

$$M_{c,s} = 0.0168qL^2 . \tag{47}$$

This moment is smaller than the connection moment capacity:

$$M_{cp} = 0.95M_{c,r} = 0.0177qL^2 . \tag{48}$$

This implies that the full connection moment capacity cannot be utilised in the design because of the limitation of the low connection stiffness.

On the other hand, if the connection stiffness is sufficiently high (say $K = 36EI/L$), but with a low moment capacity of

$$qL^2/96 ,$$

the connection will be classified as rigid according to EC3. A frame with rigid connections may be analysed by any traditional rigid frame analysis method. However, this may not be acceptable if the connection strength is not adequate. For example, the connection moment from traditional rigid frame analysis is

$$M_{c,s} = \frac{qL_b^2}{12}\frac{18\alpha}{36+19\alpha} . \tag{49}$$

If the column/beam stiffness ratio $\alpha$ is greater than 0.29, which is almost always the case in practice, the moment developed at the connection will be at least:

$$M_{c,s} = 0.0105qL_b^2 . \tag{50}$$

This moment is greater than the connection moment capacity of

$$qL_b^2/96 ,$$

hence it can never be developed in practice. Therefore, it is not appropriate to classify this kind of connection as rigid. Because of the low connection moment capacity, the full negative moment capacity of the beam cannot be developed. These findings clearly conflict with the traditional concept of a rigidly jointed frame.

The above analyses demonstrate the drawbacks in any classification system that considers the stiffness and strength separately. The method should simultaneously take into account both connection stiffness and connection strength. This will lead to a unique result consistent with the traditional concepts of rigid and pinned connections. Since structures are required to meet two separate criteria in design – strength and stiffness – corresponding to the ultimate and serviceability limit states (at which states the connection responses may well be different), the classification system should be able to distinguish these two limit states.

**Classification of Connections corresponding to Ultimate Limit State.** The connection classification with regard to ultimate limit state can be realised as follows.

*(i) Fully-connected connections.* Fully-connected connections should permit traditional rigid frame analysis to be used in the design. This requires that:

1. the connection moment capacity should be at least equal to the connected beam moment capacity;
2. the connection stiffness should be high enough to enable the connection moment capacity

to be developed.

*(ii) Pin-connected connections.* Since it is assumed that a pin-connected connection will not develop a significant moment, the connection stiffness or the moment capacity should be limited. Following the assumption of other classification systems, if the moment capacity of a connection is smaller than 25% of the moment derived from rigid frame analysis, the connection is considered as a pin connection.

*(iii) Partially-connected connections.* As a partially-connected connection also has to undergo a certain amount of rotation to develop its moment capacity, the connection should satisfy an appropriate rotational requirement.

*(iv) Non-structural connections.* If a connection cannot meet the requirements for any of the above connection types (mainly because of lack of rotation capacity), it should be classified as a non-structural connection. Non-structural connections cannot be regarded as a structural element because they lack ductility and are likely to fail prematurely before the design condition is achieved. In practice, non-structural connections should be prevented in the design..

Using the classification system, a diagram showing the classification of connections has been drawn in Figure 36a. The required connection rotation capacity shown in Figure 36a is obtained for the assumption that the hogging and sagging moment capacities of the beam are the same and the beam span section moment capacity is fully utilised. It can be seen that the actual moment–rotation curve of the connection is not required to determine the category of the connection. The required parameters are the connection design moment, the secant rotational stiffness and the rotation capacity. Some examples showing the application of this approach are given in Figure 36b.

**Classification of Connections corresponding to Serviceability Limit State.** Since the design requirement at the serviceability limit state is one of limiting deformation rather than attaining a certain strength, classification at serviceability will be affected mainly by the connection stiffness. The stiffness required for fully-connected connections should be determined such that the beam deformations with fully-connected connections are similar to those obtained assuming ideal rigid connections. The stiffness required for partially-connected connections should be high enough to considerably reduce the beam deflections compared with those obtained from ideal pin connections. If the connection stiffness is so small that the beam deflection is almost the same as that calculated assuming ideal pin connections, the connection should be regarded as a pin-connected connection.

*(i) Fully-connected connections.*

*(ii) Pin-connected connections.* If the connection stiffness at the serviceability limit state is so small that the beam deflection is almost as large as that for an ideal simply supported beam this connection should be treated as a pin-connected connection.

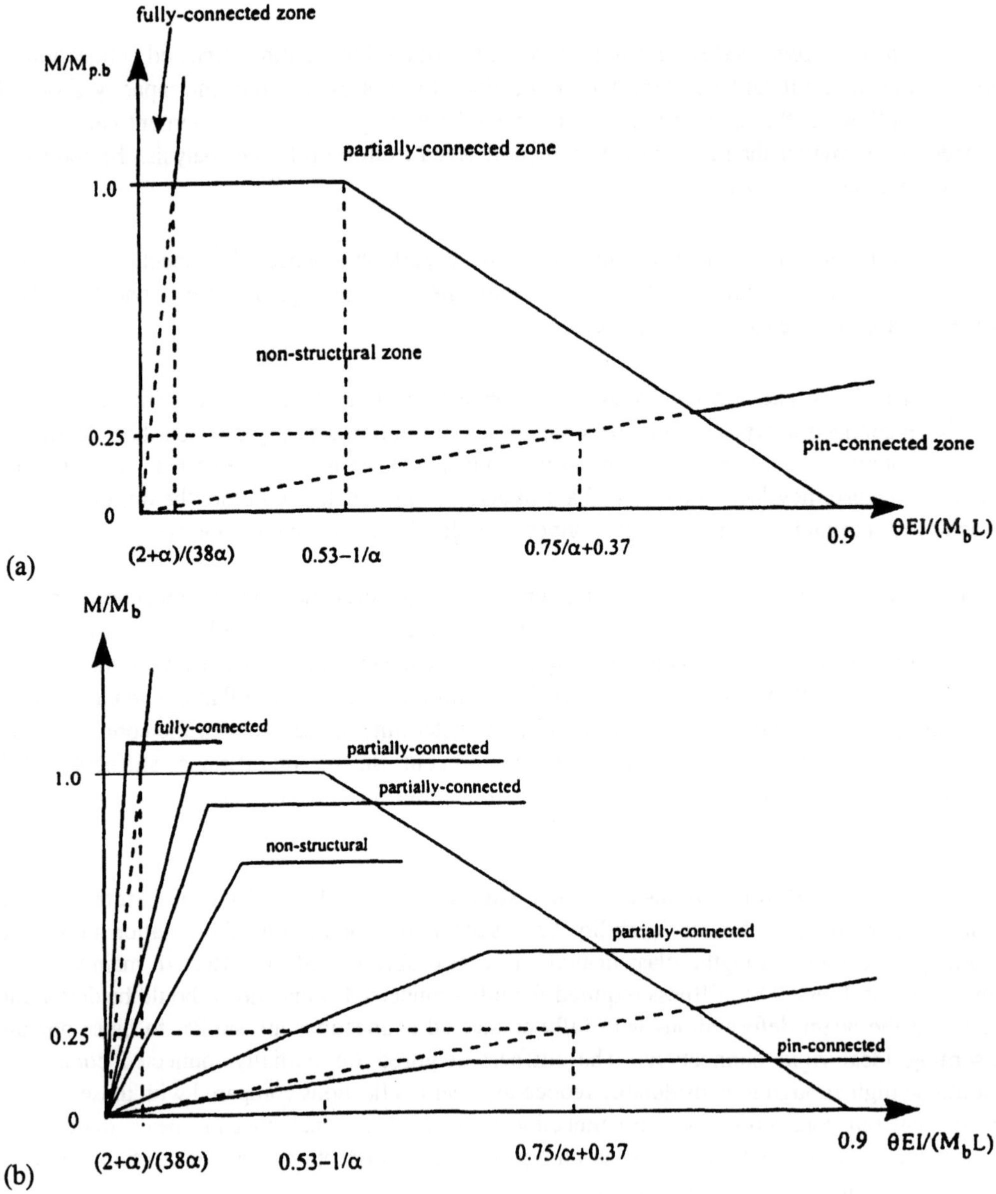


**Figure 36.** Single span subframe.
(a) unified connection classification system at ultimate limit state;
(b) examples of the application

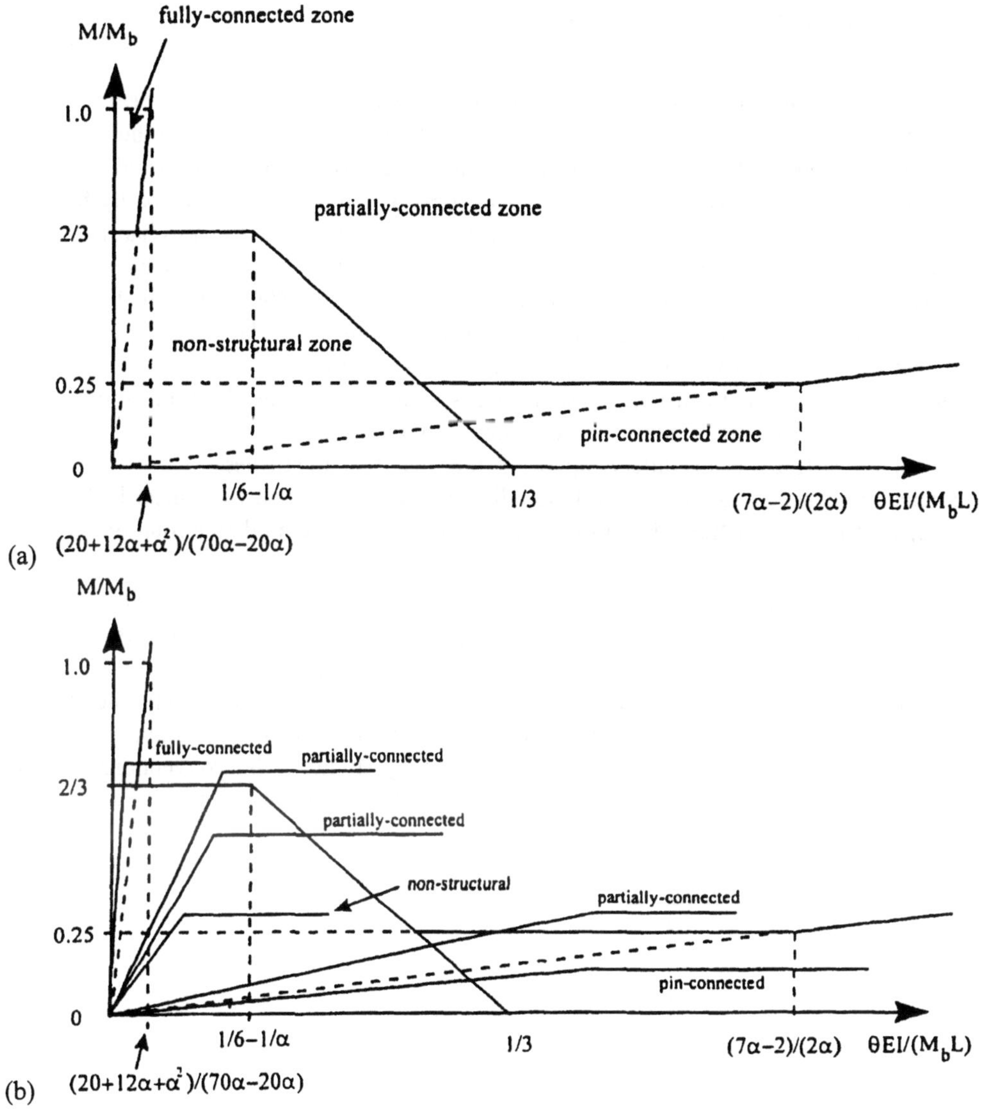


**Figure 37.** Unified connection classification system at serviceability limit state.
(a) unified connection classification system at serviceability limit state;
(b) examples of the application

*(iii) Partially-connected connections.* A diagram showing the classification system at the serviceability limit state is given in Figure 37a. Examples illustrating the use of this diagram are presented in Figure 37b.

### 3.4 Non-linear Connection Classification System

**Introduction.** Incorporating the concept of rotational flexibility of steel beam-to-column connections, modern codes (AISC, 1989) (AISC, 1994) (EC3, 1993) recognise the use of semi-rigid connections in addition to the two conventional and idealistic connections: rigid and flexible. Application of semi-rigid connections in practical construction remains remote partly because of its indistinct separation from the other two idealistic connections. The scenario had largely been improved in the recent past with the introduction of two conspicuous nondimensional classification systems (EC3, 1993) (Bjorhovde, Colson, Brozzetti, 1990) in which clear boundary lines among the three types of connections are provided, but these classification systems seem inadequate to match the real non-linear behaviour of connections.

Even though the nondimensional classification systems seem plausible, applicability of these classification systems remains questionable because of three flaws:

1. connection stiffness is absolutely expressed in terms of beam stiffness;
2. connection zones in the moment–rotation diagram are demarcated with linear lines;
3. tangent connection stiffness has a stepped distribution along the relative rotation.

**Proposed Connection Classification System.** (Hasan, Kishi, Chen, 1998) The above discussion shows the obvious necessity of having a non-linear classification system with no reliance on beam stiffness. This part aims to devise a new classification system for beam-to-column connections in an unbraced frame which will be free from the above-mentioned flaws. To that end, as opposed to previous nondimensional approaches, a dimensional one is pursued. The envisioned classification system is therefore devised by dividing the conventional moment–rotation ($M$–$\theta_r$) diagram into three connection zones with two smooth non-linear lines, as shown in Figure 38.

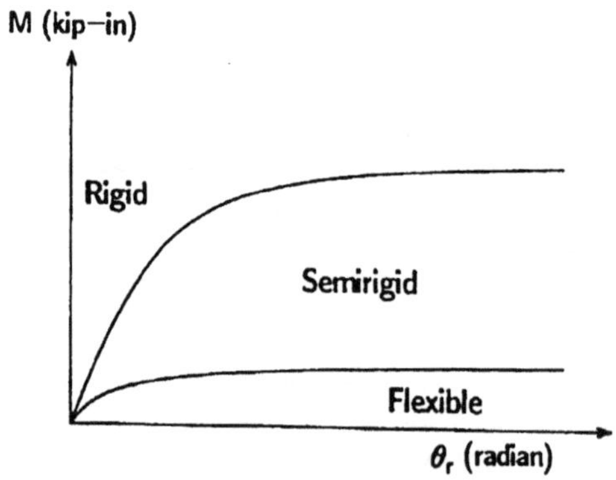


**Figure 38.** Envisioned classification system.

In search of a mathematical model to describe these two non-linear lines, the three-parameter power model of (Kishi, Chen, 1990) is found to be most suitable. The model is composed of three parameters: initial stiffness $R_{ki}$; ultimate moment capacity $M_u$ and a shape parameter *n*, and can be expressed as in Equation 48, taking the shape as shown in Figure 39:

$$M = \frac{R_{ki}\theta_r}{\left[1+\left(\frac{\theta_r}{\theta_0}\right)^n\right]^{1/n}}, \tag{51}$$

where $\theta_0 = M_u/R_{ki}$ is the reference plastic rotation. To fit the three-parameter power model into the envisioned classification system, it is therefore necessary to find the exact values of the parameters for the two demarcation lines.

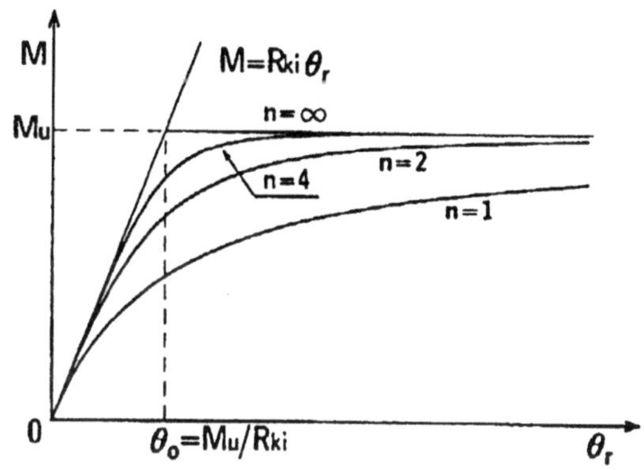


**Figure 39.** Three-parameter power model.

There are three parameters for boundary lines:

*(i) Initial stiffness $R_{ki}$.* From extensive frame analyses conducted in (Hasan, Kishi, Chen, Komuro, 1995) and (Kishi, Hasan, Goto, Komuro, 1996), it was shown that the maximum and minimum initial stiffness of a semi-rigid connection in an unbraced frame can be taken as $10^6$ and $10^{4.5}$ kip-in/rad, respectively.

*(ii) Ultimate moment capacity $M_u$.* Similar to the (EC3, 1993) classification, ultimate strength levels to define the horizontal plateaus between the rigid–semi-rigid and semi-rigid–flexible zones are taken as full plastic moment $M_p$ and 1/4 of the $M_p$ of the connecting beam, respectively.

*(iii) Shape parameter n.* The value of the shape parameter is adopted by calibration. It was observed that $n = 1$ serves the purpose of sufficient accuracy and simplicity. Figure 40 shows the appearance of the proposed classification system in US customary units.

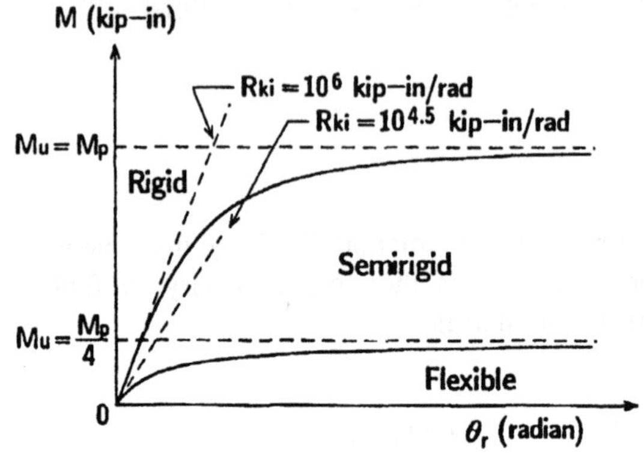


**Figure 40.** Proposed classification system.

**Conclusion.** A non-linear moment–rotation connection classification system for beam-to-column connections in an unbraced frame, consisting of these components: *(i)* initial connection stiffness; *(ii)* ultimate moment capacity; and *(iii)* a shape parameter, is proposed. The stiffness limits for the boundary lines demarcating the connection zones in the classification system have been explicitly set in numerical figures. They have been obtained from extensive numerical analyses. The strength limits are kept unchanged from those of the (EC3, 1993) classification system. Since these limits are unit-dependent, it will be necessary to adjust the classification system depending upon the type of unit to be used in the design.

This classification system can be used in the initial phase of connection/frame design to *(i)* clearly categorise a real connection into any of three types: rigid, semi-rigid or flexible; and *(ii)* select a proper analysis/design method such as type FR or type PR construction (AISC, 1994). Confusions are likely to occur in cases of mixed category connections (e.g. semi-rigid at the stiffness limit check but flexible at the strength limit check). The designers can avoid that type of situation by adjusting the connection so that the moment–rotation curve of the connection clearly falls within the boundary limits of a connection zone.

The proposed classification system is found to be simpler and more accurate than the existing systems (Hasan, Kishi, Chen, 1998).

# 4 Failure Tests of Two-Storey Steel Frames

by M. Iványi[1], L. Hegedűs[1], M. Iványi Jr.[2] and G. Varga[1]

## 4.1 Introduction

A great number of tests have been performed on isolated semi-rigid connections and flexibly connected subframes.

Tests on full-scale three-dimensional frames are somewhat less numerous. However, experimental data on full-scale frame behaviour is important. Firstly it enables the effect of column continuity through a loading level to be investigated – a parameter not present in many subframe tests – and secondly, it confirms whether the experimentally observed performance of isolated joints and subframes is indeed representative of their behaviour when they form part of an extensive frame. This latter point is of particular importance if the extensive work on isolated specimens is to be incorporated into universally accepted methods of semi-rigid and partial strength frame design.

The frame tests, presented in this contribution, were carried out in the structure testing hall of the Building Research Laboratory, Department of Steel Structures, Technical University of Budapest.

## 4.2 General arrangement of experiments

**Test Program and Description of Test Frames.** The test programme included tests of three complete frames:

- Frame COST 2: proportional loading process, horizontal load ratio: $R=H_{min}/H_{max}=1$;
- Frame COST 3. pulsating loading process, horizontal load ratio: $R=H_{min}/H_{max}=0$;
- Frame COST 4: alternating loading process, horizontal load ratio: $R=H_{min}/H_{max}=-1$.

The test frames were partially two-storey and partially two-bay ones. In the first storey the frames had two bays, and the larger one was continued in the second storey. The frames were built up from European hot-rolled sections (Figure 41).

Typical flush end plate connections have been used between the main members of the frame. There were a number of reasons for selecting this particular type of connections. Firstly, cleated connections are very often susceptible to bolt slip at relatively low moment levels. Once bolt slip has occurred, the precise moment–rotation characteristics of the connection are usually irreversibly changed. The flush end plate connection, however, is not affected by bolt slip to the same degree and therefore, exhibits similar moment–rotation behaviour under moderated repeated loading. Using such a connection, small levels of load could be applied to the frame without irreversibly deforming either the connections or the frame members. This was an important facility which allowed the loading devices and load control system to be fully commissioned prior to carrying out a test to failure.

---

[1] Department of Steel Structures, Budapest University of Technology and Economics, Hungary
[2] UVATERV Engineering Ltd., Hungary

The test frames were supported by separated rigid concrete blocks, which were connected to the testing floor. The connections of columns to the "separated" concrete blocks were realized as "hinged" bolted connections between the base plate and the concrete block. Concrete blocks were poured into a steel formwork. The formwork contained four tubes for those bolts, which were used later to connect the concrete blocks to the testing floor. Concrete blocks are reinforced, main dimensions of the blocks and the amount and form of reinforcing bars was adopted from the similar blocks of the tests of Penserini and Colson (1989). Anchor bolts were jointed to the reinforcement of the blocks before concreting. Between the top of the blocks and the bottom of baseplates, according to the usual practice, app. 1 cm thick cement layer was used.

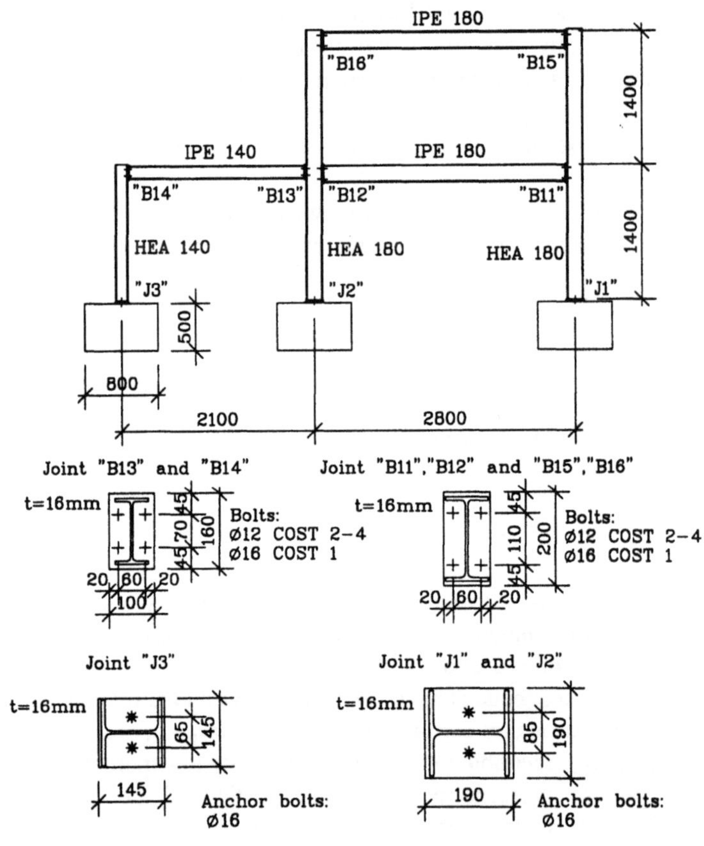


**Figure 41.** Overall view of test frames.

**Loading System.** The test frames were loaded by combined (vertical and horizontal) loads (Figure 42). Vertical loads were applied to the upper flanges of beams, this way the webs and lower flanges were not restrained laterally. Loading was exerted by means of hydraulic jacks.

To make horizontal displacements (sidesway) unrestricted, jacks were fastened not directly to the floor slab, but through a so-called gravity load simulator (Halász and Iványi 1979). This latter consisted of three elements: two "rigid" bars and one "rigid" triangle. The two bars had pin joints at both ends, resulting in a one-degree-of-freedom mechanism. Hydraulic jacks joined the third corner of the rigid triangles. This mechanism produced a vertical load acting in the intersection of the two bar axes. The simulators were designed and built in the Laboratory of Department of Steel Structures, Technical University of Budapest. Horizontal loads were applied at the levels of beams, and they were exerted by means of hydraulic jacks driven by an oil-pressure circuit independent of the vertical ones.

The original structure consists of frames and a "perpendicular" system of purlins, side-rails and wind bracing. The effect of this latter system was simulated by a "back-ground" construction, located at a distance of 3 m behind the test frames and connecting rods at the top of columns and the middle of beams.

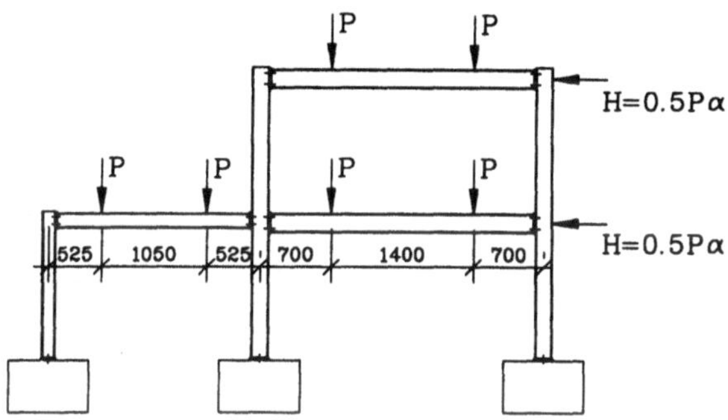


**Figure 42.** Loading arrangement.

**Measuring Techniques.** Forces were measured by means of a pressure transducer built in the oil circuit of the hydraulic system. The intensities of the forces were calculated from the pressure and the effective cross sectional area of the jacks.

Main important horizontal and vertical displacements of the frames were measured on the beam levels (a and b in Figure 43) and in the midspans of the beams (c, d and e) by inductive transducers.

Strains were measured by means of strain gauges. Altogether 16 cross sections of the frame elements were chosen and one cross section contained four strain gages on the four edges of the section. Relative rotations at joints were measured in similar sections as strains, their distribution. and notation are given in Figure 44. For this purpose two simple scales were inserted in every section in a known distance (app. 200 mm) from each other, they were hinged connected and always kept the vertical direction because of gravity. Measurements were made by geodetic devices and from the measured differences the rotation was calculated.

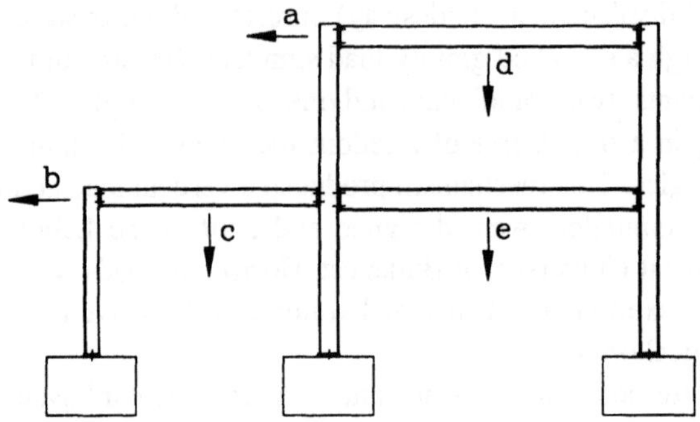


**Figure 43.** Measured displacements.

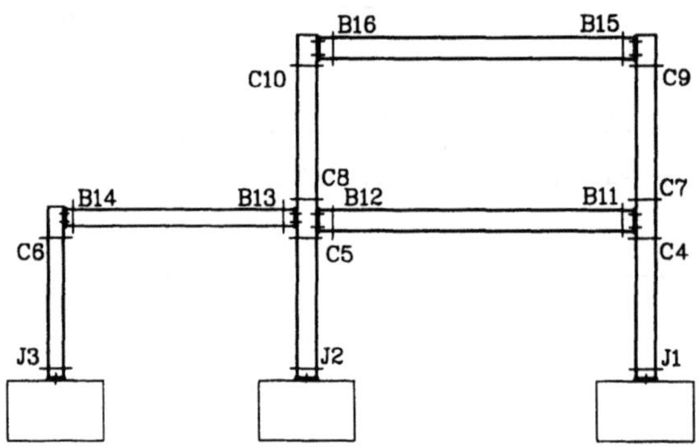


**Figure 44.** Cross-sections with relative rotation measuring scales.

### 4.3 Some results of the experiments

**Test Frame COST 2: Proportional Loading Process.** For frame COST 2 the load-to-displacement curves are presented in Figure 45. Notation of displacements can be seen in Figure 43. The load axis contains the total load produced by the hydraulic system (that is 7 x P).

Load – relative rotation and displacement curves at column bases are shown in Figure 46. The further load – relative rotation curves of other sections are given in Figure 47. The curves are giving both the load – rotation characteristics of the different sections and the differences between the neighbouring sections. These differences in case of columns (as between 4 and 7 or between 5 and 8) are not remarkable, while on two sides of a beam-to-column connection (as between 4 and 11 or between 5 and 13, etc.) they are non-negligible.

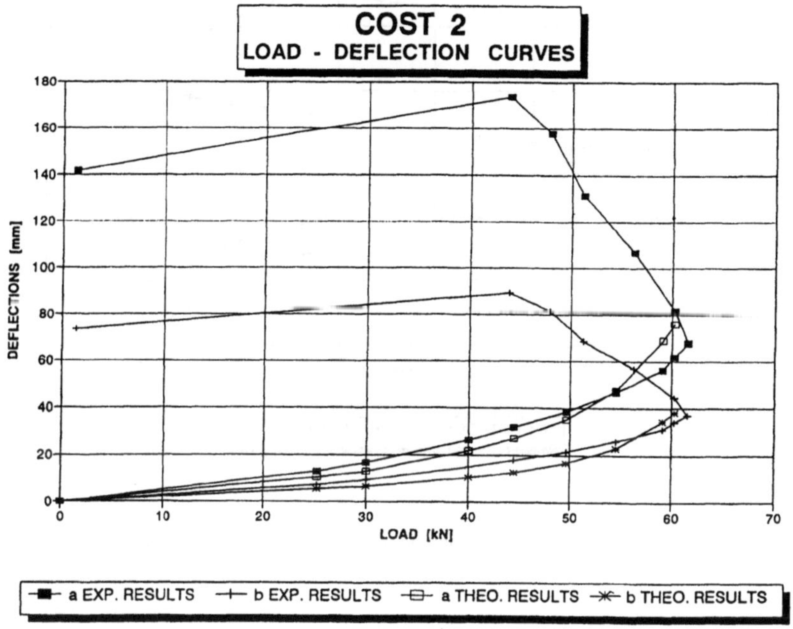


**Figure 45.** Load – displacement curves.

The loading was increased step by step. After the load level No. 6 an unloading took place. At this stage the whole structure and its connections were elastic, only the column bases showed some changes. On the compression zone there were some cracks in the cement layer, while on the tension side there was a gap. During this unloading process the rigidity of the structure caused some further cracks on the opposite side of the cement layer.

**Test Frame COST 3B: Pulsating Loading Process.** Test frame COST 3 was subjected to a pulsating loading process: the horizontal load varied between 0 and αP at every vertical load level. In test COST 3A, a low ratio for the horizontal load was chosen (α=0.07), thus failure was not encountered; while in test COST 3B, ultimate failure occurred with bolt failures, lateral-torsional buckling of beam B15 - B16, and crushing of cement layer under base plates.

The horizontal displacement of the top frame level is shown in Figure 48, as the function of vertical loading. Significant differences in joint displacements between cross-sections C4 - B11 are shown in Figure 49. Performance of column bases is illustrated on Figure 50. At the maximum level of vertical loads, the corresponding horizontal load was not applicable due to bolt failure.

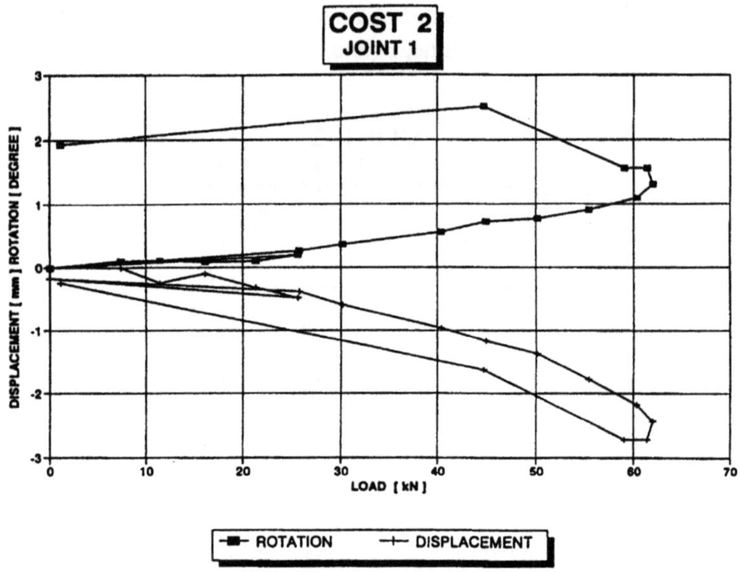


**Figure 46.** Load – relative displacement curves at the column bases.

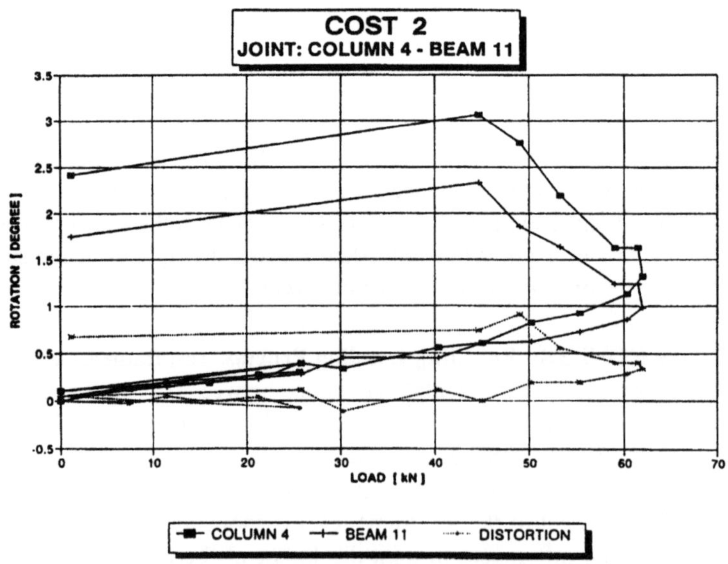


**Figure 47.** Load – joint displacement curve.

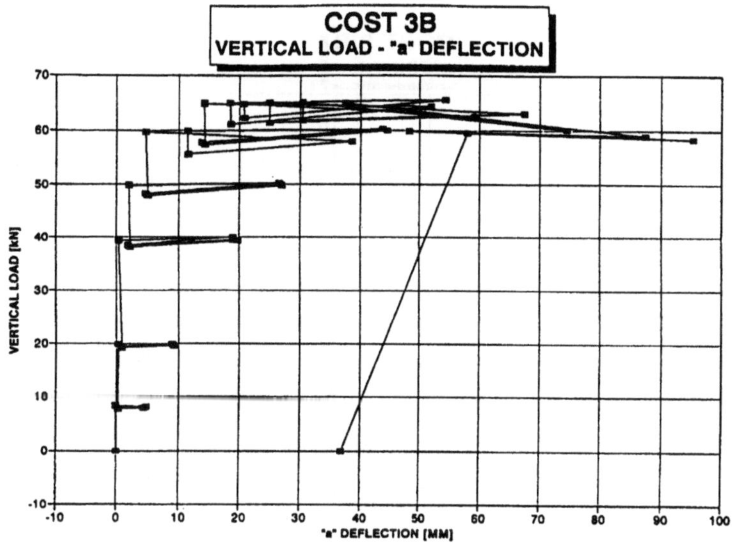


**Figure 48.** Vertical load – deflection curve.

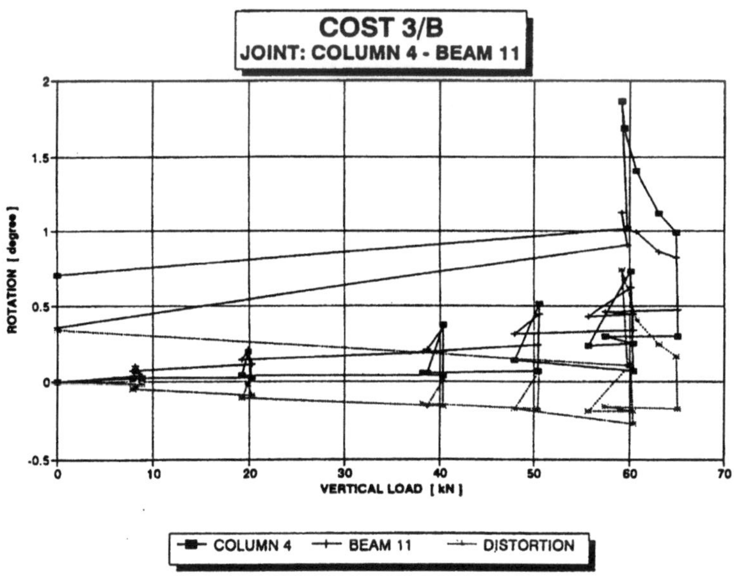


**Figure 49.** Vertical load – rotation curves.

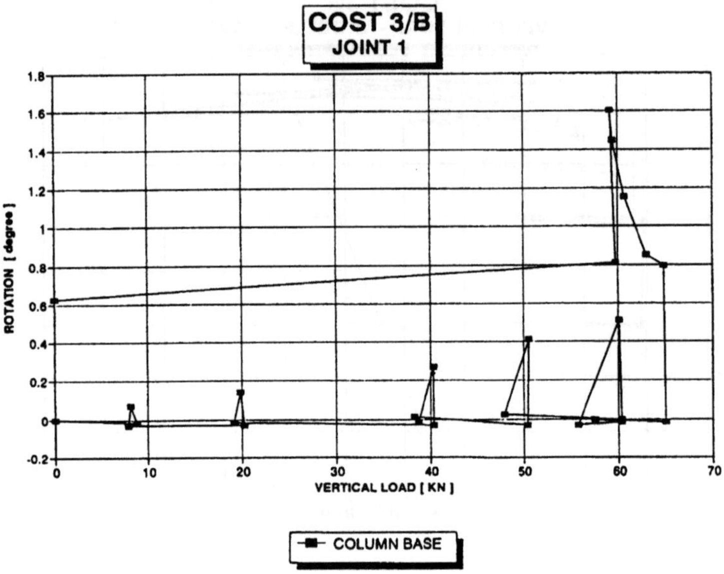


**Figure 50.** Vertical load – rotation curve.

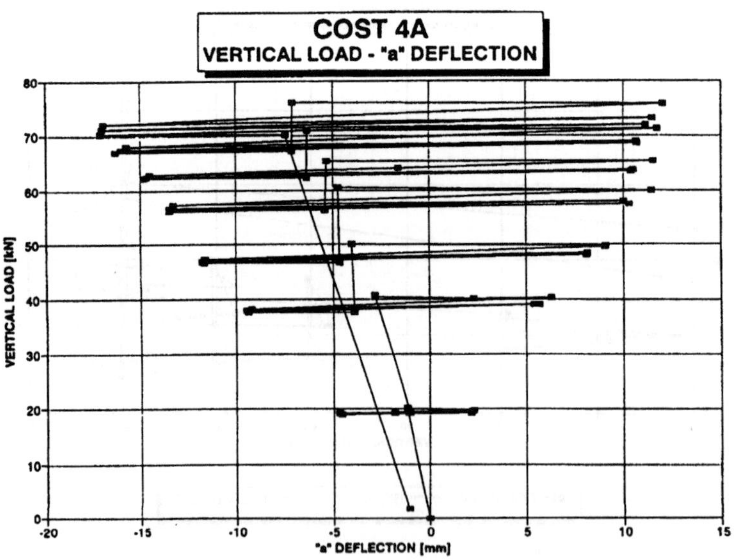


**Figure 51.** Vertical load – deflection curve.

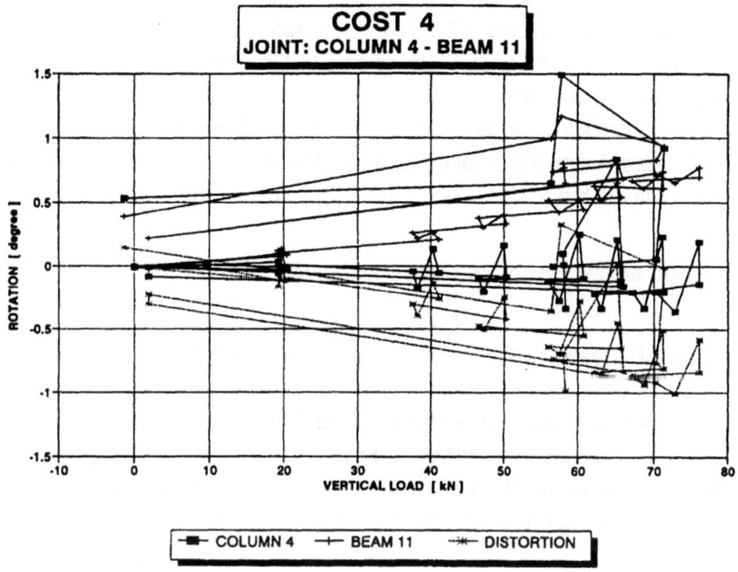


**Figure 52.** Vertical load – rotation curves.

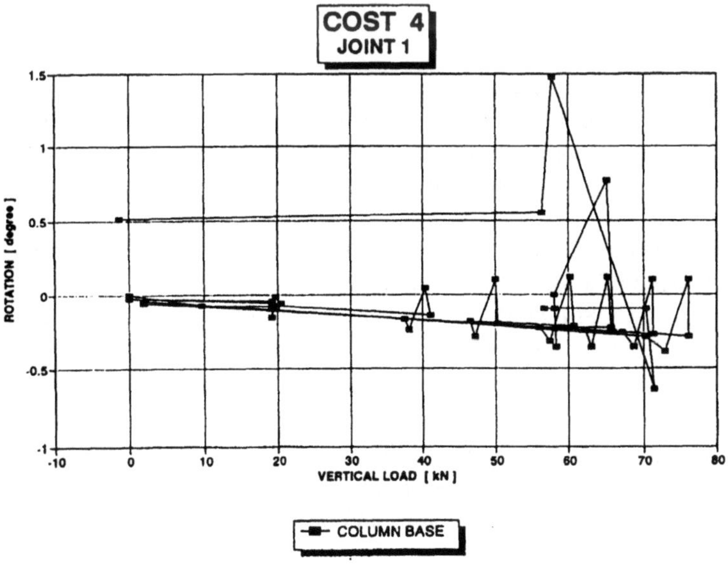


**Figure 53.** Vertical load – rotation curve.

**Test Frame COST 4: Alternating Loading Process.** Test frame COST 4 was subjected to an alternating loading process: the horizontal load varied between αP and –αP at every vertical load level. During test COST 4A, the ratio of horizontal and vertical loads was taken as α=0.16, while during test COST 4B, as α=1.0. Ultimate failure of the frames was caused by bolt failure in joints B12 and B16.

The relationship of vertical load and top level horizontal displacement is shown in Figure 51 for test COST 4A. Significant differences between the displacements of cross-sections C4 and B11 are caused by a plastic hinge (Figure 52). Behaviour of column base is shown in Figure 53.

### 4.4 Comparison of test results to theoretical calculations

**Modelling of Beam-to-Column Connections and Column Bases.** The *moment–rotation characteristics of beam-to-column connections* were calculated according to Eurocode 3 Annex J, using a computer program developed by Katula L. (Horváth and Katula 1993).

The layout of the connections was given in Figure 41. The general shape of moment–rotation characteristics that were used in the analytical study is shown in Figure 54a. In general, three types of connections were applied for the tested frames denoted by B11, B13 and B14 in Figure 41.

The moment–rotation curves can be characterized by their two points, the first associated with the 2/3 of the moment resistance of the connection $M_{Rd}$ and the initial rotational stiffness $S_{j,ini}$, and the second, with the full value of the moment resistance $M_{Rd}$ and a rotational stiffness 1/3 of $S_{j,ini}$. The first part of the curve is linear; the second one is described by a power function of the bending moment as given in Eqn. 52,

$$S_j = S_{j,ini} \cdot \left[ \frac{M_{Rd}}{1.5 \cdot M} \right]^{2.7} \qquad (52)$$

and the third part is constant at value $M=M_{Rd}$.

The connections B12 and B13 (see Figure 41) were treated as independent of each other; their characteristic curves were simply determined using transformation parameter β (expressing the influence of the common web panel in shear on the two connections) equals to 1, as recommended in EC 3.

The moment–rotation characteristics of column bases were incorporated using those determined on the basis of a parametric study by Iványi and Balogh (1995). The study involved FEM simulations using COSMOS/M software. The results for elasto-plastic constitutive law were applied. The actual curve can be seen in Figure 54b (darker line), along with a curve used in design practice.

**Effect of Proportional Loading.** In the case of test frame COST 2, with the appropriate application of lateral supports, the failure of the structure was forced to occur via developing plastic hinges at the location of column bases and beam-to-column connections, and under the concentrated loads.

The analysis was performed using the computer program PEP-Micro (developed by CTICM, France). The calculations were based on second order plastic analysis. The results are shown in

Figure 45, as theoretical ones. As for the ultimate behaviour, the results show that, the calculations were close enough to the test results considering both ultimate load and displacements. However, in the elastic range, the calculations predicted "more rigid" behaviour compared to the experimental results. On the contrary, in the ultimate range, calculations predicted "softer" behaviour.

The calculated load bearing capacity of test frame COST 2 was P=60.2kN; its load bearing capacity in tests as P=61.5kN. The load bearing capacity according to the first order theory is P=69.2 kN, which shows that the second order effects are significant.

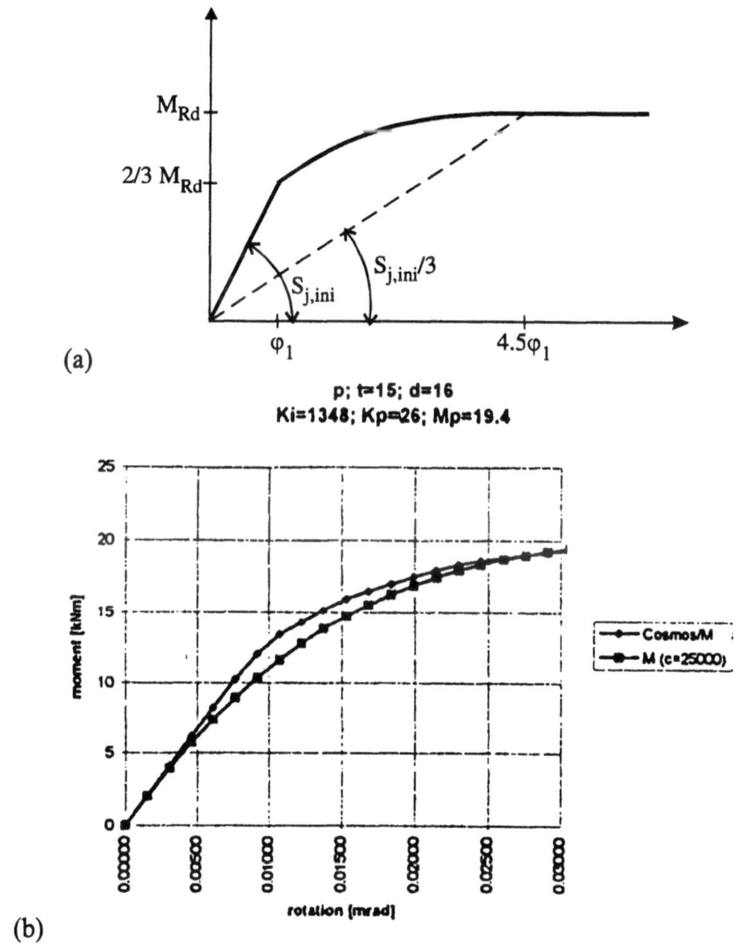


**Figure 54.** Moment – rotation curves.
(a) for bolted end-plate connections according to Eurocode 3 and (b) for column bases

**Effect of Variable Loading.** The analysis performed was based on the upper bound method for shakedown analysis (Kaliszky 1989). The method assumes that the plastic mechanism that corresponds to the shakedown state is preliminarily known. In this state, positive plastic hinge rotations of magnitude $|\vartheta_i^+|$ can develop only when the bending moment reaches the positive ultimate moment $M_{ti}$. Similarly, if at a plastic hinge location, negative rotation of magnitude $-|\vartheta_i^-|$ occurs, then the bending moment at the same location must reach its negative ultimate value $-M_{ti}$. The general condition of shakedown can then be written for the ultimate state by Eqns. 53.

$$m_b \cdot M_{res,i}^{\max} + M_i^s = M_{ti} \quad \text{and} \quad m_b \cdot M_{res,i}^{\min} + M_i^s = -M_{ti} \tag{53}$$

In these equations, $m_b$ denote the shakedown load parameter,

$$M_{res,i}^{\max} \text{ and } M_{res,i}^{\min},$$

the respective values of the maximal bending moment diagram, $M_i^s$, a value of bending moment corresponding to an equilibrium set, while $M_{ti}$ denotes the ultimate bending moment of the cross-section.

Multiplying the above equations (53) with the rotations $|\vartheta_i^+|$ and $-|\vartheta_i^-|$ respectively, considering all plastic hinges of the given plastic mechanism, and summing up the respective equations, the relationship in Eqn. 54 will be obtained.

$$m_b \cdot (\Sigma M_{res,i}^{\max} \cdot |\vartheta_i^+| - \Sigma M_{res,i}^{\min} \cdot |\vartheta_i^-|) + \Sigma M_i^s \cdot \vartheta_i = \Sigma M_{ti} \cdot |\vartheta_i| \tag{54}$$

In Eqn. (54) the first and second summation between the brackets involve all the plastic hinges where the considered hinge rotation is positive and negative, respectively. $\vartheta_i$ denote all plastic hinge rotations, thus the two other summations in the above equation involve all the considered plastic hinge locations. The right hand side of the equation gives the work of ultimate bending moments, being thus each of its members positive. The last member of the sum on the left hand side of the equation is the work of the residual moment distribution $M_i^s$ upon the plastic mechanism $\vartheta_i$. Since $M_i^s$ is an equilibrium set, the principle of virtual work gives:

$$\Sigma M_i^s \cdot \vartheta_i = 0\,. \tag{55}$$

If the plastic mechanism which corresponds to shakedown state is known, load multiplier $mp$ can be expressed from the above equation as

$$m_b = \frac{\Sigma M_{ti} \cdot |\vartheta_i|}{\Sigma M_{res,i}^{\max} \cdot |\vartheta_i^+| - \Sigma M_{res,i}^{\min} \cdot |\vartheta_i^-|} \tag{56}$$

For the case when the plastic mechanism at shakedown is not preliminarily known, it can be justified that the above load multiplier $m_b^k$ calculated with an arbitrary plastic mechanism

$$\overline{\vartheta}_i$$

is always greater than or equal to the load multiplier $m_b$. This yields that the load multiplier $m_b$ can be determined as the minimum of values $m_b^k$ corresponding to all possible plastic mechanisms.

This analysis was performed based on the plastic mechanism observed in the experiments. Thus plastic hinges were either "conventional" plastic hinges (occurring at concentrated force locations), or hinges developing at beam-to-column connections or column bases.

For test frame COST 3B, the calculated load bearing capacity (shakedown load) is P=62.4kN, while the test showed P=65kN. For test frame COST 4B, the calculated load bearing capacity (shakedown load) is P=68.3kN, while the test showed P=70.2kN.

## 4.5 Conclusions

An extensive experimental study has been carried out to analyse the effects of semi-rigid connections. Four full-scale, three-dimensional multi-storey frames have been tested. A condensed overview of the features of the experimental set-up has been given and it has been briefly explained how the complexities of the 3D nature of the tests were addressed. Comparison of theoretical and experimental results showed reasonable agreement.

# 5 Use of approximate engineering methods

by M. Iványi[1] and L. Hegedűs[1]

## 5.1 Introduction

The purpose of frame analysis is to determine the limits of structural usefulness and to compare the predicted behaviour with the required one. Such an analysis is always part of a design process, wherein adjustments and new analyses are carried out until the design requirements are met.

Frame behaviour could be best characterised by the relationship between the applied loading and characteristic deformations, the structural response.

To construct load-deflection curves for each structure would be desirable, because it would contain all the information that is necessary for checking the structural behaviour. But such curves are more or less non-linear from the very beginning because of second order geometrical effects and material non-linearities. Later the slope of the curve is further reduced because of local plastification or of some instability phenomena.

The situation becomes more complicated, if structures contain such joints, which are non-negligibly different from the ideal ones, as hinged and rigid.

In case of frames connections have special importance, as their load carrying and deformation characteristics can modify the overall behaviour of the frame in a very wide range. Between the theoretically clear, two extreme cases, as hinged and rigid, the palette of practically used connections is very wide and allowance should be taken for it during the design process.

In case of semi-rigid joints the importance of using approximate engineering methods is increasing, as they can produce upper bound envelope curves for the real behaviour in a relatively simple way. The accuracy of the results is depending on the simplifications involved.

## 5.2 Modelling joints in frames

In Annex JJ of Eurocode 3 joints are built up from two main parts, as sheared column web and components of connection (endplate, bolt, etc.) (Eurocode, 1989, 1992, 1993). In certain cases it would be evident two divide the flexibilities into two springs, that is for a web panel spring and a connection spring (Figure 55).

The web panel spring would have the task to take into account the shear deformation of the web panel, this way it could be a translational spring, as it has been treated in several papers before. This translational spring could be taken into the diagonal line connecting the tension flanges of the beam and column.

The connection spring (including all other flexibilities) could be represented typically by a rotational spring, which could be situated to the end section of the beam.

Calculation results have showed, that using these two kinds of spring models for characterising the behaviour of the joint have advantages.

[1] Department of Steel Structures, Budapest University of Technology and Economics, Hungary

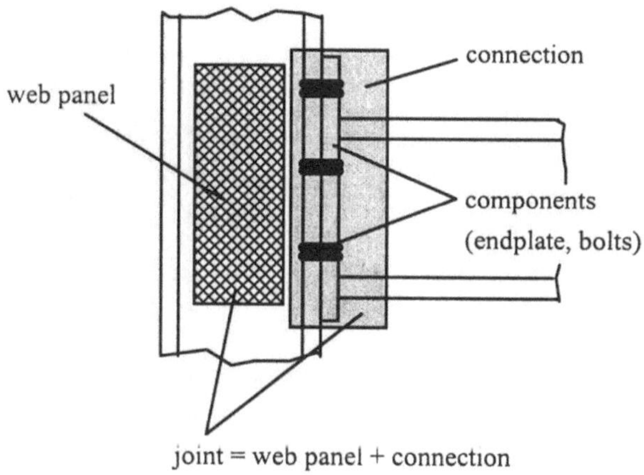


**Figure 55.** Joint model of Annex JJ.

These two springs are more suitable to simulate the joint characteristics and, which can be more important, they produce lower load carrying capacities and lower elastic critical loads for the numerically tested frame, than the calculation, which strictly follows the connection model of Eurocode 3.

Taking into account the second order effects and using these separated springs produce a relatively complicated calculation. Later on another possibility for handling these influences will be summarised in this subdivision.

## 5.3 Influences at column bases

In case of column bases there are some similarities. The typically bolted connection between the base plate and the foundation can have such flexibilities, which are comparable with those of the bolted beam-to-column connections.

If some attention is paid for the "connection" between the soil and the foundation, rather similar displacements to those of the frame joints can be found.

The position of the rotation centre of the concrete block is depending on the soil characteristics. Between this point and the lower surface of the base plate the relatively rigid concrete block is situated, which acts as a continuation of the column. But between the baseplate and the top surface of the concrete block in most practical cases a grout is situated, having different behaviour as the concrete foundation block (Figure 56).

These similarities and differences gave the idea to develop a generalised bar model, which can take "connection" and "joint" flexibilities into account.

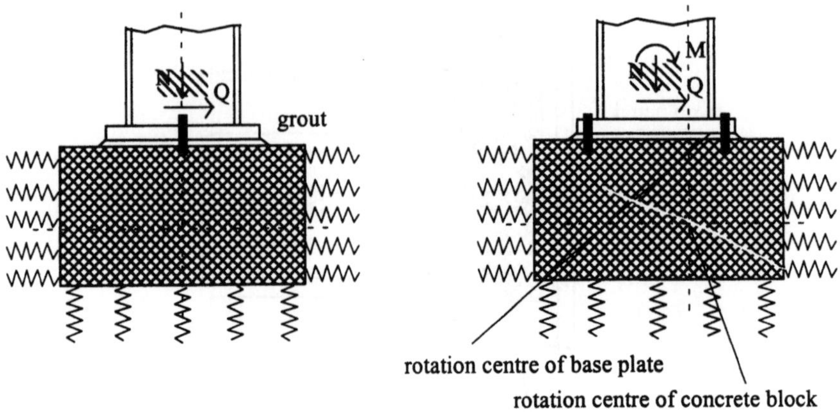


**Figure 56.** Column bases.

### 5.4 Use of stability functions

Stability functions can be practical tools to carry out an elastic second order analysis. These functions have been developed in different forms. In this paper those of Horne will be used (Horne and Merchant, 1965) (Horne and Morris, 1981) (Majid, 1972).

For the elementary case, i. e. when a member having fixed ends is influenced to a rotation on the left hand side end (Figure 57.), the end forces and moments can be expressed in the form given under the figure.

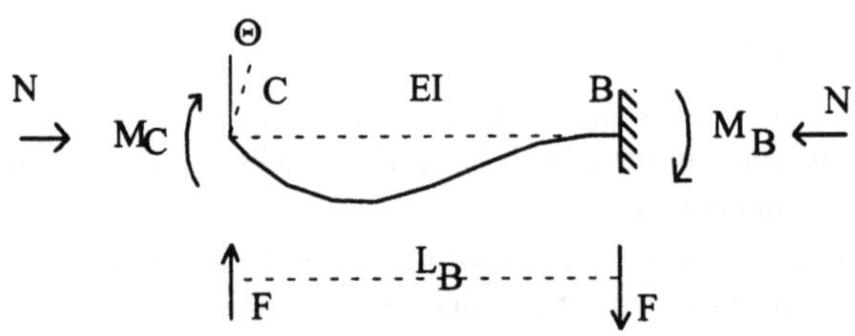


$$M_A = s.k.\Theta = S.\Theta; \quad M_B = s.c.k.\Theta = T.\Theta; \quad F = -s(1+c).k.\Theta/L = -\,U.\Theta\,/L$$

**Figure 57.** Stability functions: rotation at left end.

From this case all of the others (sway of fixed end: Figure 58., hinged cases, etc.), when there is no distributed load along the length of the member, can be developed with superposition, supposing the same normal force for the added cases.

The $s$ (stiffness) and $c$ (carry-over) functions are depending on the relative intensity of the normal force, $\rho = N/N_E$, where $N_E = \pi^2 k/L$ (and $N$ is positive in compression).

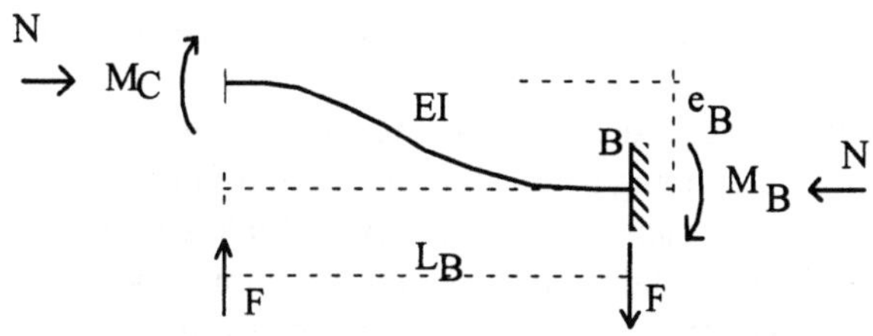


$$M_A = M_B = -u.k.e_B/L = -\,U.e_B/L; \quad F = (2.u-\pi^2\rho).k.e_B/L^2 = V.e_B/L^2$$

**Figure 58.** Stability functions: sway at right end.

These functions let to take into account the influence of shear deformations of the web panel. If we consider that the left end of the member is connected to the joint of a frame, at the joint there is a $\gamma$ shear deformation (Figure 59) and the centre of the panel remains in the same place, then the shear deformation $\gamma/2$ causes at the left end of the member: *(i)* a $\Theta$ rotation equal to $\gamma/2$ and *(ii)* an upraise vertical sway, which is equal to $L_A \bullet \gamma/2$.

Combining the cases in Figures 57 and 58 the influence of web shear deformation can be derived.

From the right part of Figure 59 it can be also clearly seen that shear deformation of the web panel causes opposite effects for beam and column respectively. This fact should be taken into account when compiling the stiffness matrix for the construction.

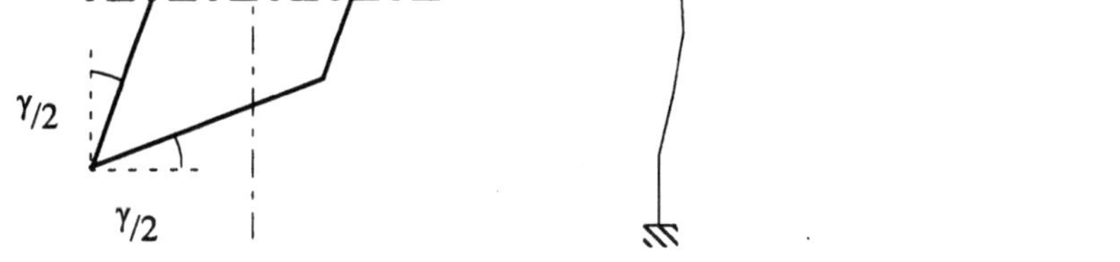


**Figure 59.** Shear deformation of web panel and its influence for deformation of members.

Before carrying out this combination process in details, because of its practical advantages, the bar model of Figures 57 and 58 is extended. A rigid part of the length of $L_A$ is joined to its left hand side, replacing the web panel between the theoretical centres of the joint (point *A*) and the beam ends (point *C*).

From simple equilibrium equations, written for *A-C*, *C-B* and *A-B* members, some new functions can be compiled, describing the relation among end forces and displacements.

These functions are given in Table 1. for fixed bar, in Table 2 for hinged bar.

If a relative rotation should be taken into account at section *C* between the column flange and the beam end, it can be done by adding a new degree of freedom there and using functions in Figure 57.

In case of displacement method the equilibrium condition can be written as

$$\left[\underline{\mathbf{q}}^*\right] = \left[\underline{\underline{\mathbf{K}}}^*\right]\left[\underline{\mathbf{u}}^*\right] \tag{57}$$

in which $\underline{\mathbf{q}}^*$ is the load vector, $\underline{\underline{\mathbf{K}}}^*$ is the stiffness matrix and $\underline{\mathbf{u}}^*$ is the displacement vector.

The displacement vector is built up from two main parts, from the group of regular nodal displacements (absolute rotations and deflections) and from the group of relative ones, caused by local flexibilities as shear deformations, connection components deformations, etc. According to the regular nodal displacements the corresponding equilibrium equations can be compiled. The second part of equilibrium equations are simply expressing the moment - relative rotation correlation at the local flexibilities, that is the $M$–$\phi$ functions.

It should be also taken into consideration that the relative rotations are always of opposite sense as the moments.

The functions for the shear or the rotation rigidity of a joint can be of any such shapes, which are in correlation (given by any mathematical function) with the change of loading.

The corresponding literature – for example (Chen, Fielding, 1972), (Chen, 1987), (Chen, Lui, 1991), (Ermopoulos, Vayas, 1991), (Fielding, Huang, 1971), (Iványi, 1992), (Lui, 1985), (Mazzolani, 1990) – illustrate that wide palette, which can be compiled either from experimental results or from modelling the above functions in different ways.

The function of $M$–$\phi$, for example (Figure 60.), has a non-linear shape in case of loading, while in case of unloading it follows the straight line determined by the initial slope of the curve (by the initial rigidity) and the $M_0$ and $\phi_0$ values achieved.

In general form the $M$–$\phi$ functions can be written, as:

$$M = R \cdot \phi + M_1 \tag{58}$$

where in case of loading:

$$R = S_{\text{sec}}(\phi) \quad \text{and} \quad M_1 = 0\,;$$

in case of unloading:

$$R = S_{\text{sec,0}} \quad \text{and} \quad M_1 = M_0 - S_{\text{sec,0}} \cdot \phi_0\,;$$

If the external load on the structure is increasing in a monotonic way and there are no joint simulating springs, which are unloading, the actual secant stiffness, $M/\phi=S_{\text{sec}}(\phi)$ should be substituted into the stiffness matrix.

**Table 1.** Stability functions including shear deformation of web panel – fixed bar.

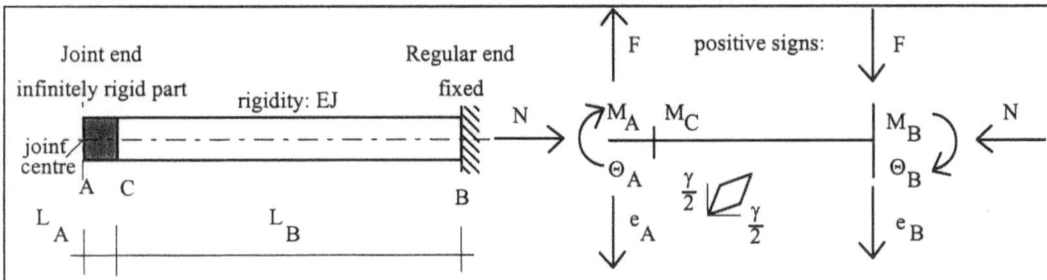


Basic notation: $s.k = S$ $\quad$ $s.(1+c).k = U$ $\quad$ $k = EJ / L_B$

$s.c.k = T$ $\quad$ $(2.s.(1+c) - \pi^2\rho).k = V$ $\quad$ $\rho = N.L_B / (\pi^2.k)$

| | $M_A$ | $M_B$ | $M_C$ | $F$ |
|---|---|---|---|---|
| $\Theta_A$ | $\left[S+\frac{L_A}{L_B}\left(1+\frac{L_A}{L_B}\right)V\right]=S_k$ | $\left[T+\frac{L_A}{L_B}U\right]=T_k$ | $\left[S+\frac{L_A}{L_B}U\right]=S_c$ | $-\left[U+\frac{L_A}{L_B}V\right]\frac{1}{L_B}=-U_k\frac{1}{L_B}$ |
| $\Theta_B$ | $\left[T+\frac{L_A}{L_B}U\right]=T_k$ | $S$ | $T$ | $-U\frac{1}{L_B}$ |
| $e_A$ | $\left[U+\frac{L_A}{L_B}V\right]\frac{1}{L_B}=U_k\frac{1}{L_B}$ | $U\frac{1}{L_B}$ | $U\frac{1}{L_B}$ | $-V\frac{1}{L_B^2}$ |
| $e_B$ | $-\left[U+\frac{L_A}{L_B}V\right]\frac{1}{L_B}=-U_k\frac{1}{L_B}$ | $-U\frac{1}{L_B}$ | $-U\frac{1}{L_B}$ | $V\frac{1}{L_B^2}$ |
| $\frac{\gamma}{2}$ | $\left[S-2U\left(\frac{L_A}{L_B}\right)^2+\pi^2\rho k\frac{L_A}{L_B}\left(1+\frac{L_A}{L_B}\right)\right]=S_\gamma$ | $\left[T-\frac{L_A}{L_B}U\right]=T_\gamma$ | $\left[S-\frac{L_A}{L_B}U\right]=S_{\gamma c}$ | $-\left[U-\frac{L_A}{L_B}V\right]\frac{1}{L_B}=-U_\gamma\frac{1}{L_B}$ |

**Table 2.** Stability functions including shear deformation of web panel – hinged bar.

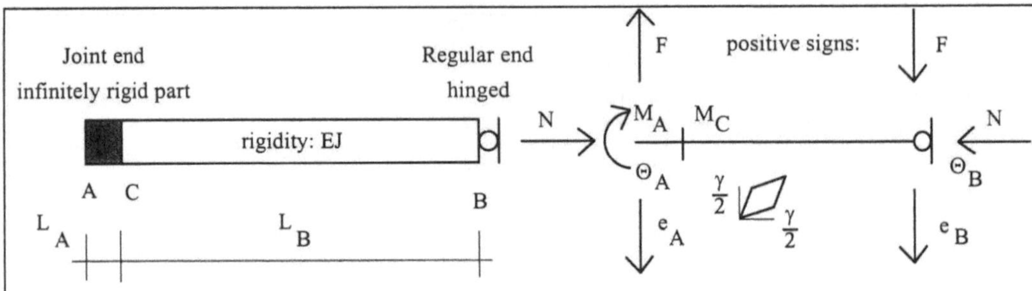


Basic notation: $s.k = S$ $s.c.k = T$ $s.(1+c).k = U$ $(2.s.(1+c) - \pi^2\rho).k = V$ $s.(1-c^2).k = S''$ $S'' - \pi^2\rho.k = V''$ $k = EJ / L_B$ $\rho = N.L_B / (\pi^2.k)$

| | $M_A$ | $M_C$ | $F$ |
|---|---|---|---|
| $\Theta_A$ | $\left[S_k - \frac{T_k^2}{S}\right] = S_k''$ | $\left[S_c - \frac{T}{S}T_k\right] = S_c''$ | $-\left[U_k - \frac{U}{S}T_k\right]\frac{1}{L_B} = -U_k''\frac{1}{L_B}$ |
| $e_A$ | $\left[U_k - \frac{U}{S}T_k\right]\frac{1}{L_B} = U_k''\frac{1}{L_B}$ | $U\left(1 - \frac{T}{S}\right)\frac{1}{L_B} = S''\frac{1}{L_B}$ | $-\left[V - \frac{U^2}{S}\right]\frac{1}{L_B^2} = -V''\frac{1}{L_B^2}$ |
| $e_B$ | $-\left[U_k - \frac{U}{S}T_k\right]\frac{1}{L_B} = -U_k''\frac{1}{L_B}$ | $-U\left(1 - \frac{T}{S}\right)\frac{1}{L_B} = -S''\frac{1}{L_B}$ | $\left[V - \frac{U^2}{S}\right]\frac{1}{L_B^2} = V''\frac{1}{L_B^2}$ |
| $\frac{\gamma}{2}$ | $\left[S_\gamma - \frac{T_k}{S}T_\gamma\right] = S_\gamma''$ | $\left[S_{\gamma c} - \frac{T}{S}T_\gamma\right] = S_{\gamma c}''$ | $-\left[U\left(1 - \frac{T_\gamma}{S}\right) - \frac{L_A}{L_B}V\right]\frac{1}{L_B} = -U_\gamma''\frac{1}{L_B}$ |

### 5.5 Extended use of stability functions and springs

This above detailed method is suitable to analyse elastic members, taking allowance for geometric non-linearities (second order effects). At the same time, as the spring rigidity functions theoretically can be of any shape, with this set of expressions those cases also can be treated, when the spring rigidities involve some plastic, hardening or softening phenomena.

If any of the rotational springs at member ends is handled this way, it can include even the plastic behaviour of that section. The function used for representing the moment-rotation relation can be either bi-linear or piecewise linear or continuous and preferably smooth.

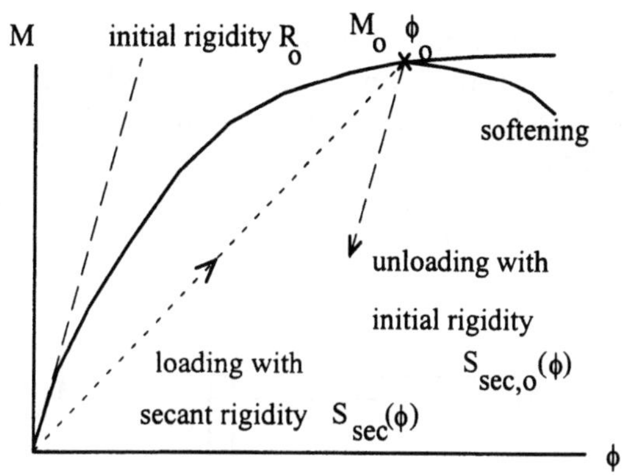


**Figure 60.** Spring characteristic.

If a spring is inserted in any section of a member, where a real connection does not exist and the spring characteristics are including the properties of a plastic hinge, this so-called pseudo connection section is able to model plastic hinges, as well.

This way the use the stability functions together with all of those possibilities, which were detailed before, can be extended to carry out an elastic - plastic hinge analysis. In this method all plastification is gathered into the plastic hinge and the complete member between two nearby hinges is supposed to be perfectly elastic.

The more exact plastic zone method, which took into consideration the reality, that plastification is a progressing process both in the fibres of a section and in a portion of a member, is more complicated and generally gives slightly different results.

In the case, when the frame to be analysed is carrying distributed loads, it is preferable to substitute it with concentrated ones, because those sections, in which plastic hinges develop in a member of distributed loading, are generally not known in advance.

This modification has two advantages:

- there is no change in shear force between two neighbouring points and
- possible sections of plastic hinges are only these division points of the members.

One very important remark, which should be taken into account in any case, is the following: either non-linear behaviour of connection or plastic behaviour of the material is the reason of the relative deformations, it should be always kept in mind, that loading (increase of deformations accompanied by either increase or decrease of load intensity) follows the mentioned curves, while unloading (decrease of deformation and loading together) is always elastic (follows the initial rigidity). Therefore, if at any increment of the loading process a decrease of a certain displacement is seen, this fact should be noticed and it must be allowed for the compilation of the stiffness matrix.

### 5.6 Demonstrative example

The frame in Figure 61 was realised from hot-rolled I sections (section height: 80mm, span of beam: 2m, height of column: 1m), having different welded beam-to-column connections, nominally pinned at supports and tested in the framework of a diploma project in the Laboratory of the Department of Steel Structures, BUTE (Iványi jr., 1993).

Three types of frame knees were tested: *(i)* no stiffeners, *(ii)* horizontal stiffeners, *(iii)* horizontal + diagonal stiffeners. The ratio of $H{:}V$=0.5:1.

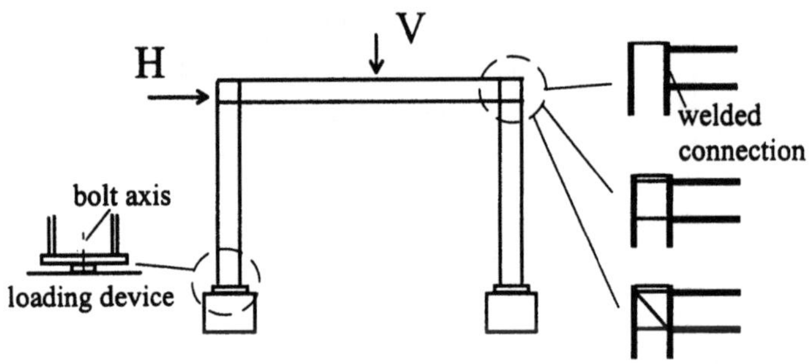


**Figure 61.** Test model built up from hot rolled sections.

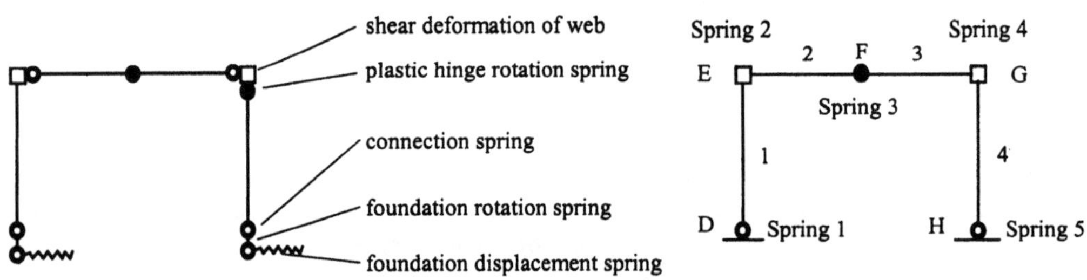


**Figure 62.** Frame model built up from springs and elastic members.

Left part of the Figure 62 shows those sections, where springs should be inserted to take account for local flexibilities, plastic hinge rotations, etc. Allowing for the real conditions, some simplifications were made, as:

1. a strong steel device was constructed to transfer the loading and to support the test frame – there were no foundation displacements;
2. flexibility of welded connection on the interface of beam-to-column is small – there is no need for connection springs at beam ends,
3. beam-to-column connections are not stronger than the sections themselves – no need for plastic hinge springs around the joints.

**Table 3.** Stiffness matrix and load vector for the model shown in Figure 62.

**Stiffness matrix**

| | 1 | 2 | 3 | 4 | 5 | 6 | 7 | 8 | 9 | 10 |
|---|---|---|---|---|---|---|---|---|---|---|
| 1 | $(S_k)_1$<br>$(S_k)_2$ | $(T_k)_2$ | 0 | $-\left(\frac{U_k}{L_B}\right)_1$ | $-\left(\frac{U_k}{L_B}\right)_2$ | $(T_k)_1$ | $-(S_\gamma)_1$<br>$(S_\gamma)_2$ | 0 | 0 | 0 |
| 2 | $(T_k)_2$ | $(S_k)_2$<br>$(S_k)_3$ | $(T_k)_3$ | 0 | $-\left(\frac{U}{L_B}\right)_2$<br>$\left(\frac{U}{L_B}\right)_3$ | 0 | $(T_\gamma)_2$ | $(S)_3$ | $(T_\gamma)_3$ | 0 |
| 3 | 0 | $(T_k)_3$ | $(S_k)_3$<br>$(S_k)_4$ | $-\left(\frac{U_k}{L_B}\right)_4$ | $\left(\frac{U_k}{L_B}\right)_3$ | 0 | 0 | $(T_k)_3$ | $(S_\gamma)_3$<br>$-(S_\gamma)_4$ | $(T_k)_4$ |
| 4 | $-\left(\frac{U_k}{L_B}\right)_1$ | 0 | $-\left(\frac{U_k}{L_B}\right)_4$ | $\left(\frac{V}{L_B^2}\right)_1$<br>$\left(\frac{V}{L_B^2}\right)_4$ | 0 | $-\left(\frac{U}{L_B}\right)_1$ | $\left(\frac{U_\gamma}{L_B}\right)_1$ | 0 | $\left(\frac{U_\gamma}{L_B}\right)_4$ | $-\left(\frac{U}{L_B}\right)_4$ |
| 5 | $-\left(\frac{U_k}{L_B}\right)_2$ | $-\left(\frac{U}{L_B}\right)_2$<br>$\left(\frac{U}{L_B}\right)_3$ | $\left(\frac{U_k}{L_B}\right)_3$ | 0 | $\left(\frac{V}{L_B^2}\right)_2$<br>$\left(\frac{V}{L_B^2}\right)_3$ | 0 | $-\left(\frac{U_\gamma}{L_B}\right)_2$ | $\left(\frac{U}{L_B}\right)_3$ | $\left(\frac{U_\gamma}{L_B}\right)_3$ | 0 |
| 6 | $(T_k)_1$ | 0 | 0 | $-\left(\frac{U}{L_B}\right)_1$ | 0 | $(S)_1$<br>$\text{Spring}_1$ | $-(T_\gamma)_1$ | 0 | 0 | 0 |
| 7 | $(S_k)_2$ | $(T_k)_2$ | 0 | 0 | $-\left(\frac{U_k}{L_B}\right)_2$ | 0 | $(S_\gamma)_2$<br>$\text{Spring}_2$ | 0 | 0 | 0 |
| 8 | 0 | $(S)_3$ | $(T_k)_3$ | 0 | $\left(\frac{U}{L_B}\right)_3$ | 0 | 0 | $(S)_3$<br>$\text{Spring}_3$ | $(T_\gamma)_3$ | 0 |
| 9 | 0 | $(T_k)_3$ | $(S_k)_3$ | 0 | $\left(\frac{U_k}{L_B}\right)_3$ | 0 | 0 | $(T_k)_3$ | $(S_\gamma)_3$<br>$\text{Spring}_4$ | 0 |
| 10 | 0 | 0 | $(T_k)_4$ | $-\left(\frac{U}{L_B}\right)_4$ | 0 | 0 | 0 | 0 | $-(T_\gamma)_4$ | $(S)_4$<br>$\text{Spring}_5$ |

**Load vector**

| | | | | | | | | | |
|---|---|---|---|---|---|---|---|---|---|
| 0 | 0 | 0 | $H$ | $V$ | 0 | 0 | 0 | 0 | 0 |

The simplified model of the frame is shown on the right of Figure 62. The model consists of four elastic bars (1-4), five nodes (D-H) and springs, Spring 2 and 4 for shear deformation, Spring 1 and 5 for nominally pinned supports, Spring 3 for plastic hinge under the vertical force. Axial deformation of bars is neglected.

Table 3 shows the stiffness matrix and load vector (In the table $Spring_1$–$Spring_5$ denote the actual secant stiffness of the springs). The first three columns of stiffness matrix correspond to absolute rotations of nodes *E-F-G*, the fourth to horizontal sway at *E-G*, the fifth to vertical deflection at *F*. Columns 6-10 are relating to spring rotations.

Lines 1-5 are giving the equilibrium equations for $M_E$, $M_F$, $M_G$, $H_E$, $V_F$, respectively.

Loading of the computational model can be increased step-by-step. Control of the loading process can be made by force or by displacement. Load control is suitable, while the structural model is stable, in the unstable stable control of the process is made by horizontal deflection at node E. In this latter case column 4 of the stiffness matrix and the load vector are exchanged and the system of equations is solved for a given horizontal deflection.

Iteration at a new loading step starts with the previous normal force distribution in the bars, the normal forces are recalculated from the new displacements, while the given tolerance in the compatibility of results is fulfilled.

Some illustrative outputs of the program are given in appendix. Systematic comparison of computer simulation and experimental results are to be published.

## 5.7 Summary

This subdivision gave a short summary about the possibilities of constructing more precise, but not too difficult connection models for frame analysis.

It has been dealing with those simple methods, which can be useful tools for

- either to construct envelop curves for the elastic and for the plastic behaviour separately,
- or to determine the elastic-plastic response of the frame by concentrating all of non-linearities into real or pseudo connection springs, which are connecting the elastic members.

## 5.8 Appendix: Parametric study on the influence of semi-rigid frame knees and support conditions

The relative moment–relative rotation curves, according to Eurocode 3, of frame knees "KS1" and "KS2" (Figure 63) are shown in Figure 64. "KS1" is the non-stiffened version, which is really semi-rigid. "KS2", containing horizontal stiffeners at flange levels, has a moment-rotation curve just beneath the limit curve in EC3 for sway frames.

The horizontally and diagonally stiffened "KS3" is a rigid and full strength joint in EC3.

The column base was structurally solved as pinned. Neither calculations nor measurements have been made to check the reality of this supposition.

Separated tests have been carried out on knee models (Figure 65.).

During the experimental work special emphasis was due to the measurement of relative rotation of joints. The absolute rotation of sections *A* and *B* in the figure were measured by means of optical methods. From the difference of the absolute rotations at *A* and *B*, the relative rotation of joint has been calculated. This way the moment-rotation curves for the joints can be constructed.

Similar measurements have been carried out during the tests on the frames themselves. In those cases the load-rotation diagrams may be determined.

From the numerical point of view there are differences among the $M$–$\phi$ curve properties, if those by EC 3, by separated knee models and by frame measurements are compared.

Therefore, more serious and systematic analysis is necessary (Figure 66).

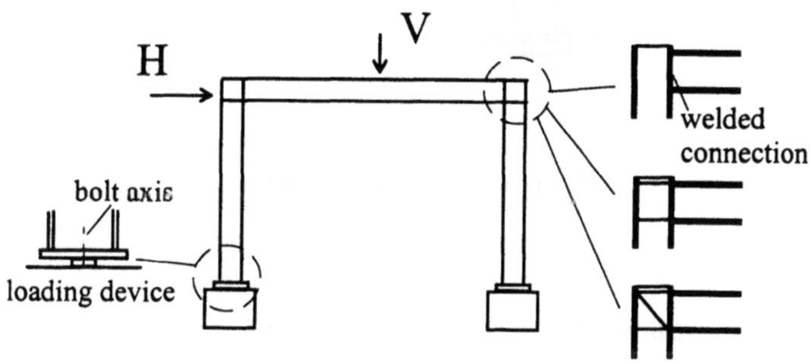


**Figure 63.** Experimental frame model.

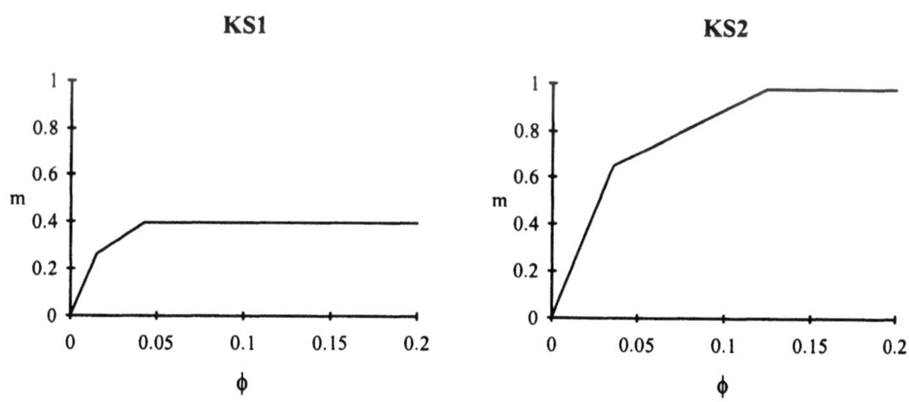


**Figure 64.** Moment – rotation curves by EC 3.

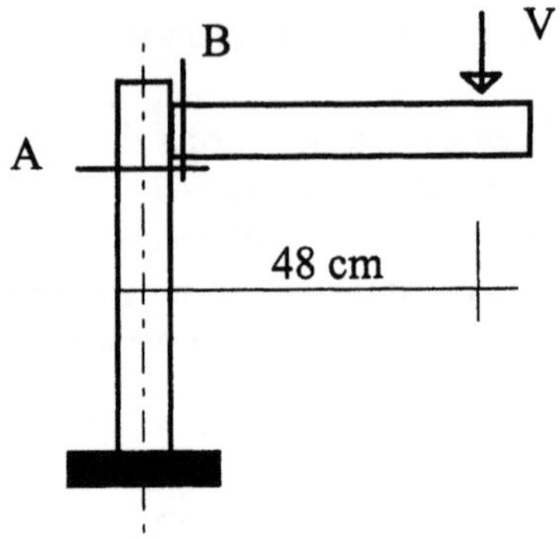


**Figure 65.** Frame knee models.

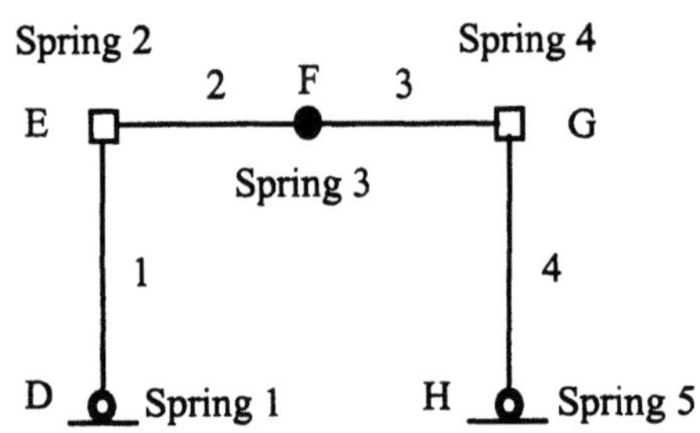


**Figure 66.** Frame model built up from springs and elastic members.

For illustration purposes the **spring properties** are supposed in the parametric study, as follows:

**Supports** (spring 1 and spring 5):

HINGED
- $S_{sec,0}$=450 kNm $M_{max}=M_{pl}/100$=0.069 kNm $\phi(M_{max}) = 0.0050$

SEMI-RIGID1
- $S_{sec,0}$=650 kNm $M_{max}=M_{pl}/10$=0.69 kNm $\phi(M_{max}) = 0.0050$

SEMI-RIGID2
- $S_{sec,0}$=850 kNm $M_{max}=M_{pl}/3$=2.3 kNm $\phi(M_{max}) = 0.0050$

SEMI-RIGID3
- $S_{sec,0}$=1050 kNm $M_{max}=2M_{pl}/3$=4.6 kNm $\phi(M_{max}) = 0.0075$

RIGID

- $S_{sec,0}$=1250 kNm  $M_{max}$=$M_{pl}$=6.9 kNm  $\phi(M_{max}) = 0.0100$

**Beam midspan** (spring 3): calculated moment capacity $M_{pl}$=6.9 kNm
plastic hinge properties are approximated as,

- $S_{sec,0}$=3000 kNm  $M_{max}$=$M_{pl}$/3=6.9 kNm  $\phi(M_{max}) = 0.0040$

**Frame knees** (spring 2 and spring 4)

/1

- $S_{sec,0}$=850 kNm  $M_{max}$=$M_{pl}$/3=2.3 kNm  $\phi(M_{max}) = 0.0050$

/2

- $S_{sec,0}$=1050 kNm  $M_{max}$=2$M_{pl}$/3=4.6 kNm  $\phi(M_{max}) = 0.0075$

/3

- $S_{sec,0}$=1250 kNm  $M_{max}$=$M_{pl}$/3=6.9 kNm  $\phi(M_{max}) = 0.0100$

In the present study the $M$–$\phi$ curve above $\phi(M_{max})$ is constant, beneath this value the curve is approximated as:

$$M = S_{\sec}.\phi = S_{\sec,o}.(1-a.\phi^n)\ ; \tag{59}$$

$$\frac{S_{\sec}(M_{\max})}{S_{\sec,o}} = \frac{n}{(n+1)}\ ; \tag{60}$$

$$a = \frac{1}{(n+1)\cdot\phi\cdot(M_{\max})^n}\ . \tag{61}$$

If the above expressions remain valid above $\phi(M_{max})$, instead of unlimited plastic behaviour the influence of descending moment-rotation characteristics can be analysed.

According to the second expression, choosing a value for $n$, the ratio of secant stiffnesses can be suitably changed.

Results of parametric study are summarized on Graphs 1-7. as follows.

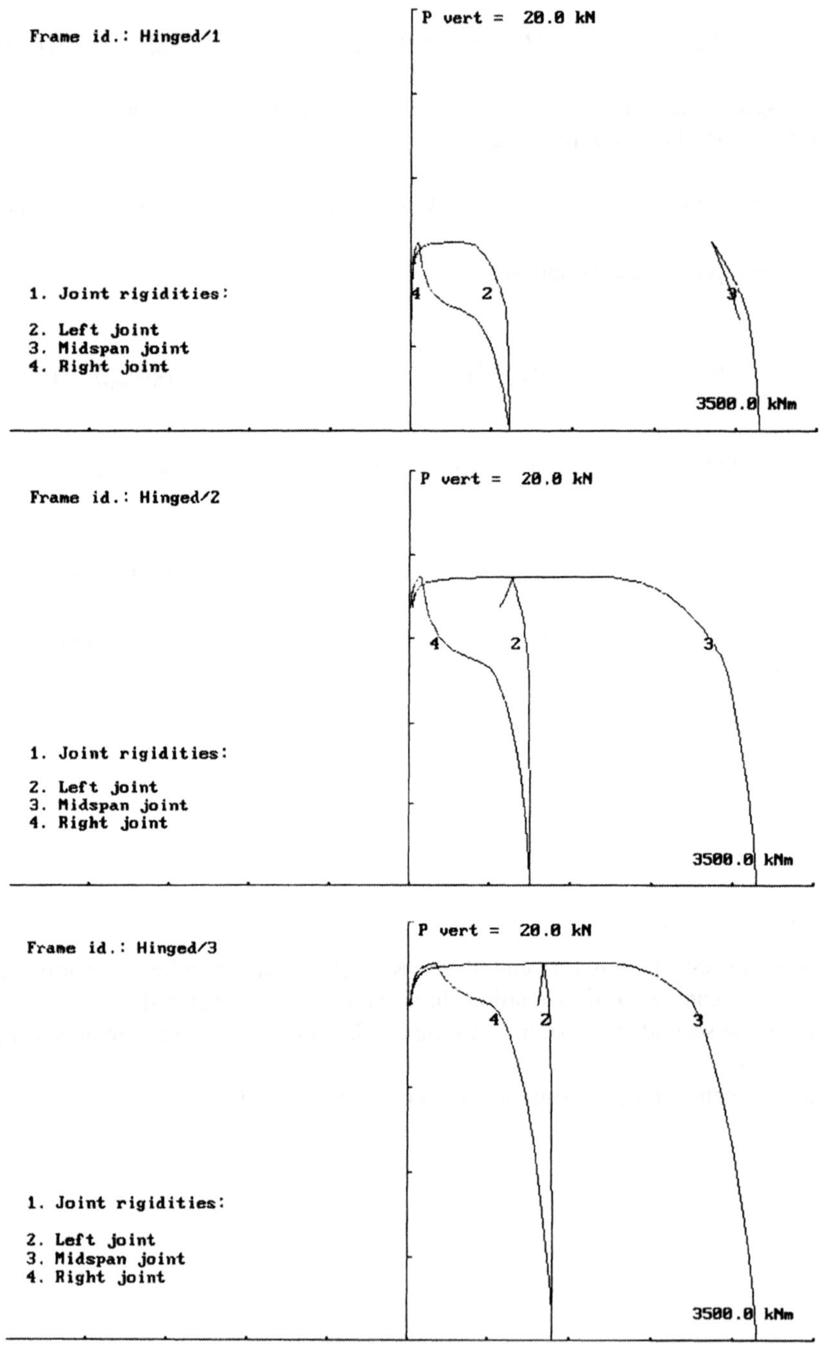


**Graph 1.** Vertical load and joint rigidity curves.

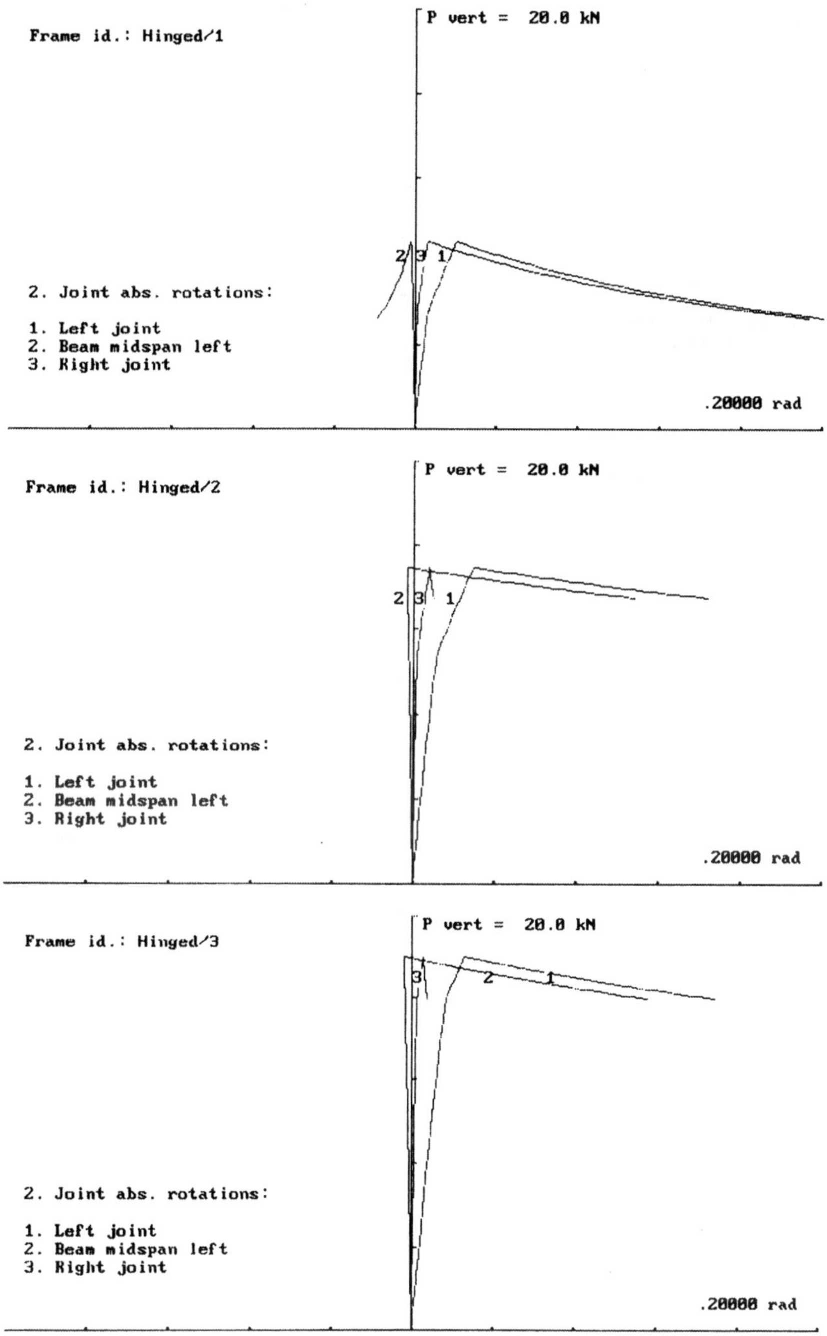


**Graph 2.** Vertical load and joint absolute rotation curves.

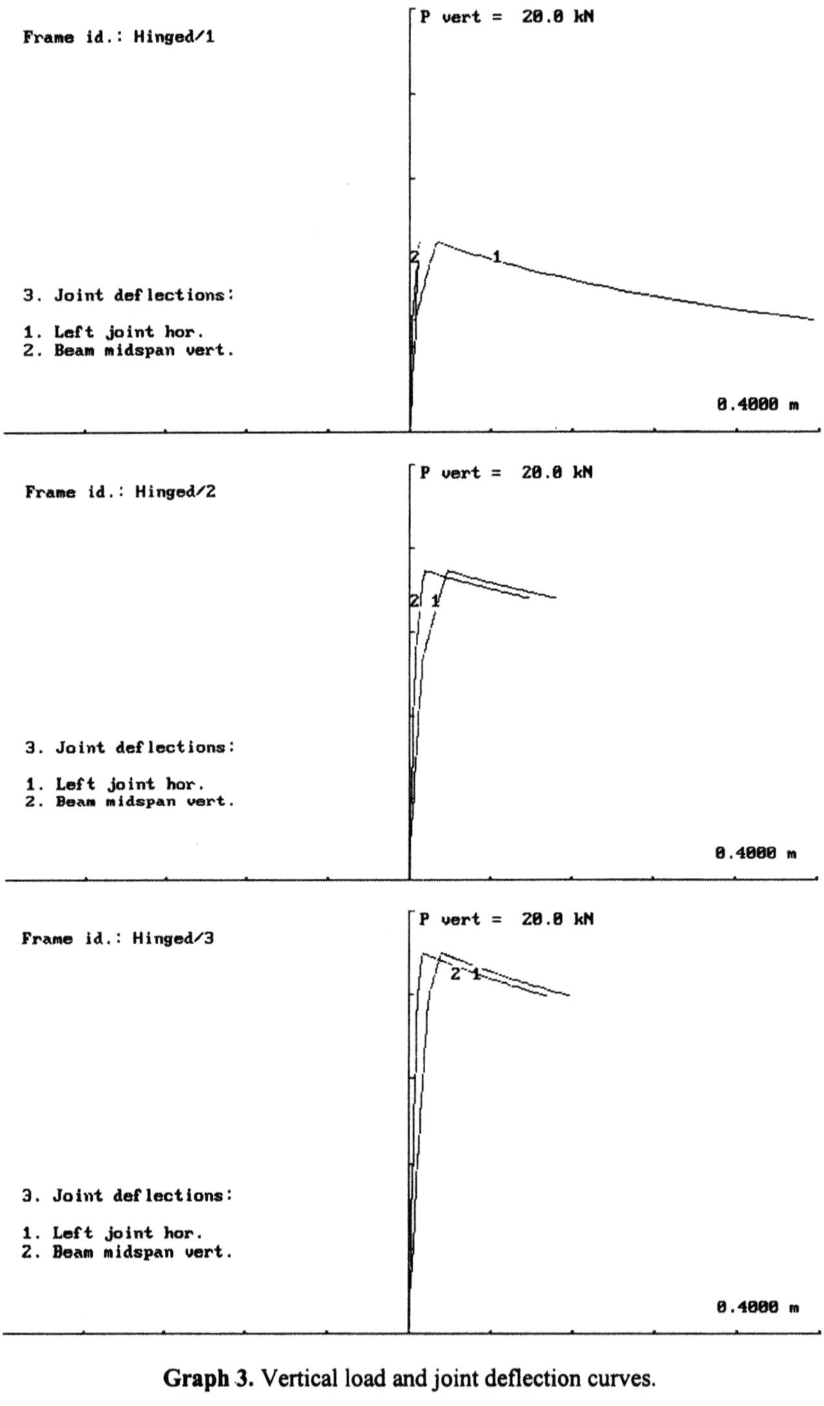


**Graph 3.** Vertical load and joint deflection curves.

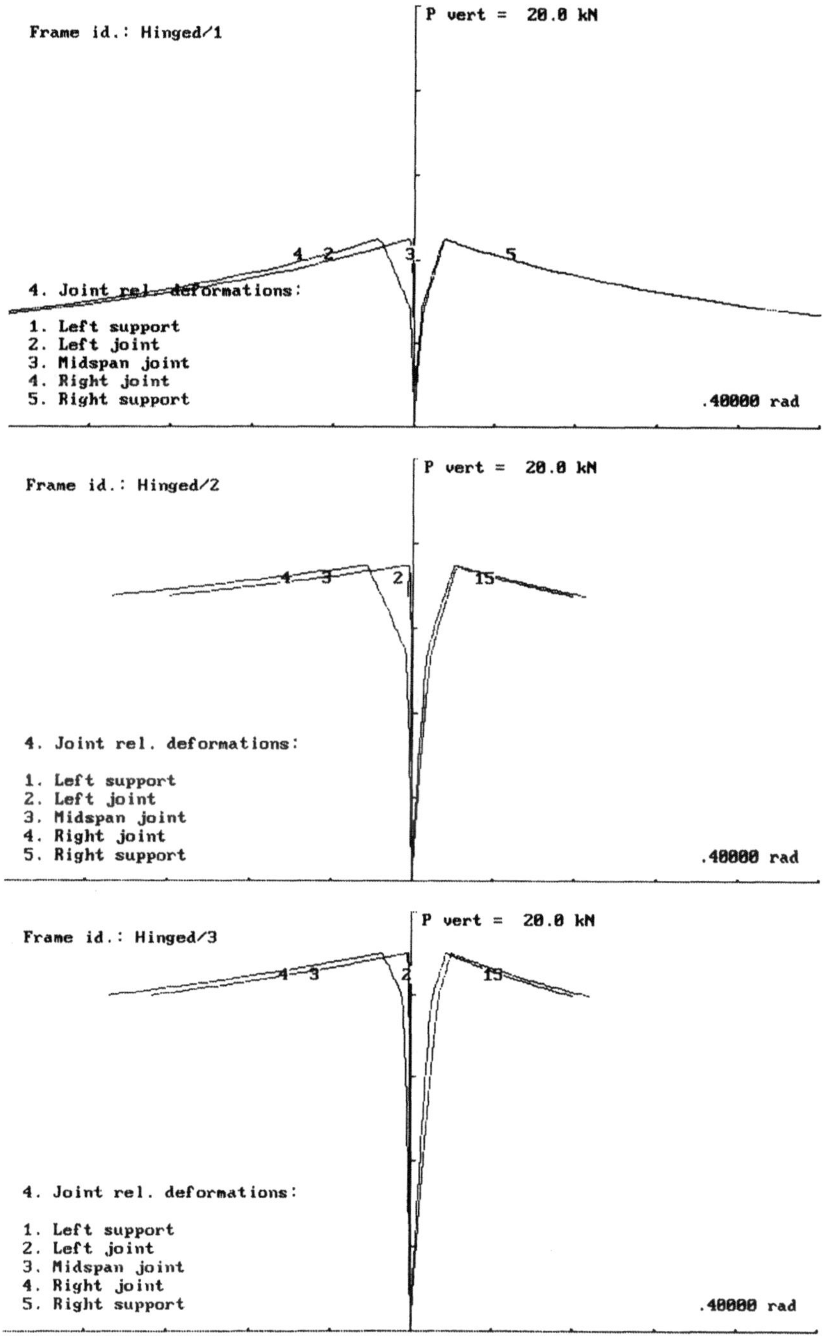


**Graph 4.** Vertical load and joint relative deformation curves.

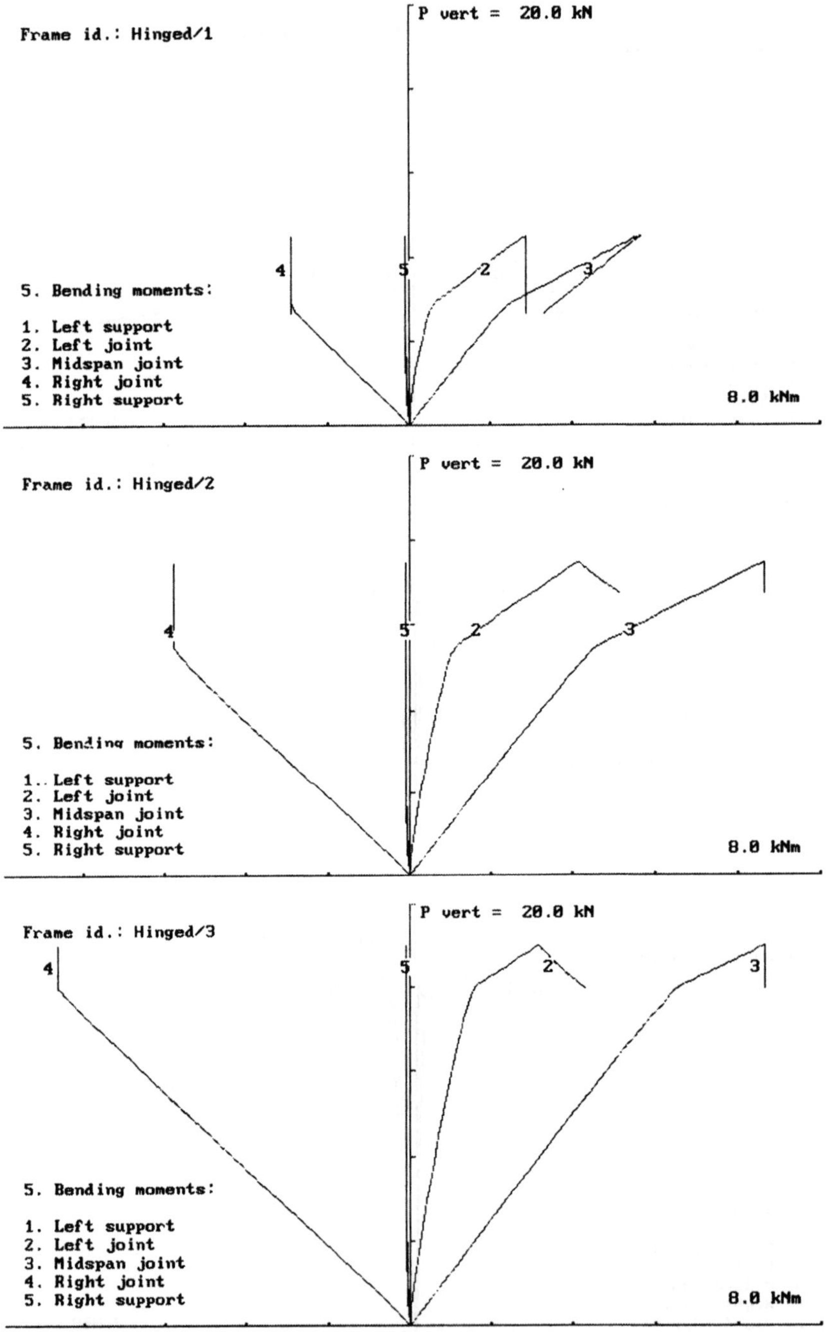


**Graph 5.** Vertical load and bending moment curves.

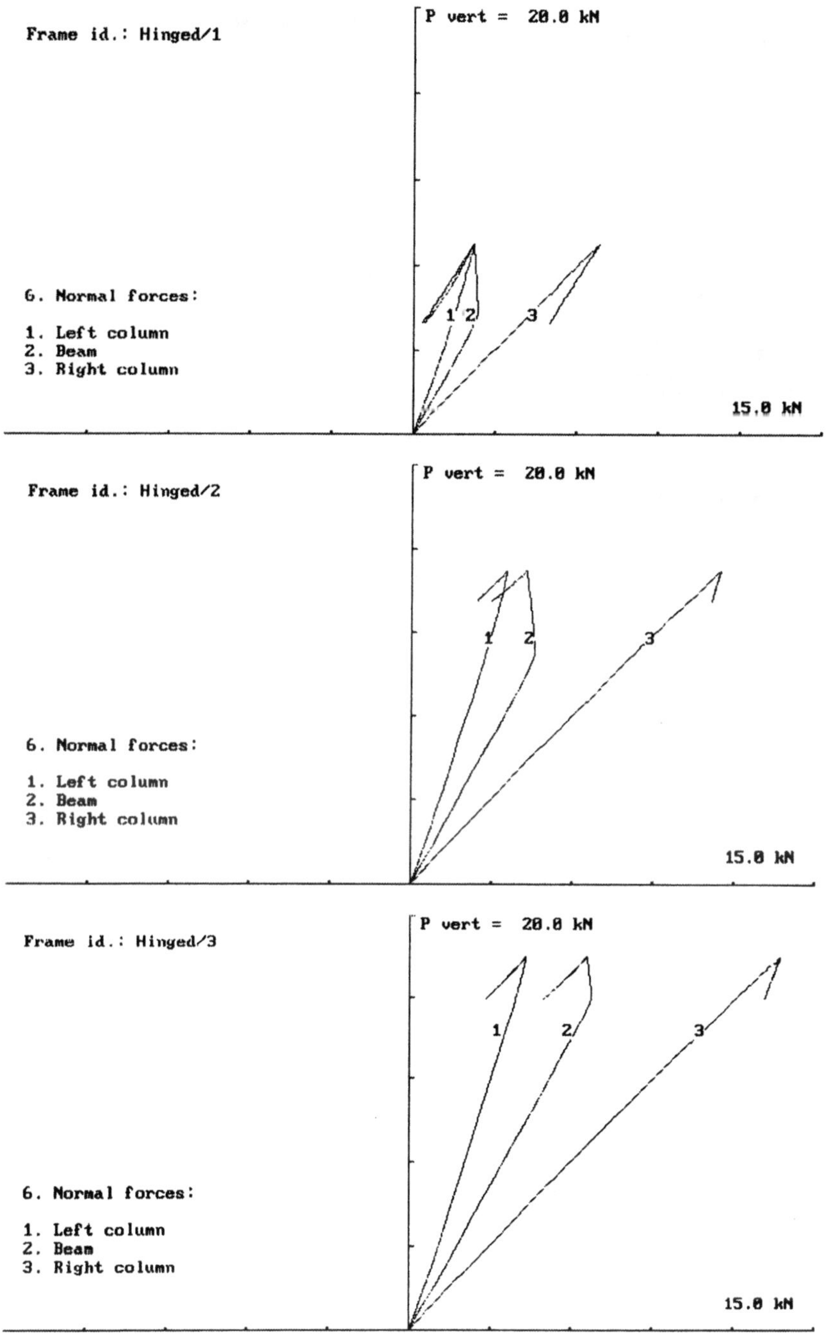


**Graph 6.** Vertical load and normal force curves.

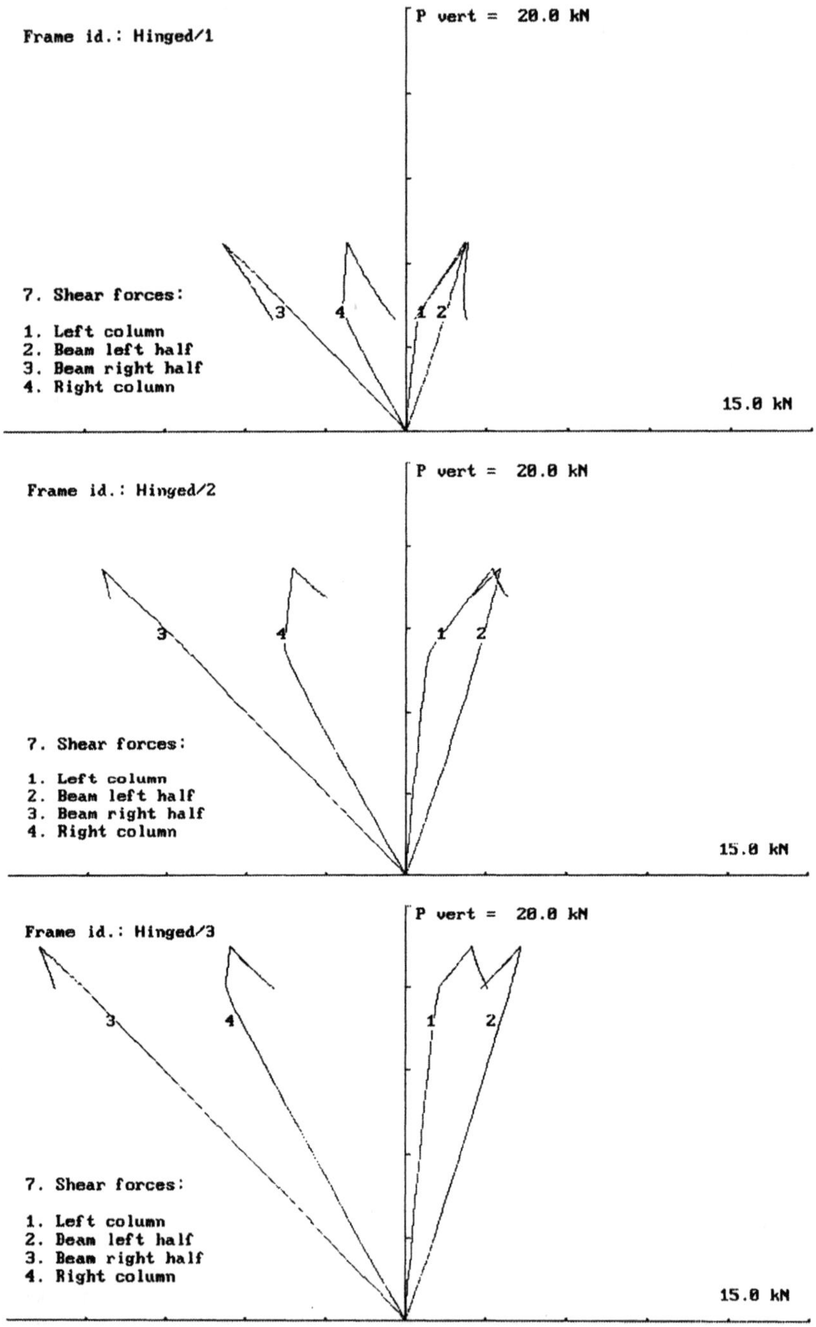


**Graph 7.** Vertical load and shear force curves.

# 6 Direct Design Method of Steel Frames with Semi-rigid Connections

## 6.1 Introduction

The most important parts in steel frames are the beam-to-column connections and the connections between columns and foundation, since their behaviour greatly influences the whole structural behaviour (distribution of moments and forces, displacements, overall stability, etc.).

## 6.2 Behaviour of beam-to-column connections and column bases

The limit analysis of steel frames starts from an idealistic elasto-platic stress-strain diagram, from this, between bending moment $M$ acting on an element of the beam, and the rotation $\phi$ due to this latter, on the basis of the hypothesis of Bernoulli and Navier, the relation shown in Figure 67 can be obtained; at the moment $M_y$ yielding starts in the extreme fibre extending more and more across the cross section, while at the ultimate moment $M_{pl}$, the whole cross section reaches the plastic stage, and the rotation increases infinitely. The concept of the classical plastic hinge theory was first introduced by Kazinczy (1914).

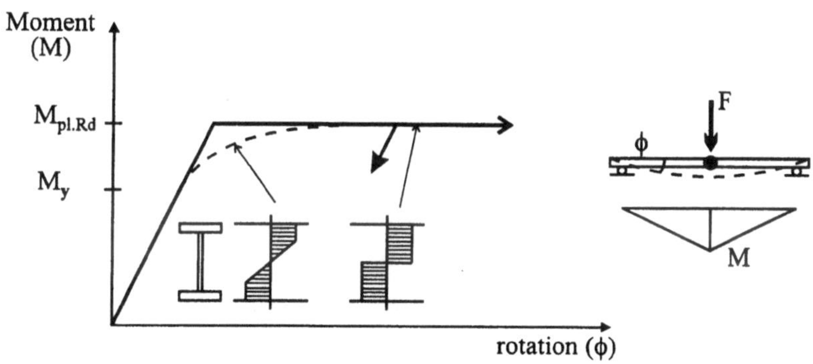


**Figure 67.** The classical plastic hinge.

The behaviour of beam-to-column connections and column bases is similar to that of the classical plastic hinge of the steel cross section (Figure 68).

Joining into the international research in this field, we tried to introduce the concept of "generalized plastic hinge", which can substitute the classical concept of plastic hinge in the traditional methods of limit design, but can reflect the effects of phenomena like concrete foundation, local plate buckling, anchorage behaviour, etc. (Iványi, 1983).

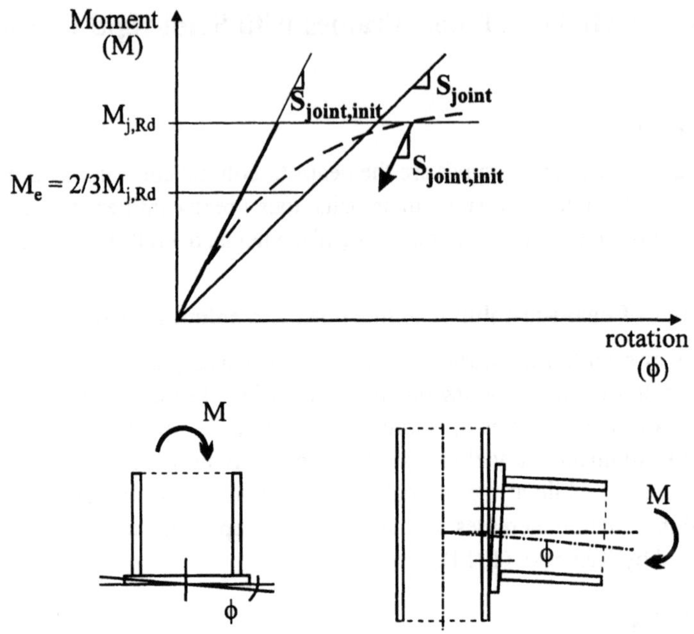


**Figure 68.** The beam-to-column connection and the column base.

## 6.3 Effect of beam-to-column connections and column bases on the behaviour of steel frames

The analysis of the behaviour of elastic-plastic frames neglecting the change in geometry of the structures while setting up the equations of equilibrium is referred to as *modified first order approach.* The steel frame has normal plastic hinges, semi-rigid connections and column bases at the respective locations. The load-deflection diagram of the frame in Figure 69 according to the first order approach $F_0^{(I)}$ can be expressed as

$$F_0^{(I)} = f_I(M_{pl}; M_{Rd})$$

with $M_{pl.Rd}$ the plastic moment of the cross section,
$M_{j.Rd}$ the moment resistance of the semi-rigid connection or column base.

If only failure load $F_0^{(I)}$ is of interest, then the detailed analysis of the structure behaviour can be omitted as the fundamental (static and kinematic) theorems of simple plastic design directly yield the $F_0^{(I)}$ value (Prager, 1951).

The fundamental theorems of simple plastic design can be utilised for two purposes, namely, *(i)* to check the failure load of a given (previously designed) structure or *(ii)* – if the value $F_0^{(I)}$ is given – to compute the required value of the generalised plastic moment $M_{pl}$ of the cross-sections. This latter will be referred to as "direct method of design", illustrated in Figure 69c–e. Based on previous consideration an adequate yield mechanism (pattern of plastic hinges) is to be chosen

(Figure 69e). Denoting the displacements of the external forces in the generalised yield mechanism by $u_i$ and the hinge rotations by $\kappa_j$ (Figure 69d), the virtual work equation furnishes

$$\sum_i \alpha_i F_0^{(I)} u_i = \sum_j \left| M_{pl} \kappa_j \right|. \tag{62}$$

Supposing all the plastic hinges to form under a common value $M_{pl}$ of frame element moment, the required value of $M_{pl}$ will be:

$$M_{pl} = F_0^{(I)} \cdot \frac{\sum_i \alpha_i u_i}{\sum_j \left| \kappa_j \right|}. \tag{63}$$

Subsequently – using the equilibrium equations – the entire moment diagram can be determined (Figure 69e) and the structure will be safe if designed so that the bending moments due to the former moment diagram nowhere exceed the full plastic moment of the corresponding joints and cross-sections.

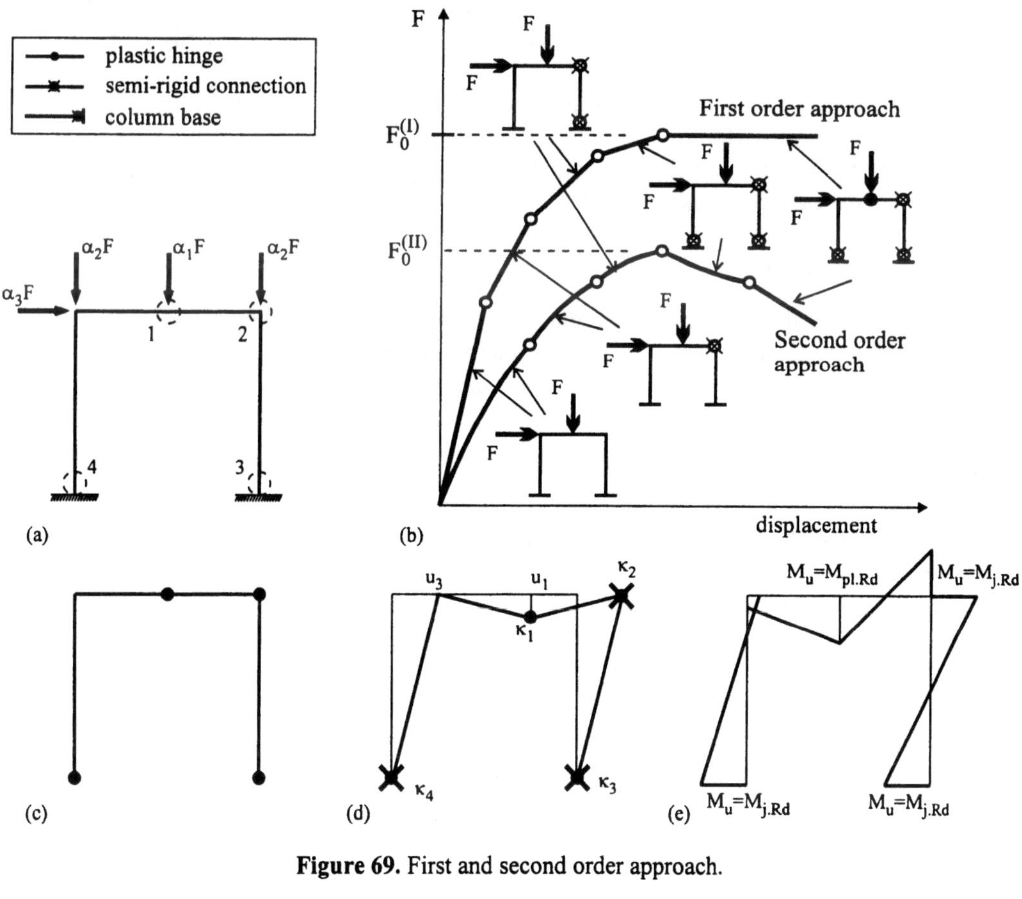


**Figure 69.** First and second order approach.

The basic assumptions in Simple Plastic Design restrict their use to cases where either axial forces or deflections are small (continuous beams, non-sway frames with stocky columns bent in double curvature etc.) (Galambos, 1968; Halász, 1967a). In other cases it may give unsafe estimate of the failure load.

*Modified second order approach* can be spoken of where the equilibrium equations are set up taking into account the deflections of the structure (Halász, 1969a). A typical load-deflection diagram according to the second order approach $F_0^{(II)}$ is illustrated in Figure 69.

$$F_0^{(II)} = f_I(M_{pl.Rd}, EI, M_{j.Rd}, S_{joint})$$

with $M_{pl.Rd}$ as above,
$EI$ the elastic stiffness of the cross sections,
$M_{j.Rd}$ as above,
$S_{joint}$ the rotational stiffness of the semi-rigid connection or column base.

The second order load-deflection curve differs basically from that based on the first order approach as follows: *(i)* the branches are curvilinear; *(ii)* the failure load (or peak load) is lower than in the case of the simple plastic (first order) theory; *(iii)* the failure may occur before the complete yield mechanism has developed and is followed by unstable behaviour. In addition, the location and sequence of occurrence of the generalised hinges do not necessarily coincide with those in the case of first order theory.

Though the elastic-plastic frame analysis based on second-order approach is dealt with in the literature (Livesley, 1959; Horne, 1961; Horne and Merchant, 1965), its practical application is cumbersome and bound to the use of a computer. This paper is to offer an approximate solution possible by manual calculation as well.

## 6.4 Assumptions

Let a frame, such as the one shown in Figure 69, be subject to monotonously increasing loads proportional to a single load factor $F$. For simplicity we confine us to cases when - up to the failure load - a good approximation can be reached, expressing the axial forces in the form

$$N_k = \beta_k \cdot F, \tag{64}$$

where the parameter $\beta_k$ is constant.

The frame may be built up of perfectly elastic members and generalised hinges at certain locations only. The load-deflection curve can be characterised by the diagram shown in Figure 70b. The subsequent branches of the curve indicate the behaviour of the frame containing an increasing number of generalised hinges. Each branch continued beyond its range of validity (dashed lines in Figure 70) approaches asymptotically a certain value $F_{cr,n}$. These values are referred to in the literature as "deteriorated critical loads" (Horne and Merchant, 1965; Halász, 1969a). They represent the load factor causing buckling of an assumed completely elastic frame with semi-rigid connections loaded by axial forces $N_k$ only, with real hinges at the locations where generalised hinges developed in the actual frame.

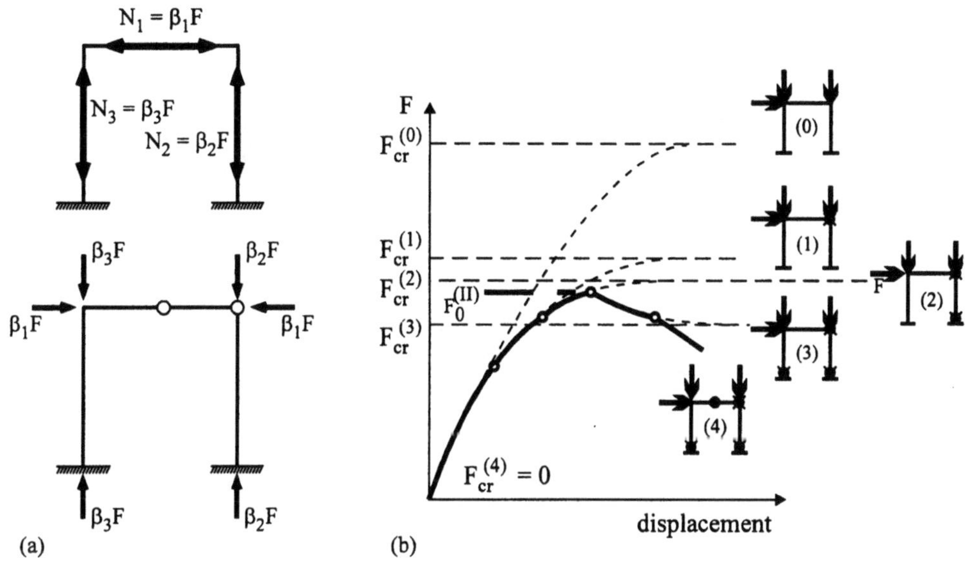


**Figure 70.** Load–displacement curve.

## 6.5 The case of the "direct method of design"

According to the first order approach, the failure load depends upon the value of the generalised moment only. In a second order approach, however, failure load depends upon two types of quantities: the first type is the moment resistance of the cross sections or the semi-rigid connections (i.e. the "strength" of the structure), and the second type is the flexural stiffness of the cross sections (*EI*) or the rotational stiffness of the semi-rigid connections or column bases (i.e. the "rigidity" of the structure).

Assuming *EI* and $S_{joint}$ to increase infinitely, the concept of rigid-plastic material is arrived at, failure will only occur after the formation of a complete yield mechanism (the "mechanism curve" in Figure 71). With decreasing stiffness the load-deflection curve may reach its peak value after the formation of generalised hinges less than needed for the complete mechanism to develop (the other curves in Figure 71).

Looking at Figure 72 one can distinguish two frame types:

- "relatively rigid" frame: the failure should take place when the number of generalized hinges has reached the number $n$ of hinges necessary for the complete mechanism, thus $i = n$;
- "relatively flexible" frame: the failure load is reached in the presence of a lower number of generalized hinges than that transforming the structure into a complete mechanism ($i < n$).

A special but easy to handle case of the "direct method of design" is to design a structure where the predetermined failure load $F_F$ coincides with one of the "deteriorated critical loads" $F_{cr,n}$. Figure 73 represents a case with $n = 3$, i.e. the structure fails as soon as the third generalised hinge has developed.

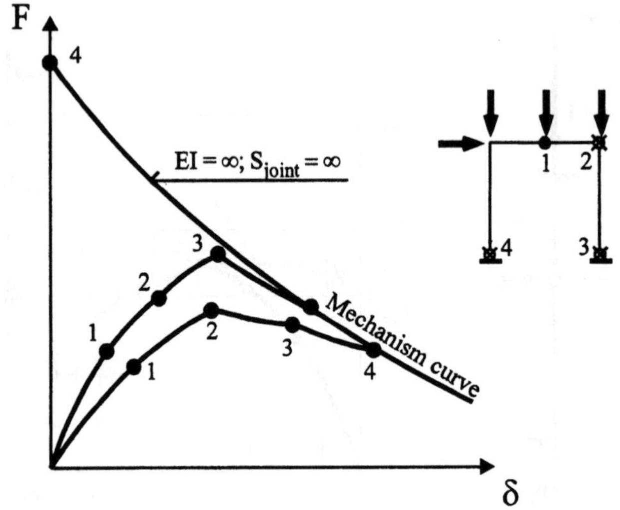


**Figure 71.** Load–deflection curves for different stiffness values.

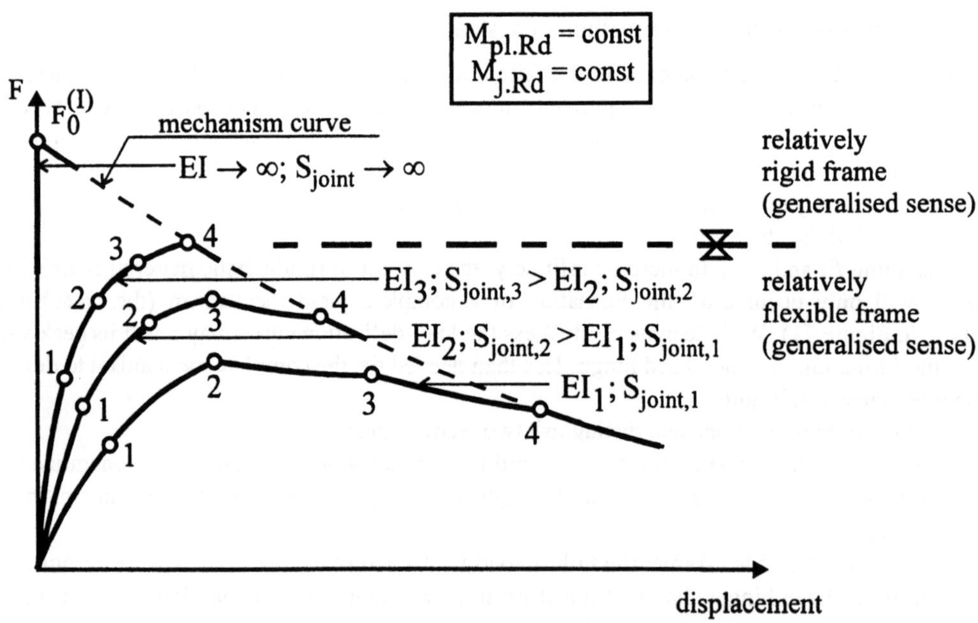


**Figure 72.** Relatively rigid and relatively flexible frames in the generalised sense.

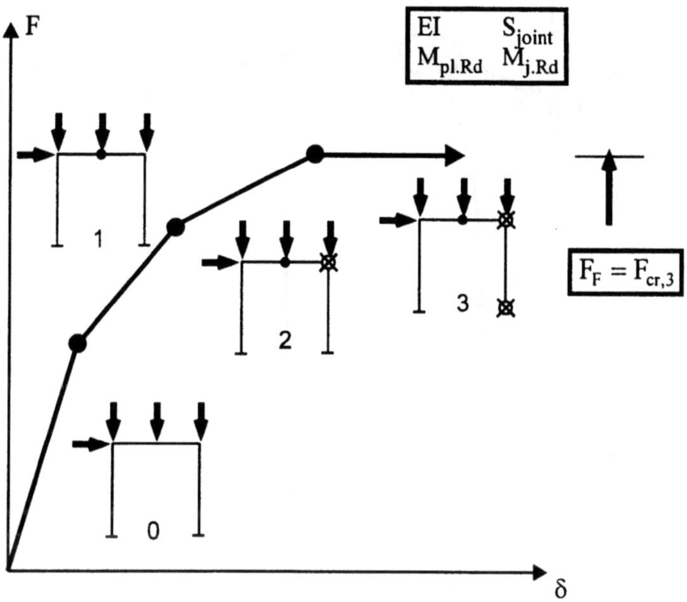


**Figure 73.** Illustration of the direct method of design.

The "deteriorated" critical load $F_{cr,3}$ (the buckling load of a completely elastic frame with three real hinges, with elements of stiffness $S_{joint}$ applied at the locations of semi-rigid beam-to-column connections and column bases, and subject to a given set of axial forces) is a function of geometry ($L$) and rigidity ($EI$) data:

$$F_{cr,3} = \frac{cEI}{L^2} \tag{65}$$

with $c$ being a constant. Setting

$$F_F = F_{cr,3}$$

the required value of the frame rigidity is

$$EI = \frac{L^2}{c} F_F \tag{66}$$

The next problem is to calculate the required value of the generalised plastic moment $M_{pl}$. For this purpose let us consider two structures (Figure 74a–b). The first is the actual structure just before failure: the load factor is equal to $F_F$, the third plastic hinge (at cross-section 3) is just about to develop. The displacements and the bending moments at cross-sections $j = 1$, 2 and 3 are denoted by $u$ and $M_j$, respectively.

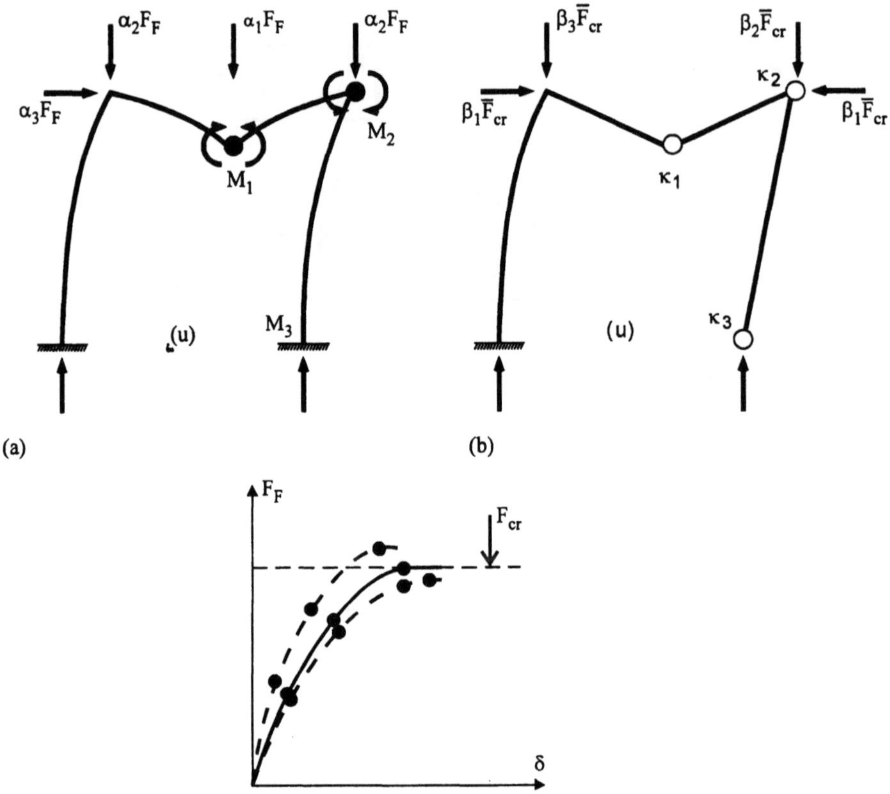


(a) (b)

(c)

**Figure 74.** Calculation of the required value of the generalised plastic moment $M_{pl}$

The second is the "deteriorated" structure with three real hinges subject to axial forces only, the load factor being $F_{cr}=F_{cr,3}$. This structure will buckle under the load $F_{cr,3}$. The displacement and the hinge rotations during buckling are denoted by

$\bar{u}$ and $\bar{\kappa}_j$,

respectively. The axial forces in both structures are defined by Eq. (64), and are supposed to keep unchanged during buckling.

Let us set up two virtual work equations, using two-way combination of loads and displacements of both systems:

$$\sum_i \alpha_i F_F \bar{u}_i - \sum_j M_j \bar{\kappa}_j + \sum_k \beta_k F_F \int_l EIu'\bar{u}' \,\mathrm{d}x = \int_l EIu''\bar{u}'' \,\mathrm{d}x$$

$$\sum_k \beta_k P_{cr,3} \int_l u'\bar{u}' \,\mathrm{d}x = \int_l EIu''\bar{u}'' \,\mathrm{d}x \qquad (67)$$

where $u'$ and $u''$ denote first and second derivatives, respectively. Note that displacements $u$ and $\overline{u}$ contain only first order terms, so

$$u_3 = \overline{u}_3 = 0 .$$

After subtraction one gets

$$\sum_i \alpha_i F_F \overline{u}_i - \sum_j M_j \overline{\kappa}_j = \sum_k \beta_k (F_{cr,3} - F_F) \int_l u' \overline{u}' \mathrm{d}x . \tag{68}$$

Considering Eq. (66)

$$F_F \sum_i \alpha_i \overline{u}_i = \sum_j M_j \overline{\kappa}_j . \tag{69}$$

Now if failure has to occur at $F_{cr}=F_{cr,3}$ (Figure 74c), bending moments $M_j$ in the generalised hinge cross-sections $j = 1$, 2 and 3 have to equal the generalised plastic moment, thus

$$F_F \sum_i \alpha_i \overline{u}_i = \sum_j \left| M_{pl} \overline{\kappa}_j \right| \tag{70}$$

Eq. (70) replaces Eq. (62) of the Modified Simple Plastic Theory and helps to compute the required value of $M_{pl}$. Supposing all the generalised plastic hinges to form under the same value of $M_{pl}$, by analogy to Eq. (63),

$$M_{pl} = F_F \frac{\sum_i \alpha_i \overline{u}_i}{\sum_j \left| \overline{\kappa}_j \right|} . \tag{71}$$

Using Eqs. (66) and (71) the flexural rigidity $EI$ of the members, the connection stiffnesses $S_{joint}$, and the generalised plastic moment in the plastic hinge cross-sections can be computed.

For designing the rest of the cross-sections the entire bending moment diagram at failure is to be known. This can only be determined indirectly, as the structure is in an indifferent state of equilibrium during failure, the failure load $\mathrm{F_u}$ being equal to the "deteriorated" critical load (Figure 75a)

$$F_F = \overline{F}_{cr} = F_{cr,3} \tag{72}$$

Without giving detailed prove of the procedure (Halász, 1969b), let the bending moment diagram $M$ at failure be composed of two parts as

$$M = M_0 + a\overline{M} \tag{73}$$

The moment diagram $M_0$ refers to the structure after removing one generalised plastic hinge chosen arbitrarily (Figure 75c). As thus the frame gets into a state of stable equilibrium and the generalised plastic hinges can be considered as real hinges with external moments equal to the generalised plastic moment acting upon them, the diagram $M_0$ can be determined by a second order elastic analysis (Horne and Merchant, 1965). The second bending moment diagram represents the bending moments arising during buckling of the "deteriorated" structure (Figure 75d) (with three real hinges and subject to axial forces only), which can be determined by known methods of second order elastic theory (Horne and Merchant, 1965), at least as far as its shape is concerned. The constant factor a is to be chosen as follows.

Like the bending moments, the displacements – among them rotations $\kappa_j$ of generalised plastic hinges – can be built up of similarly chosen components:

$$\kappa_j = \kappa_{0j} + a\bar{\kappa}_j \tag{74}$$

where

$$\kappa_{0j} \text{ and } \bar{\kappa}_j$$

are generalised hinge rotations in the two structures defined above. We will get the actual moments and displacements by selecting the value $a$ so that for all hinge rotations

$$\operatorname{sign}\kappa_j = \operatorname{sign}\bar{\kappa}_j$$

and

$$\left|\kappa_j\right|_{\min} = 0 . \tag{75}$$

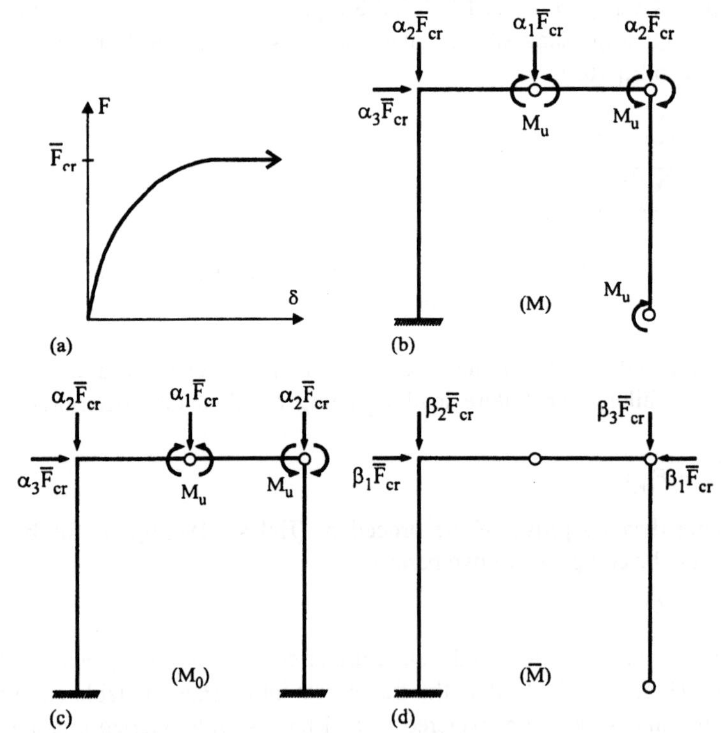


$M = M_0 + a \cdot M$;  $\kappa = \kappa_0 + a \cdot \kappa$;  $a \rightarrow \operatorname{sign}\kappa = \operatorname{sign}\kappa$;  $\kappa_{min} = 0$

**Figure 75.** Calculation of the "deteriorated" critical load

Put in words, all generalised hinge rotations should be of the same sign as hinge rotations of the deteriorated structure during buckling, but one of them (of course the last plastic hinge, just forming at failure) should be equal to zero.

Knowing the bending moment diagram $M$ at failure the frame can be designed.

The direct design method suggested above is based on the supposition that plastic hinges do not form but at predetermined cross-sections (indicated in the example by numbers 1, 2 and 3). The design according to the bending moment diagram

$$M = M_0 + a\overline{M}$$

at failure will make it certain to avoid other plastic hinges to exist at failure load, but further prove is needed that no other plastic hinges develop in previous stages of the loading process. In conformity with the condition that the cross-sections are of unit shape factor, the interaction curve between axial force and full plastic moment will be a straight line like that indicated in Figure 76. It is to prove that bending moments at any value of the load factor $F$ will not exceed the interaction curve.

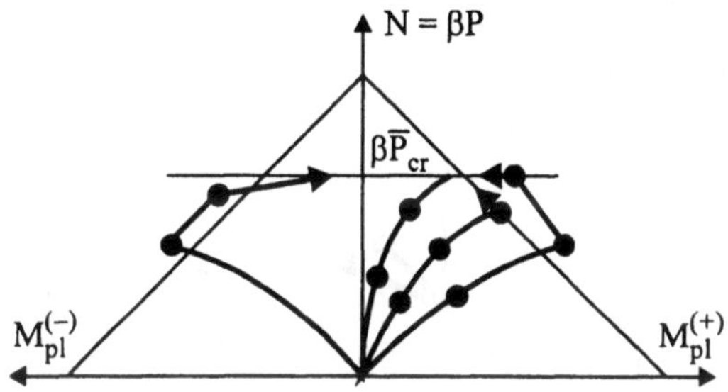


**Figure 76.** Interaction between axial force and full plastic moment

## 6.6 A generalization of Shanley's phenomenon

Attention is to be paid to the fact that the assumption of active loading process disregards the possibility bifurcation of equilibrium which can take place in stable state of equilibrium as well (Hill, 1958), due to the "two-faced" nature of generalized hinges, i.e. the loading and unloading processes are different.

As an illustrative example (see Figure 77) a symmetric and symmetrically loaded frame can be considered with only two possible locations for generalized hinges to occur. At the load level $F$, although having stable state of equilibrium, two different loading paths are possible: a symmetric and another including side-sway as well. This special case emerges if in Eq. (70)

$$\sum \alpha_i \bar{u}_i = 0 \tag{76}$$

as e.g. in case of the frame indicated in Figure 78, the actual deformations being normal to the buckling deformation. This case may lead to a bifurcation under stable conditions (Halász, 1967b).

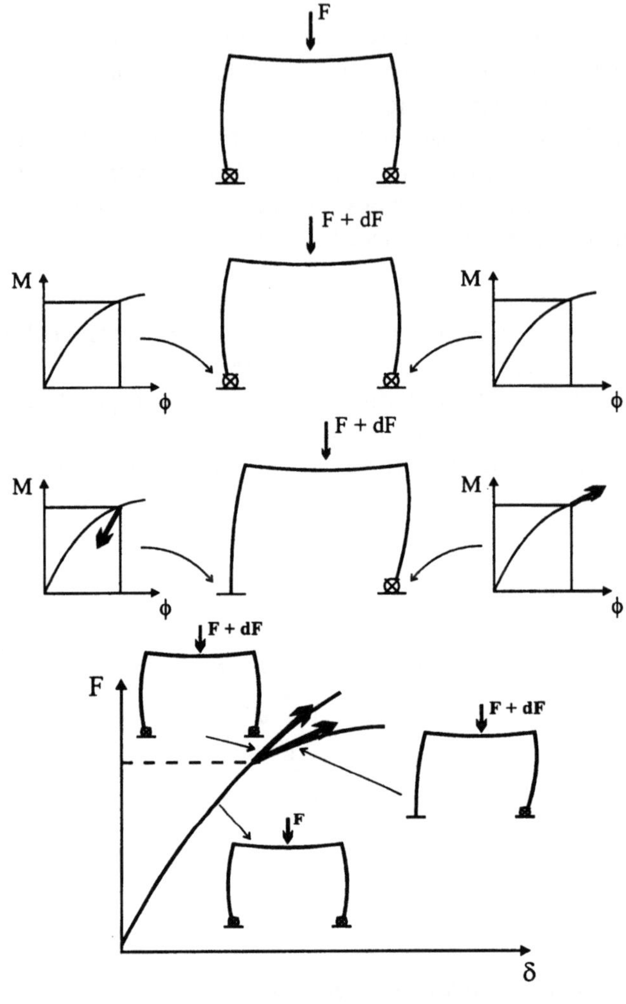


**Figure 77.** The generalized Shanley's phenomenon.

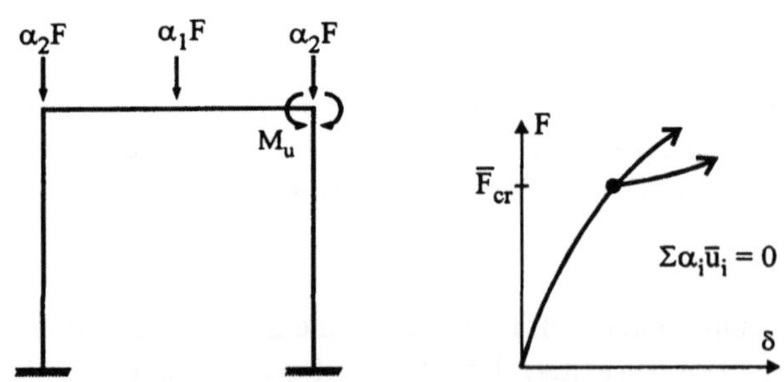


**Figure 78.** Bifurcation under stable conditions

## References


AISC (1989). *Manual of Steel Construction*. Allowable Stress Design. AISC, Chicago, IL.

AISC (1994). *Manual of Steel Construction*. Lond and Resistance Factor Design. AISC, Chicago, IL.

Bijlaard, F.S.K., and Zoetemeijer, P. (1986). Joint Characteristics and Structural Response of Frames. In *Steel Structures – Recent research advances and their application to design*, Elsevier, 109-133.

Bjorhovde, R., Brozzetti, J., and Colson, A. (1988). *Connections in Steel Structures*. Elsevier Applied Science Publishers Ltd.

Bjorhovde, R., and Colson, A. (1989). Semi-Rigid Behaviour and Classification of Bolted Connections. In *International Colloquium: Bolted and Special Structural Connections*, USSR, Moscow, May 15-20, Proceedings, Vol.3, 5-15.

Bjorhovde, R., Colson, A., and Brozzetti, J. (1990). Classification system for beam-to-column connections. In *Journal of Structural Engineering*, ASCE, 116(11):3059-76.

Chen, W.F., ed. (1987). *Joint flexibility in steel frames*. Elsevier Applied Science, London.

Chen, W.F., and Fielding, D.J. (1972). Frame analysis considering connection shear deformation, Structural design of tall steel buildings. *In Proceedings of International Conference on Planning and Design of Tall Buildings*, Lehigh University, Bethlehem, August 21-26, Vol. II., 365-370.

Chen, W.F., and Lui, E.M. (1991). *Stability design of steel frames*. CRC Press, Boca Raton.

Colson, A. (1996). Research on Connections in Europe. In *The COST C1 Action "Semi-Rigid Structural Connections"*, IABSE Colloquium, Istanbul, Report, 13-27.

Cosenza, E., De Luca, A., and Faella, C. (1989). Inelastic Buckling of Semi-Rigid Sway Frames. In *Structural Connections: Stability and Strength*. Elsevier Applied Science, London.

Cosenza, E., De Luca, A., and Faella, C. (1989a). Elastic Buckling of Semi-Rigid Sway Frames. In *Structural Connections: Stability and Strength*. Elsevier Applied Science, London.

EC3 (1992). *Design of Steel Structures*, Part 1.1. Eurocode 3, European Committee for Standardization, CEN, Brussels.

EC3 (1993). *Design of Steel Structures*, Part 1.1. Eurocode 3, European Committee for Standardization, CEN, Brussels.

Ermopoulos, J., and Vayas, I. (1991). Zum Nachweis von Rahmentragwerken mit verformbaren Knoten. In *Der Stahlbau*, Vol. 60, 326-332.

ESDEP (1994). *Structural Systems: Buildings*. Vol. 14, European Steel Design Education Programme, Steel Construction Institute.

ESDEP (1994). *Connection Design: Static Loading*. Vol. 15-16, European Steel Design Education Programme, Steel Construction Institute.

Eurocode 3. (1992). *Design of Steel Structures, Part 1 - General Rules and Rules for Buildings*. Commission of the European Communities, Chapter 6, Annex J, ENV 1993-1-1: 1992

Eurocode 3. (1989). *Design of Steel Structures, Part 1 - General Rules and Rules for Buildings*. Commission of the European Communities, Background Documentation, Chapter 6, Document 6.09, Beam to column connections, March 1989.

Eurocode 3. (1993). *Design of Steel Structures, Part 1 - General Rules and Rules for Buildings*. Commission of the European Communities, Second draft of part "Joints in Building Frames", Annex JJ/V2-02.93.

Fielding, D.J., and Huang, J.S. (1971). Shear in steel beam-to-column connections. In *Welding Research Supplement*, July 1971, 313-325.

Galambos, T.V. (1968). *Structural members and frames*. Prentice-Hall, New York.

Gomes, F.C.T. (1999). Classification of Joints. In *Proceedings of the Conference EUROSTEEL'99*, Praha, Vol.2, 535-538.

Gomes, F.C.T., Khulmann, U., De Matteis, G., and Mandara, A. (1998). Recent Developments of Classification of Joints. In *Proceedings of the International Conference: COST C1: Control of the semi-rigid behaviour of civil engineering structural connections*, Liege, 17-19 Sept., 187-198.

Gomes, F.C.T., and Neves, L.F.C. (1996). Influence of Semi-Rigid Connections on the Behaviour of Frames. Classification of Connections. In *Proceedings of the 5th International Colloquium on Structural Stability*, SSRC, Rio de Janeiro, August.

Goto, Y., and Miyashita, S. (1995). Validity of Classification System of Semi-Rigid Connections. In *Journal of Engineering Structures*, Vol.17, No.8, 544-553.

Halász, O. (1967a). Adoption of the theory of plasticity to steel structures. In *Acta Tecnica Acad. Sci. Hung.*, T59

Halász, O. (1967b). A generalisation of Shanley's phenomenon. In *Symposium of IUTAM Hungarian National Group*, Miskolc

Halász, O. (1969a). Design of elastic-plastic frames under primary bending moments. In *Periodica Polytechnica*, Budapest, Vol. 13, 3-4, 95-102.

Halász, O. (1969b). *Thesis*. Manuscript, Budapest.

Halász, O., and Iványi, M. (1979). Tests with simple elastic-plastic frames. In *Periodica Polytechnica, Civil Engineering*, Vol. 23 (3 - 4), 151-182.

Hasan, R., Kishi, N., and Chen, W.F. (1998). A New Nonlinear Connection Classification System. In *Journal of Constructional Steel Research*, Vol.47, 119-140.

Hasan, R., Kishi, N., Chen, W.F., and Komuro, M. (1995). Evaluation of rigidity of extended end-plate connections by utilizing updated data base. In *Structural Engineering*. Report, CE-STR-95-19, Purdue University, West Lafayette, IN.

Hill, R. (1958). A general theory of uniqueness and stability in elastic-plastic solids. In *Journal of the Mechanics and Physics of Solids*, Vol. 6.

Horne, M.R. (1961). The stability of elasto-plastic structures. In *Progress in Solid Mechanics*, Vol. 2, North-Holland Publishing, Amsterdam.

Horne, M.R., and Merchant, W. (1965). *The stability of frames*. Pergamon Press, Oxford.

Horne, M.R., and Morris, L.J. (1981). *Plastic Design of Low-Rise Frames*. Constrado Monographs, Granada, London.

Horváth, L., and Katula L., (1993). Design of semi-rigid knee joints according to Eurocode 3. In Muzeau, J.P., and Iványi, M., eds., *Steel design and Eurocodes*, TEMPUS JEP 2184 Technical Report, Budapest.

Iványi, M. (1983). A new concept in limit design of steel structures. In *Stability of Metal Structures, Final Report of 3rd International Colloqium*, Paris, 257-264.

Iványi, M. (1992). Prediction of ultimate load of steel frames with softening of semi-rigid connection. In *COST C1 Workshop*, Strasbourg, France, October 28-30.

Iványi, M. jr. (1993). *Experiments on frames, Influence of semi-rigid connections* (in Hungarian). Diploma thesis (supervisor: Hegedûs, L.), Budapest.

Iványi, M. jr. (1994). Tests with semi-rigid steel frames. In *Proceedings of The 17th Czech and Slovak International Conference on Steel Bridges*, Bratislava, Vol. I., 59-64.

Iványi, M., and Balogh, J. (1995). Parametric study of a column-base connection. In *Column-base strength and rigidity modeling*, PECO-COST Contract, Final Report for 1993-95.

Johnston, B.G., and Mount, E.H. (1941). *Analysis of Building Frames with Semi-Rigid Connections*. Transactions, ASCE, March, 993-1038.

Jones, S.W., Kirby, P.A., and Nethercot, D.A. (1982). Columns with Semi-Rigid Joints. In *Journal of Structural Division,* ASCE 108, 361-72.

Kaliszky, S. (1989). *Plasticity: Theory and engineering applications*, Akadémiai Kiadó, Budapest.

Kazinczy, G. (1914). Experiments with fixed-end beams (in Hungarian). In *Betonszemle*, 2.68.

Kishi, N., and Chen, W.F. (1990). Moment–rotation relations of semirigid connections with angles. In *Journal of Structural Engineering*, ASCE, 116(7):1813-34.

Kishi, N., Hasan. R., Goto, Y., and Komuro, M. (1996). Investigation on the validity of connection classification system. In *IABSE International Colloquium on Semirigid Structural Connections*, Istanbul, Turkey, 73-82.

Li, T.Q., Choo, B.S., and Nethercot, D.A. (1995). Determination of Rotation Capacity Requirements for Steel and Composite Beams. In *Journal of Constructional Steel Research*, Vol.32, 303-332.

Livesley, R.K. (1959). *Symposium on the use of electronic computers in structural engineering.* University of Southampton

Lui, E.M. (1985). *Effect of connection flexibility and panel zone deformation on the behavior of plane steel frames.* PhD dissertation, School of Engineering, Purdue University.

Majid, K.I. (1972). *Non-linear structures.* Butterworths, London.

Mazzolani, F.M. (1990). Stability of steel frames with semi-rigid joints. In *International Advanced School "Stability Problems of Steel Structures"* (lecture notes), CISM, Udine, Italy, September 24-28, 1990.

McGuire, W. (1992). Introduction to Connection Design. In Dowling, P.J., Harding, J.E., Bjorhovde, R., eds., *Constructional Steel Design, An International Guide.* Elsevier Applied Science, London.

Nethercot, D.A., Li, T.Q., and Ahmed, B. (1998). Unified Classification System for Beam-to-Column Connections. In *Journal of Constructional Steel Research*, Vol.45, No.1, 39-65

Owens, G.W., and Cheal, B.D. (1989). *Structural Steelwork Connections.* Butterworths, London.

Penserini, P., and Colson, A. (1989). Ultimate limit strength of column-base connections. In *Journal of Constructional Steel Research*, Vol. 14, 301-320.

Prager, W. (1951). *An introduction to plasticity.* Wesley Publishing Company, Reading, Massachusetts, U.S.A.

Snijder, H.H., Bijlaard, F.S.K., and Stark, J.W.B. (1983). Use of the Elastic Effective Length Theory for Stabilty Checks on Beams in Braced Frames. In Morris, L.J., ed., *Instability and Plastic Collapse of Steel Structures.* London, Granada, 152-63.

Sugimoto, H., and Chen, W.F. (1982). Small End Restraint Effects on Strength of H-Columns. In *Journal of Structural Division*, ASCE 108, 661-81.

Tautchnig, A. (1983). *Entwicklung eines neuen makromechanischen Knotenmodells and Erstellung eines darauf aufbauenden EDV-Programmes zur berechnung van Stahlskeletellragwerkwn unter Berücksischtigung nichtlinearer Nachgiebigkeiten fer Verbindungselemente ins besondere bei steifenloser Bauweise.* PhD Thesis, University of Innsbruck.

Zoetemeijer, P. (1974). *A Design Method for the Tension Side of Statically Loaded Bolted Beam-to-Column Connections.* Monogr. Heron 20 (No.1), Stevin Lab Techn. Univ. Delft/TNO Building and Construction Research.

Zoetemeijer, P. (1981a). Bolted Connections with Flush End-Plates and Haunched Beams. In *Test and Limit State Design Methods*, Rept 6-81-15, Dept. Civ. Eng. Techn. Univ. Delft.

Zoetemeijer, P. (1981b). Bolted Beam-to-Column Connections with Flush End-Plates and Haunched Beams. In *Test and Limit State Design Methods*, Rept 6-81-23, Dept. Civ. Eng. Techn. Univ. Delft.

## PART II

# INTEGRATION OF THE JOINT ACTUAL BEHAVIOUR INTO THE FRAME ANALYSIS AND DESIGN PROCESS

**J-P. Jaspart**
**University of Liège, Liège, Belgium**

**Abstract:** The present chapter describes the sources of joint deformability and gives information on how to integrate the actual joint behaviour into the frame design and analysis process. Four main aspects are particularly dealt with: the so-called joint characterisation, classification, modelling and idealisation. Finally practical procedures for the evaluation of the response of moment resisting joints are presented.

## 1 A Consistent Approach for Structural Joints

The rotational behaviour of actual joints is well recognised as being often intermediate between the two extreme situations, i.e. rigid or pinned.

Later in this lecture, the difference between *joints* and *connections* will be introduced. For the time being, examples of joints between one beam and one column only will be used. Consider now the bending moments and the related rotations at a joint (Figure 1):

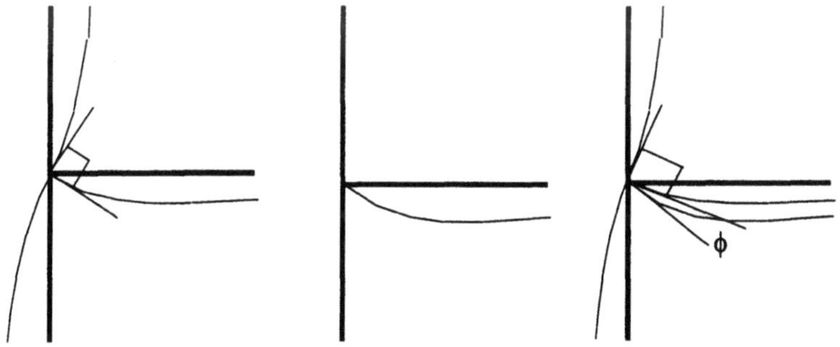


(a) Rigid joint (b) Pinned joint (c) Semi-rigid joint

**Figure 1** Classification of joints according to stiffness

When all the different parts in the joint are sufficiently stiff (i.e. ideally infinitely stiff), the joint is *rigid*, and there is no difference between the respective rotations at the ends of the members connected at this joint (Figure 1.a). The joint experiences a single global rigid-body rotation which is the nodal rotation in the commonly used analysis methods for framed structures.

Should the joint be without any stiffness, then the beam will behave just as simply supported whatever the behaviour of the other connected member(s) (Figure 1.b). This is a *pinned* joint.

For intermediate cases (non zero and non infinite stiffness), the transmitted moment will result in there being a difference $\phi$ between the absolute rotations of the two connected members (Figure 1.c). The joint is *semi-rigid* in these cases.

The simplest means for representing the concept is a rotational (spiral) spring between the ends of the two connected members. The rotational stiffness $S$ of this spring is the parameter that links the transmitted moment $M_j$ to the relative rotation $\phi$, which is the difference between the absolute rotations of the two connected members.

When this rotational stiffness $S$ is zero, or when it is relatively small, the joint falls back into the pinned joint class. In contrast, when the rotational stiffness $S$ is infinite, or when it is relatively high, the joint falls into the rigid joint class. In all the intermediate cases, the joint belongs to the semi-rigid joint class.

For semi-rigid joints the loads will result in both a bending moment $M_j$ and a relative rotation $\phi$ between the connected members. The moment and the relative rotation are related through a constitutive law which depends on the joint properties. This is illustrated in Figure 2, where, for the sake of simplicity, the global analysis is assumed to be performed with linear elastic assumptions.

At the global analysis stage, the effect of having semi-rigid joints instead of rigid or pinned joints is to modify not only the displacements, but also the distribution and magnitude of the internal forces throughout the structure.

As an example, the bending moment diagrams in a fixed-base simple portal frame subjected to a uniformly distributed load are given in

Figure 3 for two situations, where the beam-to-column joints are respectively either pinned or semi-rigid. The same kind of consideration holds for deflections.

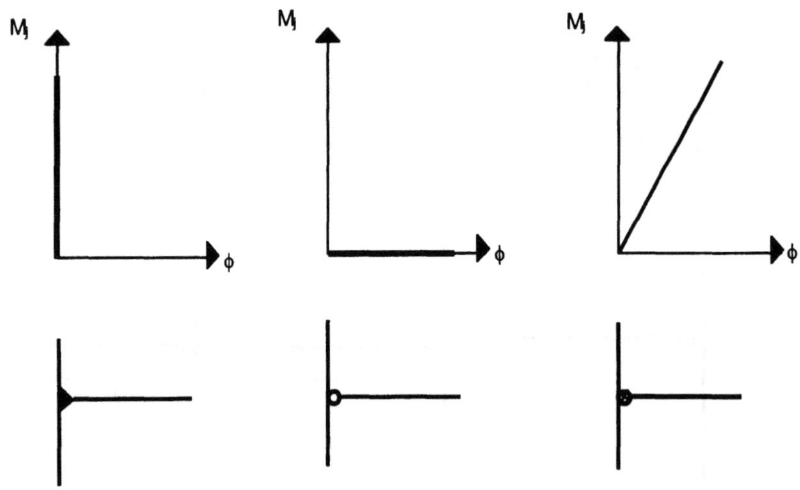


(a) Rigid joint($\phi = 0$) (b) Pinned joint ($M_j = 0$) (c) Semi-rigid joint ($M_j$ and $\phi \neq 0$)

**Figure 2** Modelling of joints (case of elastic global analysis)

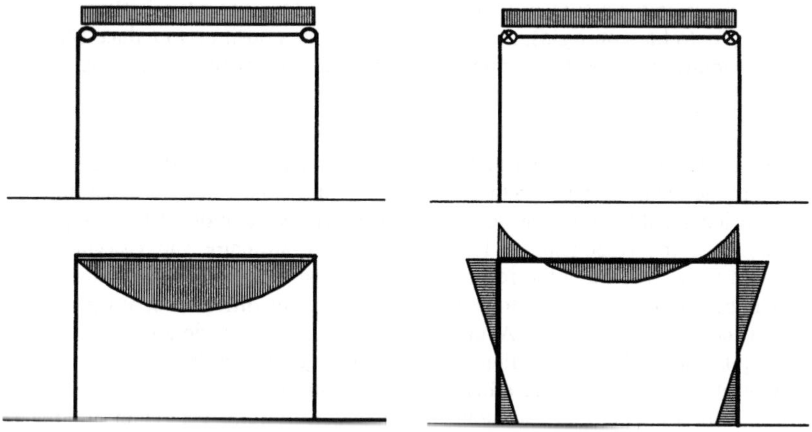

(a) Pinned joints (b) Semi-rigid joints

**Figure 3** Elastic distribution of bending moments in a simple portal frame

### 1.1 The merits of the consistent approach for structural joints

Both the *Eurocode 3* requirements and the desire to model the behaviour of the structure in a more realistic way leads to the consideration of the semi-rigid behaviour when necessary.

Many designers would stop at that basic interpretation of *Eurocode 3* and hence would be reluctant to confront the implied additional computational effort involved. Obviously a crude way to deal with this new burden will be for them to design joints that will actually continue to be classified as being either pinned or fully rigid. However such properties will have to be proven at the end of the design process and, in addition, such joints will certainly be found to be uneconomical in a number of situations.

It should be noted that the concept of rigid and pinned joints still exists in *Eurocode 3*. It is accepted that a joint which is *almost rigid*, or *almost pinned*, may still be considered as being *truly rigid* or *truly pinned* in the design process. How to judge whether a joint can be considered as rigid, semi-rigid or pinned depends on the comparison between the joint stiffness and the beam stiffness, which depends on the second moment of area and length of the beam.

The designer is strongly encouraged to go beyond this "all or nothing" attitude. Actually it is important to consider the benefits to be gained from the semi-rigid behaviour of joints. Those benefits can be brought in two ways :

1. The designer decides to continue with the practice of assuming -sometimes erroneously- that joints are either pinned or fully rigid. However, *Eurocode 3* requires that proper consideration be given to the influence that the actual behaviour of the joints has on the global behaviour of the structure, i.e. on the precision with which the distribution of forces and moments and the displacements have been determined. This may not prove to be easy when the joints are designed at a late stage in the design process since some iterations between global analysis and design checking may be required. Nevertheless, the following situations can be foreseen:

   - So that a joint can be assumed to be rigid, it is common practice to introduce web stiffeners in the column. *Eurocode 3* now provides the means to check whether such stiffeners are really necessary for the joint to be both rigid and have sufficient resistance. There are practical cases where they are not needed, thus permitting the adoption of a more economical joint design.

- When joints assumed to be pinned are later found to have fairly significant stiffness (i.e. to be semi-rigid), the designer may be in a position to reduce beam sizes. This is simply because the moments carried by the joints reduce the span moments in the beams.

2. The designer decides to give consideration, at the preliminary design stage, not only to the properties of the members but also to those of the joints. It may be shown that this new approach is not at all incompatible with the sometimes customary separation of the design tasks between those who have the responsibility for conceiving the structure and carrying out the global analysis and those who have the responsibility for designing the joints. Indeed, both tasks are very often performed by different people, or indeed, by different companies, depending on national or local industrial habits. Adopting this novel early consideration of joints in the design process requires a good understanding of the balance between, on the one hand, the costs and the complexity of joints and, on the other hand, the optimisation of the structural behaviour and performance through the more accurate consideration of joint behaviour for the design as a whole. Two examples are given to illustrate this:

- It was mentioned previously that it is possible in some situations to eliminate column web stiffeners and therefore to reduce costs. Despite the reduction in its stiffness and, possibly, in its strength, the joint can still be considered to be rigid and be found to have sufficient strength. This is shown to be possible for industrial portal frames with rafter-to-column haunch joints, in particular, but other cases can be envisaged.
- In a more general way, it is worthwhile to consider the effect of adjusting the joint stiffness so as to strike the best balance between the cost of the joints and the costs of the beams and the columns. For instance, for braced frames, the use of semi-rigid joints, which are probably more costly than the pinned joints, leads to reducing the beam sizes. For unbraced frames, the use of less costly semi-rigid joints, instead of the rigid joints, leads to increased beam sizes and possibly column sizes.

Of course the task may seem a difficult one, and this is why the present lectures are aimed at providing the reader with useful information. The whole philosophy could be termed as *"Because you must do it, take advantage of it"*.

Thus *Eurocode 3* now presents the designer with a choice between a *traditionalist attitude*, where however something may often be gained, and an *innovative attitude*, where the most economical result may best be sought.

It is important to stress the high level of similarity that exists between the member classification and the joint classification. This topic is addressed in the next section.

## 1.2 A parallel between member sections and joints

Member cross-section behaviour may be considered through an $M$-$\phi$ curve for a simply supported beam loaded at mid span ($M$ : bending-moment at mid-span ; $\phi$ : sum of rotations at the span ends). Joint behaviour will be considered through a similar relationship, but with $M = M_j$ being the bending moment transmitted by the joint and $\phi$ being the relative rotation between the connected member and the rest of the joint. Those relationships have similar shapes as illustrated in Figure 4.

To flexural stiffness EI/L and the design resistance $M_{b.Rd}$ of the member correspond the initial stiffness $S_{j,ini}$ and the design resistance $M_{j.Rd}$ of the joint.

According to *Eurocode 3* member cross-sections are divided into four classes according to their varying ability to resist local instability, when partially or totally subject to compression, and the consequences this may have on the possibility for plastic redistribution. Therefore their resistance ranges from the full plastic resistance (*class 1* and *2*) to the elastic resistance (*class 3*) or a reduced elastic resistance (*class 4*).

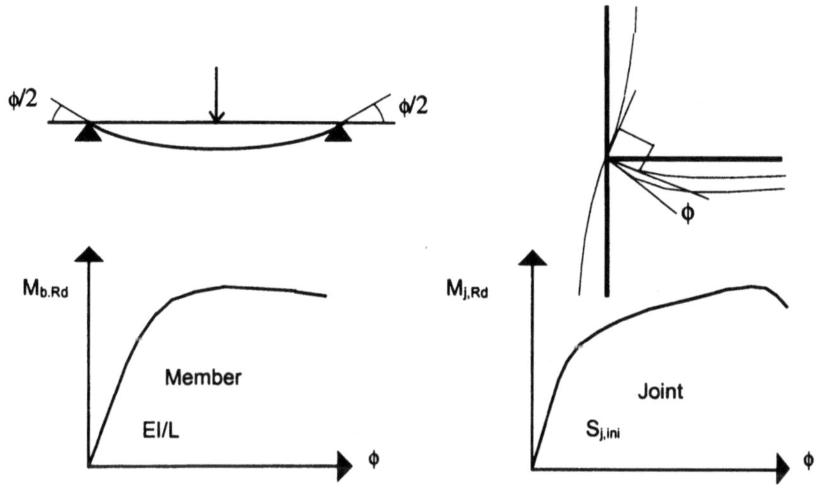


**Figure 4** $M$-$\phi$ characteristics for member cross-section and joint

The allocation of a cross-section to a specific class is governed by the assumptions on:

- The behaviour to be idealised for global analysis (i.e. *class 1* will allow the formation of a plastic hinge and permit the redistribution of internal forces in the frame as loads are increased up to or beyond the design loads);
- The behaviour to be taken into account for local design checks (i.e. *class 4* will imply that the resistance of the cross-section is based on the properties of a relevant effective cross-section rather than of the gross cross-section).

In *Eurocode 3*, the classification of a cross-section is based on the width-to-thickness ratio of the component walls of the section. Ductility is directly related to the amount of rotation during which the design bending resistance will be sustained. For joints, the *rotation capacity* concept is equivalent to the *ductility* concept for sections.

In a manner similar to that for member cross-sections, joints are classified in terms of *ductility* or *rotation capacity*. This classification is a measure of their ability to resist premature local instability and, even more likely, premature brittle failure (especially due to bolt failure) with due consequences on the type of global analysis allowed.

The practical interest of such a classification for joints is to check whether an elasto-plastic global analysis may be conducted up to the formation of a plastic collapse mechanism in the structure, which implies such hinges in at least some of the joints.

As will be shown, this classification of joints by ductility, while not explicitly stated in *Eurocode 3*, may be defined from the geometric and mechanical properties of the joint components (bolts, welds, plate thickness, etc.).

Joints may therefore be classified according to both their *stiffness* and their *ductility*. Moreover, joints may be classified according to their *strength*.

In terms of their strength, joints are classified as *full-strength* or *partial-strength* according to their resistance compared to the resistance of the connected members. For elastic design, the use of partial-strength joints is well understood. When plastic design is used, the main use of this classification is to foresee the possible need to allow a plastic hinge to form in the joint during the global analysis. In order to permit a further increase of loads beyond that corresponding to the formation of the hinge, a partial-strength joint may be required to act as a hinge from the moment when its plastic bending resistance is reached. In that case, the joint must also have sufficient ductility.

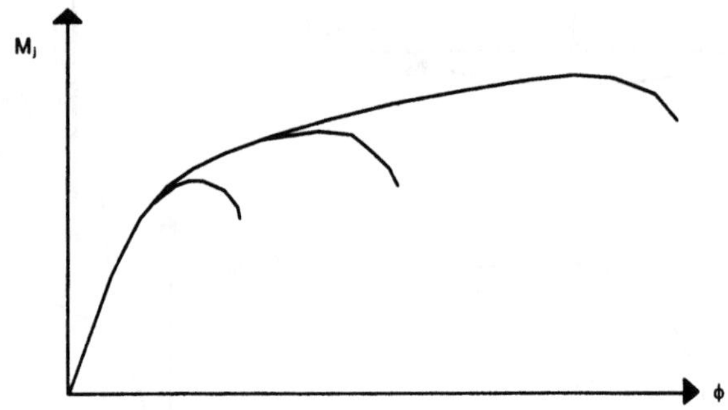


**Figure 5** Ductility or rotation capacity in joints

## 2 Definitions of Joint Configuration, Joint and Connection

Building frames consist of beams and columns, usually made of H or I shapes, that are assembled together by means of connections. These connections are between two beams, two columns, a beam and a column or a column and the foundation (Figure 6).

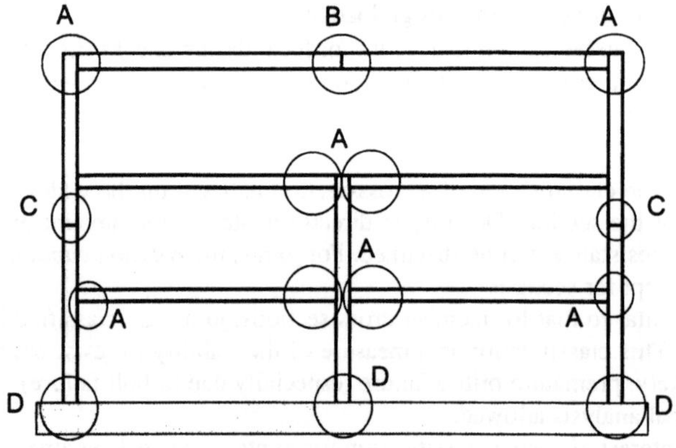


**Figure 6** Different types of connections in a building frame

A *connection* is defined as the set of the physical components, which mechanically fasten the connected elements. One considers the connection to be concentrated at the location where the fastening action occurs, for instance at the beam end/column interface in a major axis beam-to-column joint. When the connection as well as the corresponding zone of interaction between the connected members are considered together, the wording *joint* is then used (Figure 7.a).

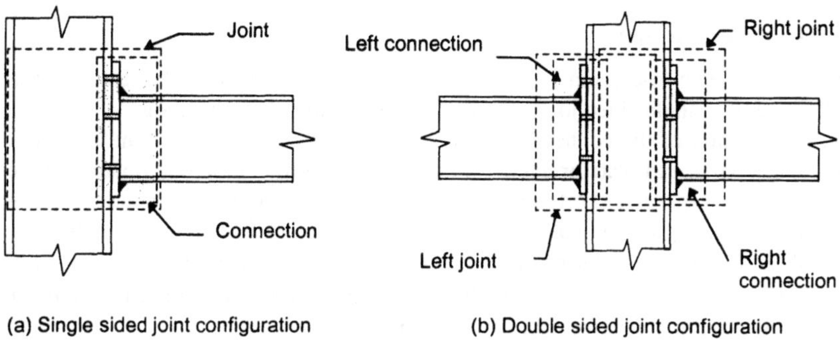


**Figure 7** Joints and connections

Depending on the number of in-plane elements connected together, single-sided and double-sided joint configurations are defined (Figure 8). In a double-sided configuration (.b), two joints - left and right - have to be considered (Figure 7.b.).

The definitions illustrated in Figure 7 and Figure 8 are valid for other joint configurations and connection types.

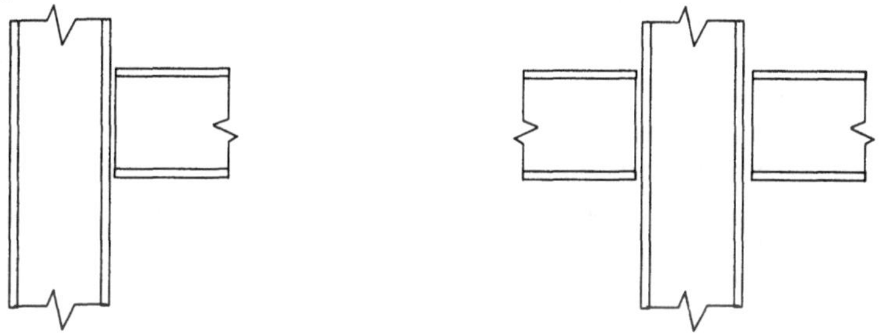

(a) Single-sided (b) Double-sided

**Figure 8** In-plane joint configurations

As explained previously, the joints which are traditionally considered as rigid or pinned and are designed accordingly, possess, in reality, their own degree of finite flexibility or finite stiffness resulting from the deformability of all the constitutive components. The next section is aimed at describing the main *joint deformability sources.*

# 3 Sources of Joint Deformability

The rotational behaviour of the joints may affect the local and/or global structural response of the frames. In this section, the sources of rotational deformability are identified for beam-to-column joints, splices and column bases.

It is worthwhile mentioning that the rotational stiffness, the joint resistance and the rotation capacity are likely to be affected by the shear force and/or the axial force acting in the joint.

These shear and axial forces may obviously have contributions to the shear and axial deformability within the connections. However it is known that these contributions do not affect significantly the frame response; therefore, the shear and axial responses of the connection, in terms of rotational deformability, are neglected.

## 3.1 Beam-to-column joints

### 3.1.1 Major axis joints

In a major axis beam-to-column joint, different sources of deformability can be identified. For the particular case of a single-sided joint (Figure 9.a and Figure 10.a), these are:

The deformation of the connection. This includes the deformation of the connection elements : column flange, bolts, end-plate or angles,... and the load-introduction deformation of the column web resulting from the transverse shortening and elongation of the column web under the compressive and tensile forces $F_b$ acting on the column web. The couple of $F_b$ forces are statically equivalent to the moment $M_b$ at the beam end. These deformations result in a relative rotation $\phi_c$ between the beam and column axes; this rotation, which is equal to $\theta_b$ - $\theta_c$ (see Figure 9.a) is concentrated mainly along edge AB and provides a flexural deformability curve $M_b$ - $\phi_c$.

The shear deformation of the column web panel associated with the shear force $V_{wp}$ acting in this panel. This leads to a relative rotation $\gamma$ between the beam and column axes; this rotation makes it possible to establish a shear deformability curve $V_{wp}$-$\gamma$.

The deformability curve of a connection may obviously be influenced by the axial and shear forces possibly acting in the connected beam.

Similar definitions apply to double-sided joint configurations (Figure 9.b and Figure 9.b). For such configurations, two connections and a sheared web panel, forming two joints, must be considered.

In short, the main sources of deformability which must be contemplated in a beam-to-column major axis joint are :

*Single-sided joint configuration :*

- the connection deformability $M_b$-$\phi_c$ characteristic;
- the column web panel shear deformability $V_{wp}$-$\gamma$ characteristic.

*Double-sided joint configuration :*

- the left hand side connection deformability $M_{b1}$-$\phi_{c1}$ characteristic;
- the right hand side connection deformability $M_{b2}$-$\phi_{c2}$ characteristic;
- the column web panel shear deformability $V_{wp}$-$\gamma$ characteristic.

The deformability of the connection (connection elements + load-introduction) is only due to the couple of forces transferred by the flanges of the beam (equivalent to the beam end moment $M_b$). The shear deformability of the column web panel results from the combined action of these equal but opposite forces and of the shear forces in the column at the level of the beam flanges. Equilibrium equations of the web panel provide the shear force $V_{wp}$ (see Figure 10 for the sign convention) :

$$V_n = \frac{M_{b1} + M_{b2}}{h_b} - \frac{V_{c1} + V_{c2}}{2} \tag{1}$$

Another formula to which it is sometimes referred, i.e. :

$$V_n = \frac{M_{b1} + M_{b2}}{h_b} \tag{2}$$

is only a rough and conservative approximation of (1).

In both formulae, $z$ is the lever arm of the resultant tensile and compressive forces in the connection(s). Guidance on the derivation of $z$ is given in *Eurocode 3*.

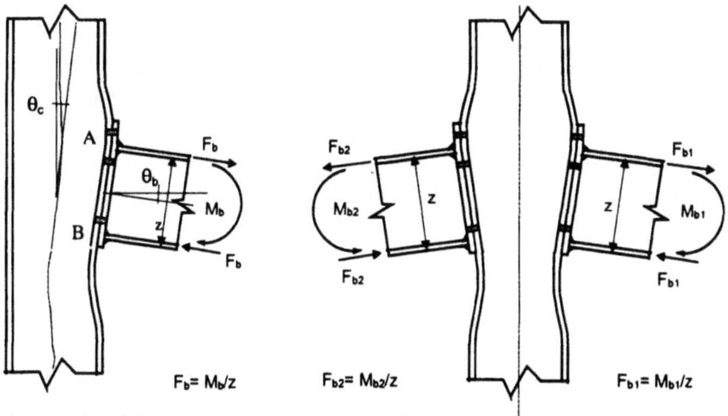


(a) Single-sided joint configuration (b) Double-sided joint configuration

**Figure 9** Sources of joint deformability

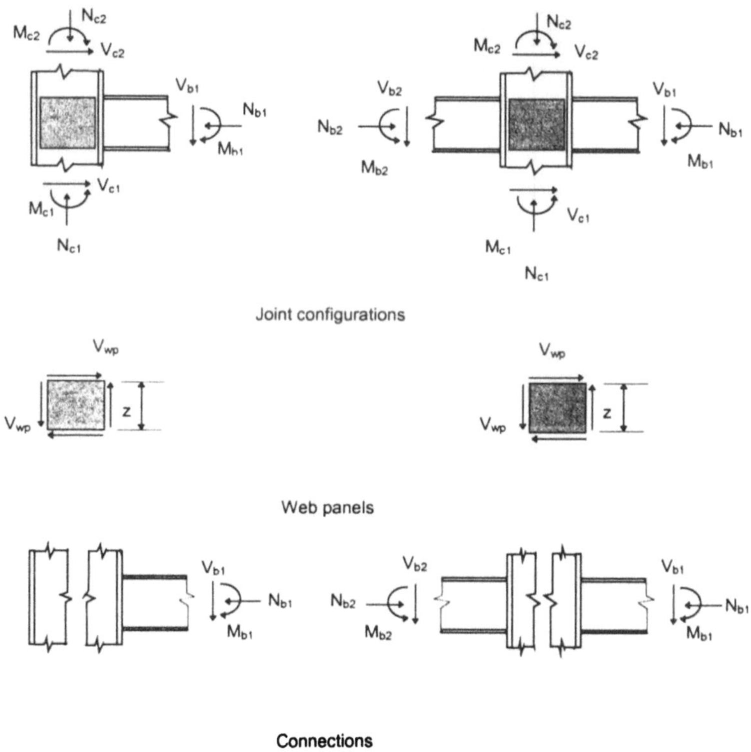


(a) Single-sided joint (b) Double-sided joint

**Figure 10** Loading of the web panel and the connections

### 3.1.2 Minor axis joints

A similar distinction between the *web panel* and the *connection* shall also be made for a minor axis joint (Figure 11). The column web exhibits a so-called out-of-plane deformability while the connection deforms in bending as it does in a major axis joint. However no load-introduction deformability is involved.

In a double-sided joint configuration, the out-of-plane deformation of the column web depends on the bending moments experienced by the right and left connections (see Figure 12):

$$\Delta M_b = M_{b1} - M_{b2} \qquad (3)$$

For a single-sided joint configuration (Figure 11), the value of $\Delta M_b$ equals that of $M_b$.

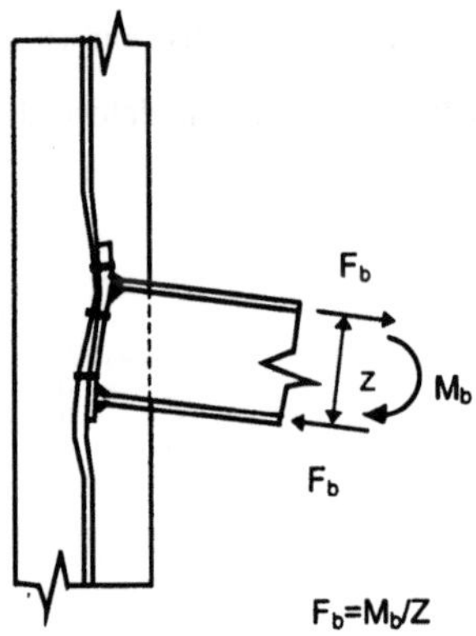


**Figure 11** Deformability of a minor axis joint

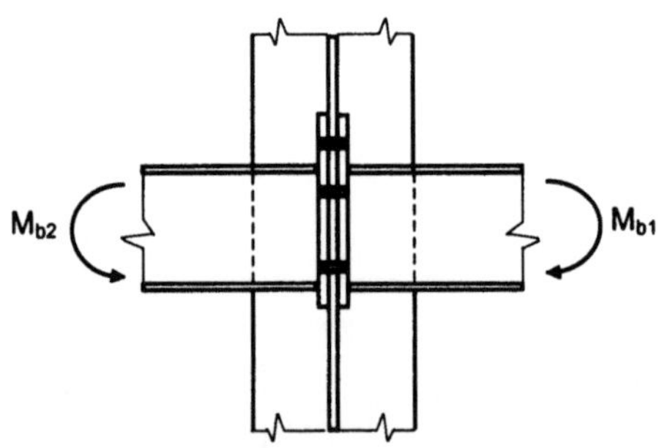


**Figure 12** Loading of a double-sided minor axis joint

### 3.1.3 Joints with beams on both major and minor column axes

A 3-D joint (Figure 13) is characterised by the presence of beams connected to both the column flange(s) and web. In such joints, a shear deformation and an out-of-plane deformation of the column web develop coincidentally.

The loading of the web panel appears therefore as the superposition of the shear loading given by formulae (1) or (2) and the out-of-plane loading given by formula (3).

The joint configuration of Figure 13 involves two beams only; configurations with three or four beams can also be met.

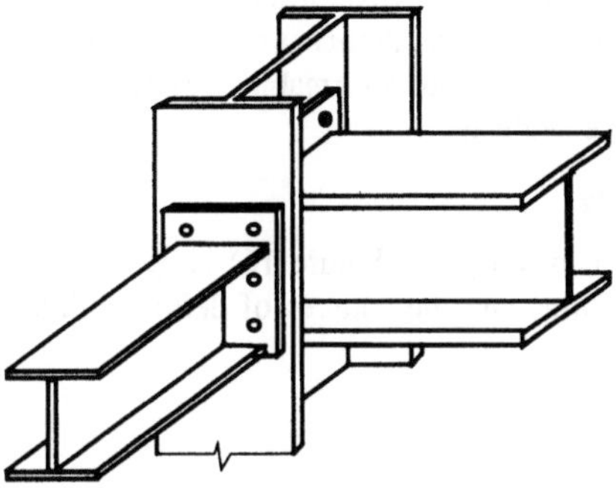

**Figure 13** Example of a 3-D joint

### 3.2 Beam splices and column splices

The sources of deformability in a beam splice (Figure 14) or in a column splice (Figure 15) are less than in a beam-to-column joint; indeed they are concerned with connections only. The deformability is depicted by the sole $M_b$-$\phi$ curve.

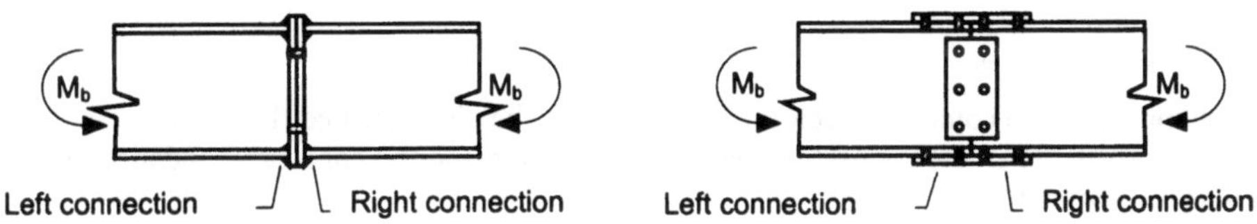


**Figure 14** Deformation of a beam splice

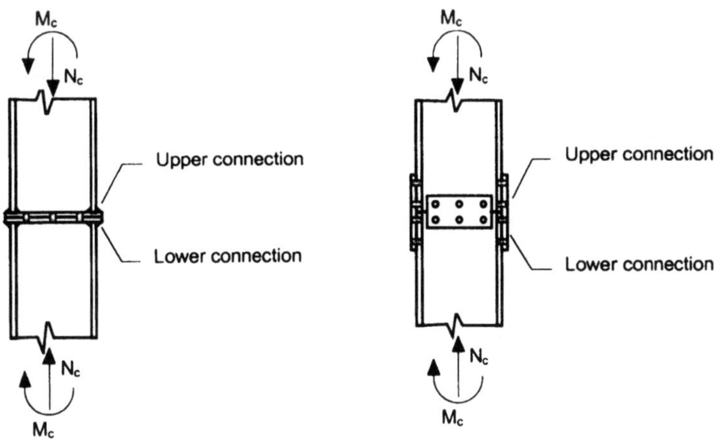


**Figure 15** Deformation of a column splice

The single $M_b$-$\phi$ curve corresponds to the deformability of the whole joint, i.e. the two constituent connections (left and right connections in a beam splice, upper and lower connections in a column splice).

In a column splice where the compressive force is predominant, the axial force affects the mechanical properties of the joint, i.e. its rotational stiffness, its strength and its rotation capacity, in a significant way. The influence of the axial deformability of splices on the global frame response is however limited and is therefore neglected.

### 3.3 Beam-to-beam joints

The deformability of a beam-to-beam joint (Figure 16) is quite similar to the one of a minor axis beam-to-column joint; the loadings and the sources of deformability are similar to those expressed earlier for minor axis joints.

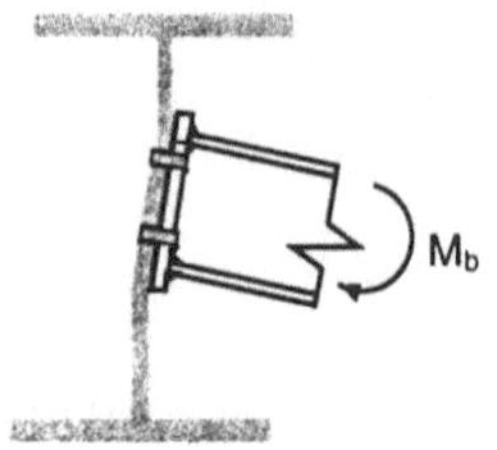


**Figure 16** Deformation of a beam-to-beam joint

### 3.4 Column bases

In a column base, two connection deformabilities need to be distinguished (Figure 17) :

the deformability of the connection between the column and the concrete foundation (*column-to-concrete connection*);

the deformability of the connection between the concrete foundation and the soil (*concrete-to-soil connection*).

For the column-to-concrete connection, the bending behaviour is represented by a $M_c$ - $\phi$ curve, the shape of which is influenced by the ratio of the bending moment to the axial load at the bottom of the column.

For the connection between the concrete foundation and the soil, two basic deformability curves are identified:

a $N_c$-$u$ curve which corresponds to the soil settlement due to the axial compressive force in the column; in contrast with the other types of joint, this deformability curve may have a significant effect on the frame behaviour;

a $M_c$ - $\phi$ curve characterising the rotation of the concrete block in the soil.

As for all the other joints described above, the deformability of the column base due to the shear force in the column may be neglected.

The column-to-concrete connection and concrete-to-soil connection $M_c$-$\phi$ characteristics are combined in order to derive the rotational stiffness at the bottom of the column and conduct the frame analysis and design accordingly.

Similar deformability sources exist in column bases subjected to biaxial bending and axial force. The connection $M_c$-$\phi$ characteristics are then defined respectively for both the major and the minor axes.

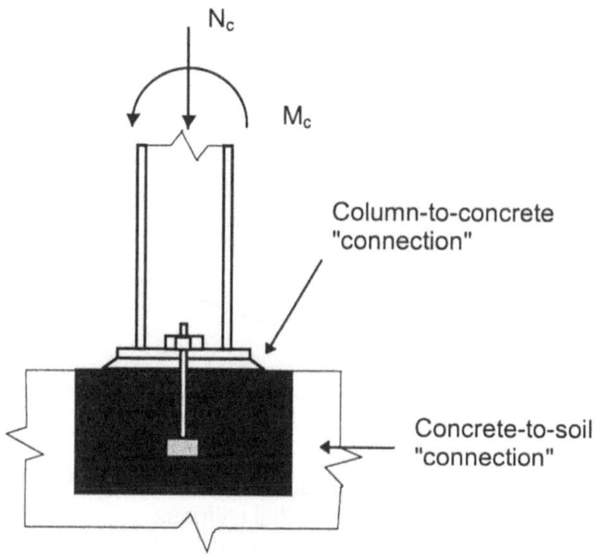


**Figure 17** Connections in a column base

# 4 Joint Classification

## 4.1 General

Later in this lecture, it is shown that the joints need to be modelled for the global frame analysis and that three different types of joint modelling are introduced : simple, semi-continuous and continuous.

It will also been explained that the type of joint modelling to be adopted is dependent both on the type of frame analysis and on the class of the joint in terms of stiffness and/or strength.

Classification criteria are used to perform this task and these are described below.

## 4.2 Stiffness classification

The *stiffness classification* into rigid, semi-rigid and pinned joints is performed by comparing simply the design joint stiffness to two stiffness boundaries (Figure 19). For sake of simplicity, the stiffness boundaries have been derived so as to allow a direct comparison with the *initial design* joint stiffness, whatever the type of joint idealisation that is used afterwards in the analysis.

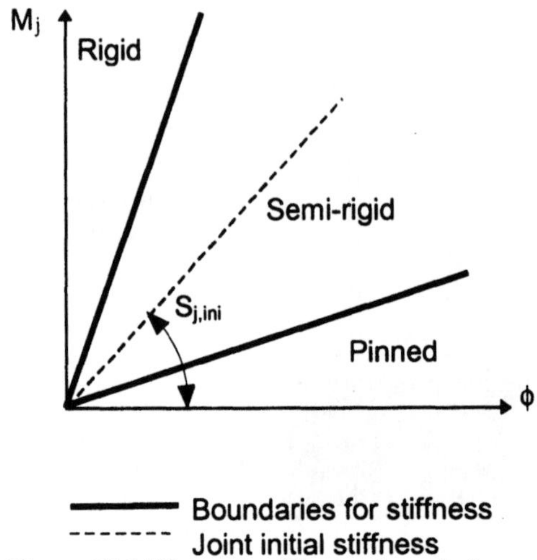


**Figure 18** Stiffness classification boundaries

## 4.3 Strength classification

The *strength classification* simply consists of comparing the joint *design* moment resistance to "ful strength" and "pinned" boundaries (Figure 19).

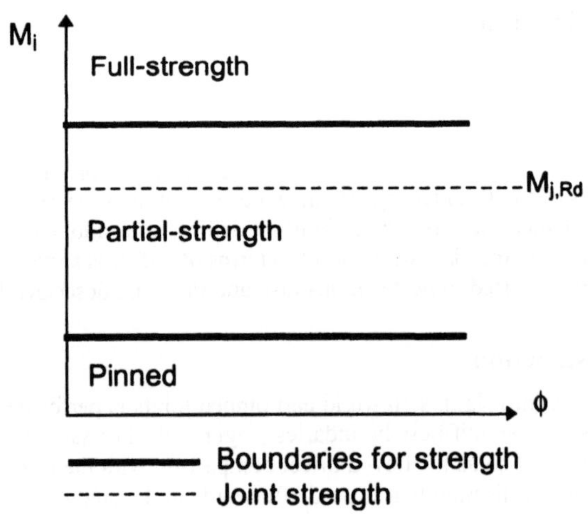


**Figure 19** Strength classification boundaries

### 4.4 Boundaries for classification

It is while stressing that a classification based on the *experimental* joint $M$-$\phi$ characteristics is not allowed, as *design* properties only are of concern.

The stiffness and strength boundaries for the joint classification are given as follows :

*Classification by stiffness*

- rigid joint — $S_{j,ini} \geq 25\ EI/L$ (unbraced frames)
  $S_{j,ini} \geq 8\ EI/L$ (braced frames)
- semi-rigid joint — $0{,}5EI/L < S_{j,ini} < 25EI/L$ (unbraced frames)
  $0{,}5EI/L < S_{j,ini} < 8EI/L$ (braced frames)
- pinned joint — $S_{j,ini} \leq 0{,}5EI/L$

*Classification by strength*

- full-strength joint — $M_{j,Rd} \geq M_{full-strength}$
- partial strength joint — $0{,}25M_{full-strength} < M_{j,Rd} < M_{full-strength}$
- pinned joint — $M_{j,Rd} \leq 0{,}25M_{full-strength}$

In this notation, EI/L and $M_{full\text{-}strength}$ designate respectively :

- the flexural stiffness of the connected beam;
- the design resistance of the weaker of the connected members.

## 5 Ductility classes

Experience and proper detailing result in so-called *pinned* joints which exhibit a sufficient rotation capacity to sustain the rotations imposed on them.

For moment resisting joints the concept of ductility classes is introduced to deal with the question of rotation capacity. This item is briefly raised in Annex A and in Section 9.2.2.

## 6 Joint modelling

### 6.1 General

Joint behaviour affects the structural frame response and shall therefore be modelled, just as for beams and columns, for the frame analysis and design. Traditionally, the following types of *joint modelling* are considered :

*For rotational stiffness :*
- rigid
- pinned

*For resistance :*
- full-strength
- partial-strength
- pinned

When the joint rotational stiffness is of concern, the wording *rigid* means that no relative rotation occurs between the connected members whatever be the applied moment. The wording *pinned* postulates the existence of a perfect (i.e. frictionless) hinge between the members. In fact these definitions may be relaxed, as explained earlier. Indeed rather flexible but not fully pinned joints and rather stiff but not fully rigid joints may be considered as effectively pinned and sufficiently rigid respectively. The stiffness boundaries allowing one to classify joints as rigid or pinned have been examined.

For joint resistance, a *full-strength joint* is stronger than the weaker of the connected members, which is in contrast to a *partial-strength joint*. In the everyday practice, partial-strength joints are used whenever the joints are designed to transfer the internal forces but not to resist the full capacity of the connected members. A *pinned joint* is considered to transfer no moment. Related classification criteria have also been presented.

Consideration of rotational stiffness and joint resistance properties leads to three significant joint models:

rigid/full-strength;
rigid/partial-strength;
pinned.

However, as far as the joint rotational stiffness is considered, joints designed for economy may be neither rigid nor pinned but semi-rigid. There are thus new possibilities for joint modelling:

semi-rigid/full-strength;
semi-rigid/partial-strength.

With a view to simplification, *Eurocode 3 - Chapter 6* and *Annex J* account for these possibilities by introducing three joint models (Table 1) :

*continuous* : covering the rigid/full-strength case only;
*semi-continuous* : covering the rigid/partial-strength, the semi-rigid/full-strength and the semi-rigid/partial-strength cases;
*simple* : covering the pinned case only.

| STIFFNESS | RESISTANCE | | |
|---|---|---|---|
| | Full-strength | Partial-strength | Pinned |
| Rigid | Continuous | Semi-continuous | * |
| Semi-rigid | Semi-continuous | Semi-continuous | * |
| Pinned | * | * | Simple |
| * : Without meaning | | | |

**Table 1** Types of joint modelling

The following meanings are given to these terms :

*continuous:* the joint ensures a full rotational continuity between the connected members;
*semi-continuous:* the joint ensures only a partial rotational continuity between the connected members;
*simple:* the joint prevents from any rotational continuity between the connected members.

The interpretation to be given to these wordings depends on the type of frame analysis to be performed. In the case of an elastic global frame analysis, only the stiffness properties of the joint are relevant for the joint modelling. In the case of a rigid-plastic analysis, the main joint feature is the resistance. In all the other cases, both the stiffness and resistance properties govern the manner in which the joints should be modelled. These possibilities are illustrated in Table 2.

| MODELLING | TYPE OF FRAME ANALYSIS | | |
|---|---|---|---|
| | Elastic analysis | Rigid-plastic analysis | Elastic-perfectly plastic and elastoplastic analysis |
| Continuous | Rigid | Full-strength | Rigid/full-strength |
| Semi-continuous | Semi-rigid | Partial-strength | Rigid/partial-strength<br>Semi-rigid/full-strength<br>Semi-rigid/partial-strength |
| Simple | Pinned | Pinned | Pinned |

**Table 2** Joint modelling and frame analysis

## 6.2 Modelling and sources of joint deformability

The difference between the loading of the connection and that of the column web in a beam-to-column joint requires, from a theoretical point of view, that account be taken separately of both deformability sources when designing a building frame.

However doing so is only feasible when the frame is analysed by means of a sophisticated computer program which enables a separate modelling of both deformability sources. For most available software, the modelling of the joints has to be simplified by concentrating the sources of deformability into a single rotational spring located at the intersection of the axes of the connected members.

## 6.3 Simplified modelling according to Eurocode 3

For most applications, the separate modelling of the connection and of the web panel behaviour is neither useful nor required; therefore only the simplified modelling of the joint behaviour will be considered in the present document. This idea is the one followed in the *ENV 1993-1-1* experimental standard (*Chapter 6* and *Annex J* ). Table 3, taken from the revised *Annex J*, shows how to relate the simplified modelling of typical joints to the basic wordings used for the joint modelling: simple, semi-continuous and continuous.

| JOINT MODELLING | BEAM-TO-COLUMN JOINTS MAJOR AXIS BENDING | BEAM SPLICES | COLUMN BASES |
|---|---|---|---|
| SIMPLE | 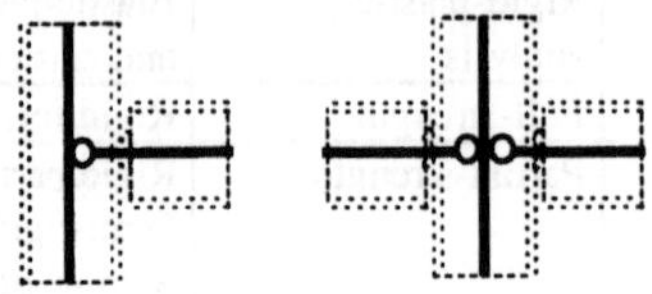 | 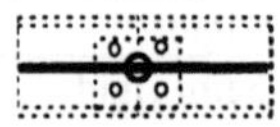 | 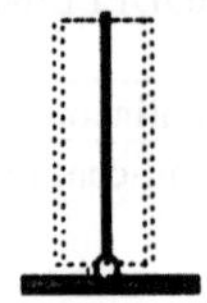 |
| SEMI-CONTINUOUS | 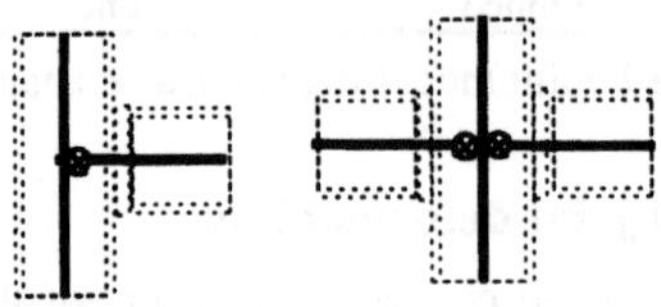 |  | 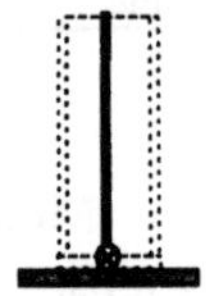 |
| CONTINUOUS | 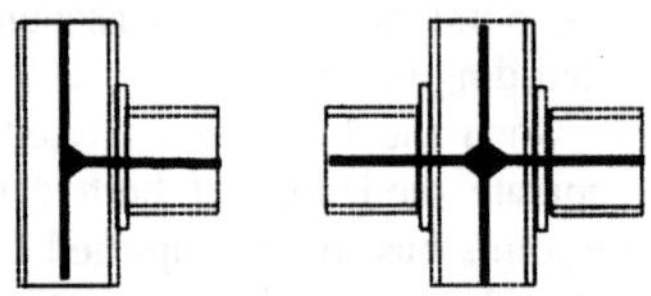 | 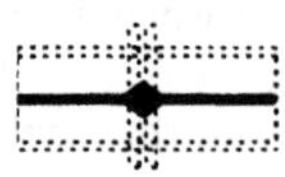 | 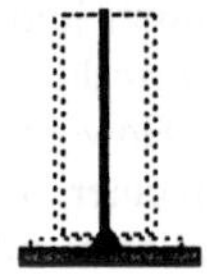 |

**Table 3** Simplified modelling for joints according *Eurocode 3*

## 6.4 Concentration of the joint deformability

For normal practice taking account of both the flexural behaviour of the connection and the shear (major axis beam-to-column joint) separately or out-of-plane behaviour of the column web panel (minor axis beam-to-column joint configurations or beam-to-beam configurations) is not feasible. This section is aimed at explaining how to concentrate the two deformabilities into a single flexural spring located at the intersection of the axes of the connected members.

### 6.4.1 Major axis beam-to-column joint configurations

In a single-sided configuration, only one joint is concerned. The characteristic shear-rotation deformability curve of the column web panel (Figure 20.b) is first transformed into a $M_b$-$\gamma$ curve through the use of the ***transformation parameter*** $\beta$. This parameter, defined in Figure 21.a, relates the web panel shear force to the (load-introduction) compressive and tensile forces connection (see also formulae 1 and 2).

The $M_b$-$\phi$ spring characteristic which represents the joint behaviour is shown in Figure 21.a; it is obtained by summing the contributions of rotation, from the connection ($\phi_c$) and from the shear panel ($\gamma$). The $M_j$-$\phi$ characteristic of the joint rotational spring located at the beam-to-column interaction is assumed to identify itself to the $M_b$-$\phi$ characteristic obtained as indicated in Figure 20.c.

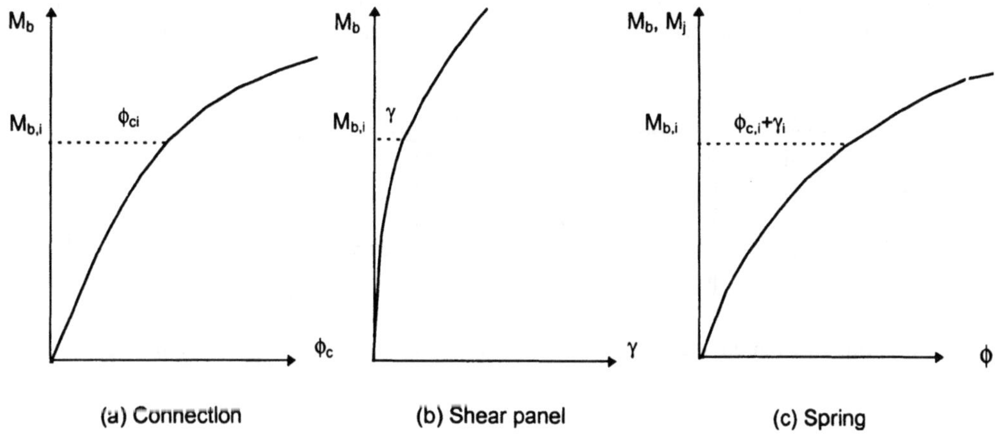


**Figure 20** Flexural characteristic of the spring

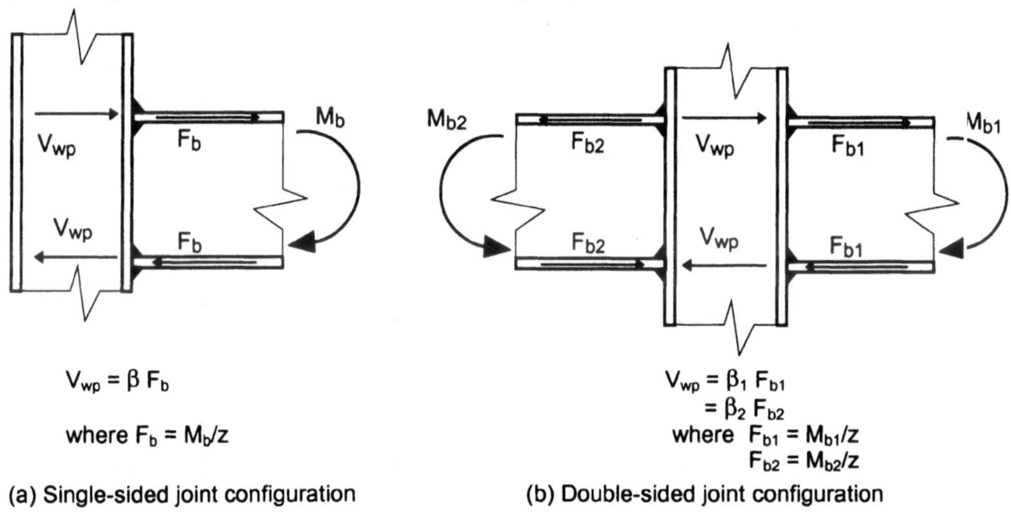


**Figure 21** Definition of the transformation parameter β for major axis joints

In a double-sided configuration, two joints - the left one and the right one – are involved. The derivation of their corresponding deformability curves is conducted similarly as in a single-sided configuration by using transformation parameters $\beta_1$ and $\beta_2$ (Figure 21.b).

Because the values of the β parameters can only be determined once the internal forces are known, their accurate determination requires an iterative process in the global analysis. For practical applications, such an iterative process is hardly acceptable provided safe β values are available. These values may be used a priori to model the joints and, on the basis of such joint modelling, the frame analysis may be performed safely in a non-iterative way.

The recommended but approximate values of β, where $\beta_1$ is taken as equal to $\beta_2$ for double-sided configurations, are given in the Lecture on "Practical procedures for the characterisation of the response of moment resisting joints". They vary from $\beta = 0$ (double-sided joint configuration with balanced moments in the beams) to $\beta = 2$ (double-sided joint configuration with equal but unbalanced moments in the beams). These two extreme cases are illustrated in Figure 22.

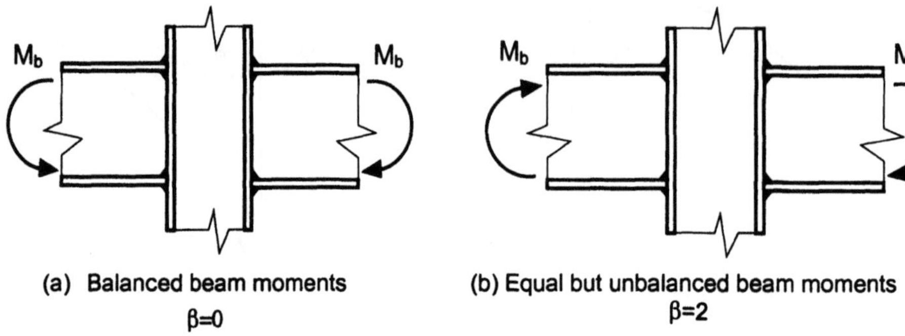


**Figure 22** Extreme cases for β values

### 6.4.2 Minor axis beam-to-column joint configurations and beam-to-beam configurations

Similar concepts as those developed in Section 6.4.1. are referred to for minor axis beam-to-column joint configurations and beam-to-beam configurations. The definition of the transformation parameter is somewhat different (Figure 23).

Approximate values of β (assuming $\beta_1 = \beta_2$) for these joint configurations may also be recommended, in a similar way than what is done for major axis beam-to-column joint configurations.

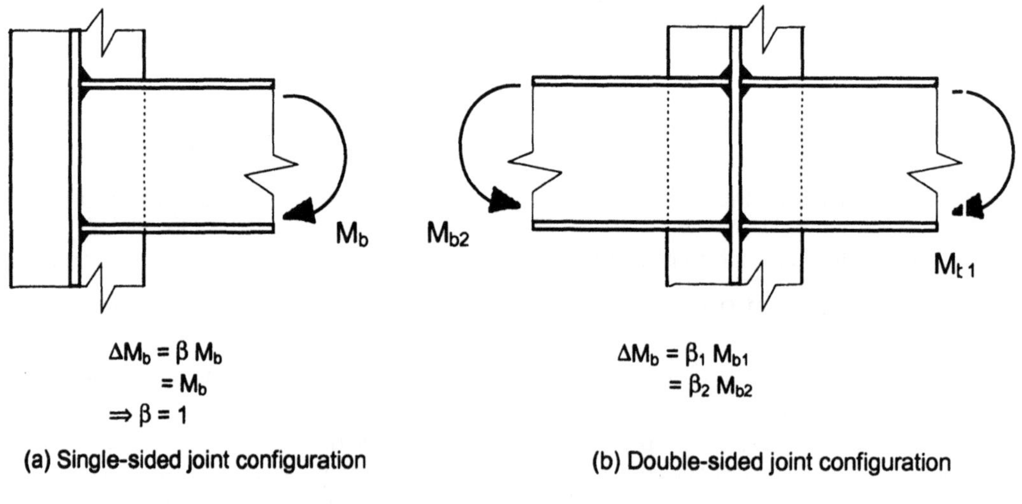


**Figure 23** Definition of the transformation parameter β for minor axis joints

# 7 Joint Characterisation

## 7.1 General

An important step when designing a frame consists of the characterisation of the rotational response of the joints, i.e. the evaluation of the mechanical properties in terms of stiffness, strength and ductility.

Three main approaches may be followed :
- experimental
- numerical
- analytical.

The only practical option for the designer is the analytical approach. Analytical procedures have been developed which enable a prediction of the joint response based on the knowledge of the mechanical and geometrical properties of the joint components.
In this section a general analytical procedure, termed *component method*, is introduced. It applies to any type of steel or composite joints, whatever the geometrical configuration, the type of loading (axial force and/or bending moment, ...) and the type of member sections.
The method is used where the mechanical properties of joints subjected to bending moment and shear force are computed.

## 7.2 Introduction to the component method

The component method considers any joint as a set of *individual basic components*. For the particular joint shown in Figure 25.b. (joint with an extended end-plate connection subject to bending), the relevant components are the following :

*Compression in zone :*
column web in compression;
beam flange and web in compression;
Tension zone :
column web in tension;
column flange in bending;
bolts in tension;
end-plate in bending;
beam web in tension;
*Shear zone :*
column web panel in shear.

Each of these basic components possesses its own strength and stiffness either in tension or in compression or in shear. The column web is subject to coincident compression, tension and shear. This coexistence of several components within the same joint element can obviously lead to stress interactions that are likely to decrease the resistance of the individual basic components.

The application of the component method requires the following steps :
- *identification* of the active components in the joint being considered;
- *evaluation of the stiffness and/or resistance characteristics* for each individual basic component (specific characteristics - initial stiffness, design resistance, ... - or the whole deformability curve) ;

- *assembly* of all the constituent components and evaluation of the stiffness and/or resistance characteristics of the whole joint (specific characteristics - initial stiffness, design resistance, ... - or the whole deformability curve).

In Figure 24, the principles of the component method are illustrated in the specific case of a beam-to-column joint with a welded connection.

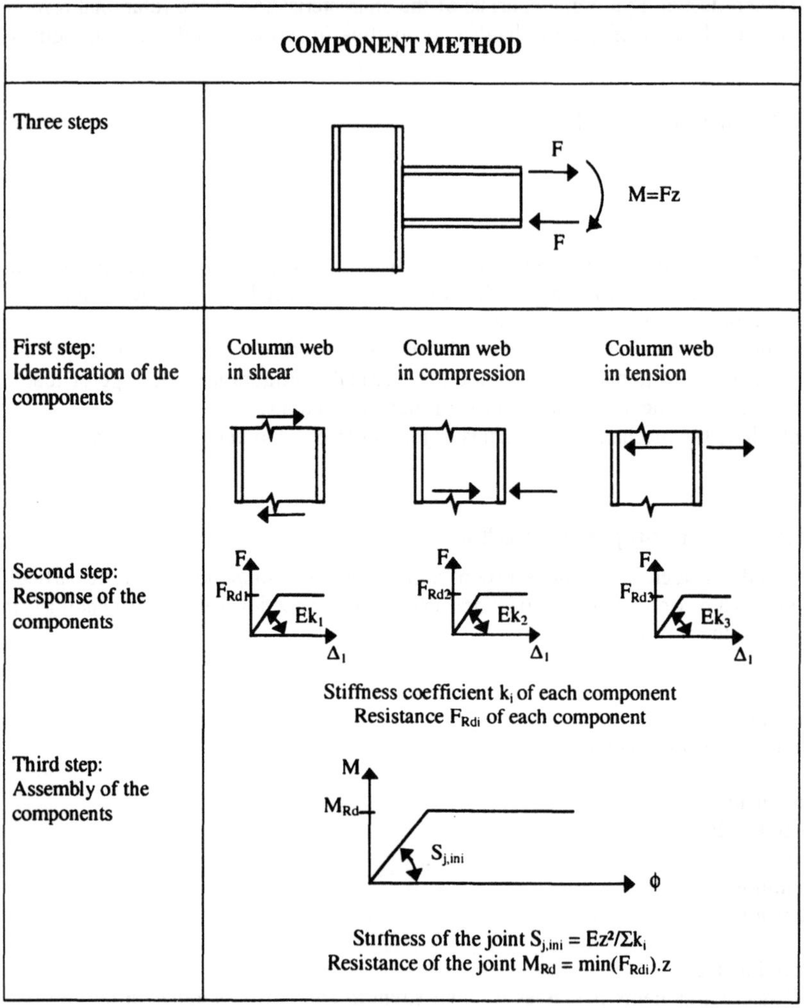


**Figure 24** Application of the component method to a welded joint

The assembly procedure consists in deriving the mechanical properties of the whole joint from those of all the individual constituent components. This requires a preliminary distribution of the forces acting on the joint into internal forces acting on the components in a way that satisfies equilibrium.

*In Eurocode 3 Annex J*, the analytical assembly procedures are described for the evaluation of the initial stiffness and the design moment resistance of the joint. These two properties enable the

| N° | Component | |
|---|---|---|
| 1 | Column web panel in shear |   |
| 2 | Column web in compression | 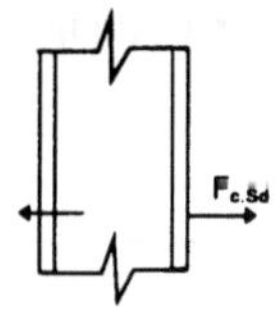  |
| 3 | Beam flange and web in compression | 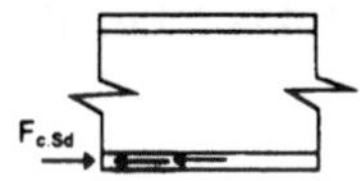  |
| 4 | Column flange in bending | 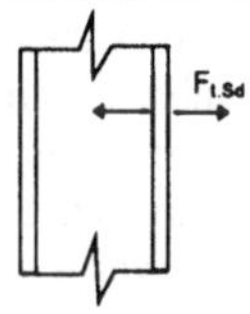  |
| 5 | Column web in tension | 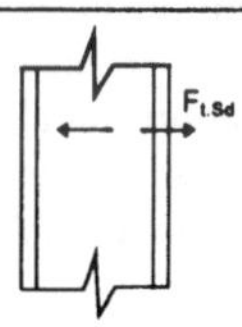  |
| 6 | End-plate in bending | 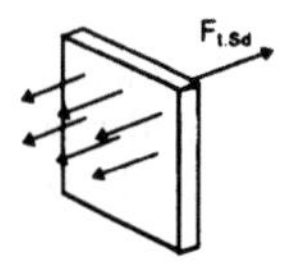  |
| 7 | Beam web in tension | 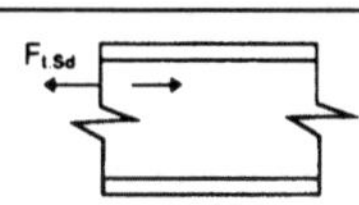  |
| 8 | Flange cleat in bending | 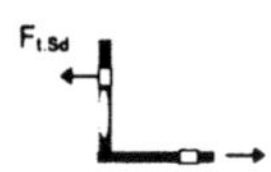  |

**Table 4** List of components covered by *Eurocode 3*

| | | |
|---|---|---|
| 9 | Bolts in tension | 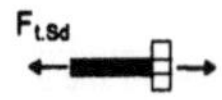  |
| 10 | Bolts in shear |   |
| 11 | Bolts in bearing (on beam flange, column flange, end-plate or cleat) | 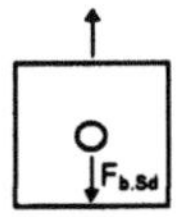  |
| 12 | Plate in tension or compression | 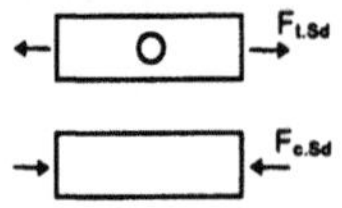  |

**Table 4** List of components covered by *Eurocode 3 (sheet 2)*

designer to determine the design joint moment-rotation characteristic whatever the type of analysis (Figure 28 to Figure 30). In Annex A of the present lecture, information is provided on how the stiffness and strength assembly is carried out.

The application of the component method requires a sufficient knowledge of the behaviour of the basic components. Those covered by *Eurocode 3* are listed in Table 1. The combination of these components allows one to cover a wide range of joint configurations, which should be sufficient to satisfy the needs of practitioners as far as beam-to-column joints and beam splices in bending are concerned. Examples of such joints are given in Figure 25.

Some fields of application can also be contemplated :

Joints subject to bending moment (and shear) and axial force;

Column bases subject to coincident bending moment, shear force and axial force where the components such as :

- concrete foundation in compression;
- end-plates with specific geometries;
- anchorages in tension;
- contact between soil and foundation,

will be activated.

These situations are however not yet covered, or only partially covered, by *Eurocode 3*.

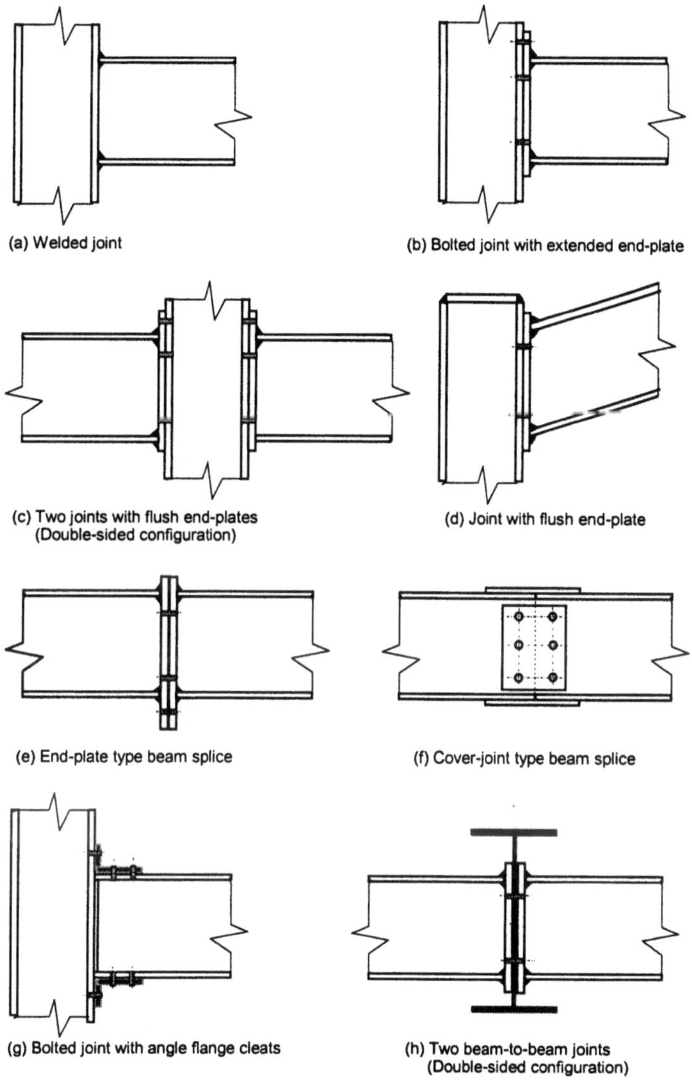


**Figure 25** Examples of joints covered *by Eurocode 3*

# 8 Joint Idealisation

The non-linear behaviour of the isolated flexural spring which characterises the actual joint response does not lend itself towards everyday design practice. However the moment-rotation characteristic curve may be idealised without significant loss of accuracy. One of the most simple idealisations possible is the elastic-perfectly plastic relationship (Figure 26.a). This modelling has the advantage of being quite similar to that used traditionally for the modelling of member cross-sections subject to bending (Figure 26.b).

The moment $M_{j.Rd}$ that corresponds to the yield plateau is termed the *design moment resistance* in *Eurocode 3*. It may be considered as the *pseudo-plastic moment resistance* of the joint. Strain-hardening effects and possible membrane effects are henceforth neglected, which explains the difference in Figure 26 between the actual *M-$\phi$* characteristic and the *yield plateau* of the idealisation.

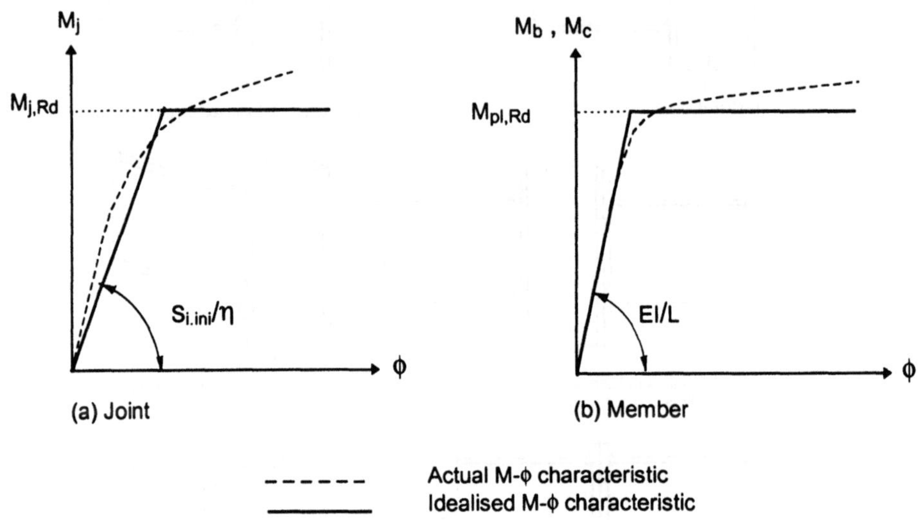


**Figure 26** Bi-linearisation of moment-rotation curves

The value of the constant stiffness $S_{j.ini}/\eta$ is discussed below.

In fact there are different possible ways to idealise a joint *M-$\phi$* characteristic. The choice of one of them is dependent upon the type of frame analysis which is contemplated:

- *Elastic idealisation for an elastic analysis* (Figure 27) :

The principal joint characteristic is the constant rotational stiffness.

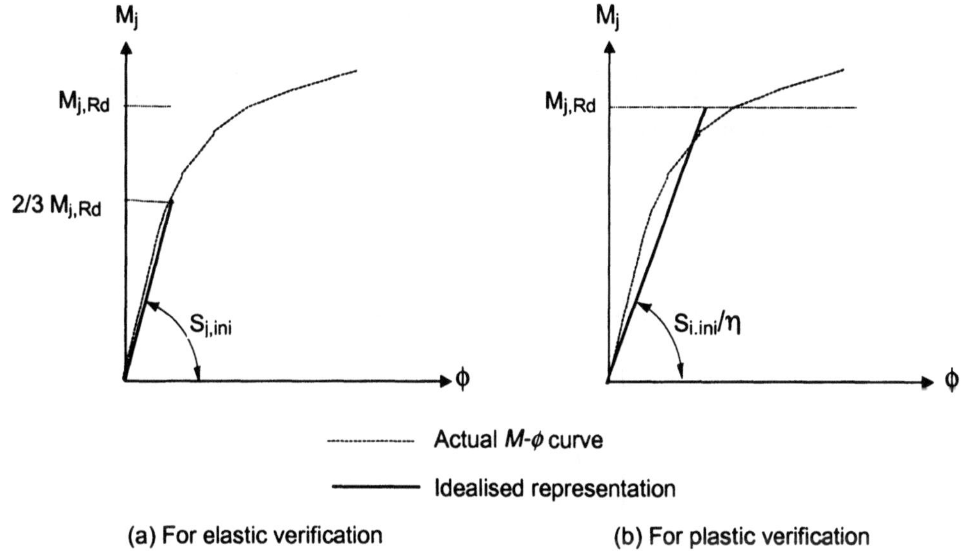


**Figure 27** Linear representation of a $M$-$\phi$ curve

Two possibilities are offered in *Eurocode 3-(revised) Annex J* :

- *Elastic verification of the joint resistance* (Figure 27.a) : the constant stiffness is taken equal to the initial stiffness $S_{j.ini}$; at the end of the frame analysis, a check that the design moment $M_{Sd}$ experienced by the joint is less than the maximum elastic joint moment resistance defined as 2/3 $M_{j,Rd}$.
- *Plastic verification of the joint resistance* (Figure 27.b) : the constant stiffness is taken equal to a fictitious stiffness, the value of which is intermediate between the initial stiffness and the secant stiffness relative to $M_{j,Rd}$; it is defined as $S_{j.ini}/\eta$. This idealisation is valid for $M_{Sd}$ values less than or equal to $M_{j,Rd}$ .

The values of $\eta$ are as follows (Table 5) :

| Type of connection | Beam-to-column joint | Other types of joint |
|---|---|---|
| Welded | 2 | 3 |
| Bolted end-plate | 2 | 3 |
| Bolted cleat | 2 | 3,5 |

**Table 5** Values of $\eta$

- *Rigid-plastic idealisation for a rigid-plastic analysis* (Figure 28).

Only the design resistance $M_{j,Rd}$ is needed. In order to allow the possible plastic hinges to form and rotate at the joint locations, it is necessary to check that the joint has a sufficient rotation capacity.

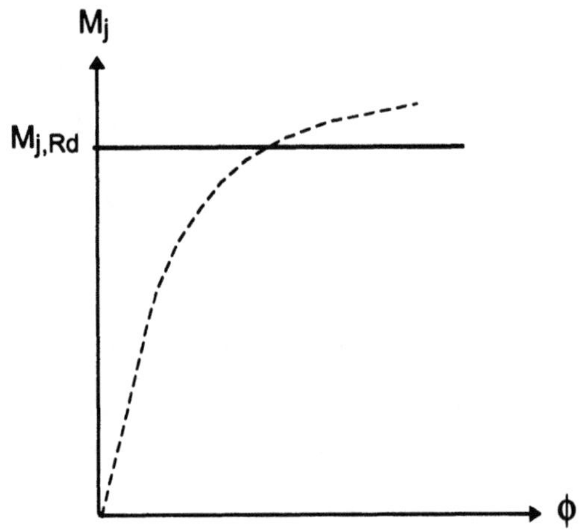


**Figure 28** Rigid-plastic representation of a $M$-$\phi$ curve

*Non-linear idealisation for an elastic-plastic analysis* (Figure 29).

The stiffness and resistance properties are of equal importance in this case. The possible idealisations range from bi-linear or tri-linear representations to the fully non-linear curve. Again rotation capacity is required in joints where plastic hinges are likely to form and rotate.

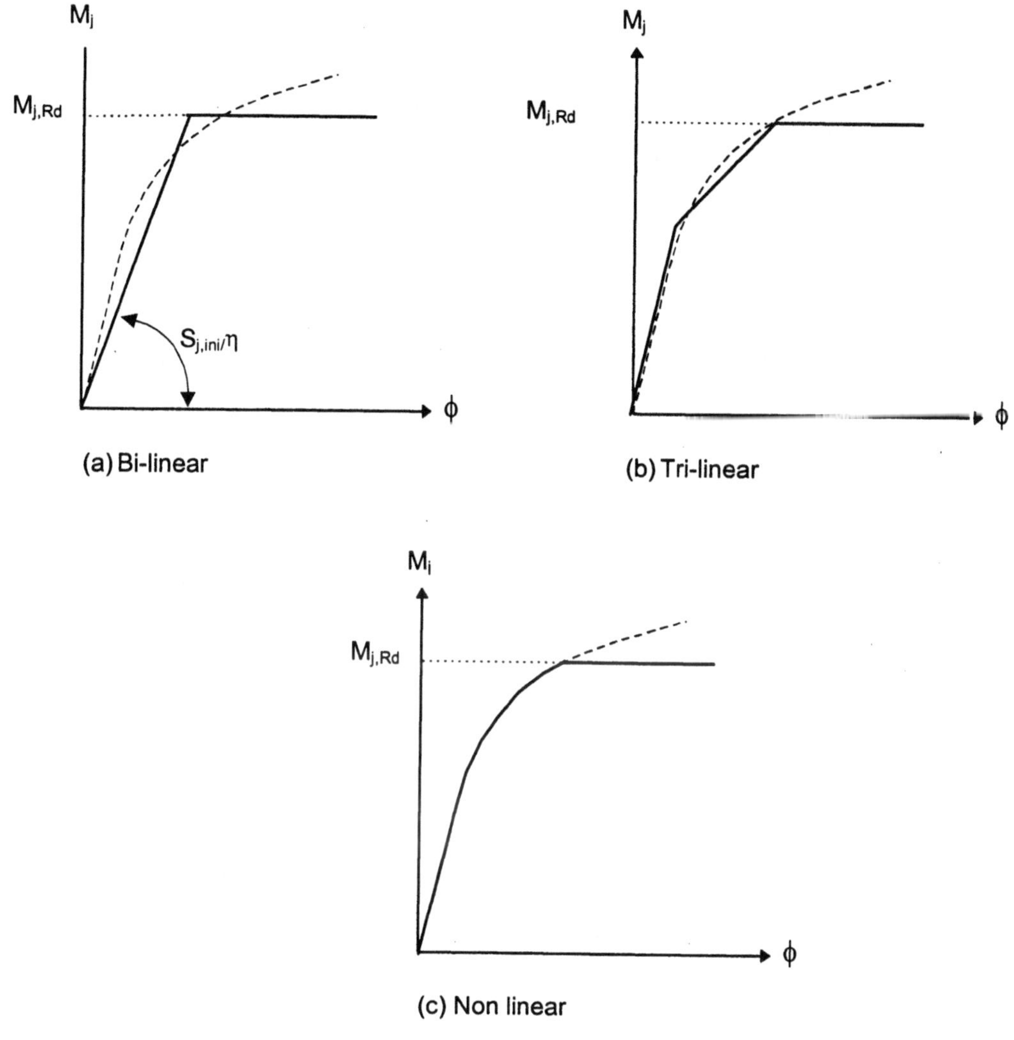


**Figure 29** Non-linear representations of a M-$\phi$ curve

## 9 Practical Tools for Joint Characterisation

For design purposes, detailed design aids are given in Annex B to the present chapter. Each of these design aids is devoted to a specific type of joint. They are composed of two parts :

a. *a calculation procedure*, presented in the format of design sheets;
b. *design tables.*

*The calculation procedure* is aimed at assisting the designer who wishes to take account of the full potential of joint semi-rigidity, without having to go through the more complex approach of *Eurocode 3-(revised)Annex J.*

The first design sheet is devoted to determining the mechanical characteristics and geometrical parameters of the joint. In the remaining sheets, the calculation procedure first provides all the expressions for both stiffness and resistance for each of the joint components in a logical order and finally shows how to derive the characteristics of the whole joint, i.e. the initial and nominal stiffnesses and the design moment resistance. The failure mode corresponds to the component whose resistance determines the design moment resistance of the joints. Additional considerations on joint design are given hereunder.

The *shear resistance of the joint (resistance to a shear force in the beam)* is an important value. However, for the sake of clarity, it *is not dealt with in the design sheets.* Relevant information is provided later in this lecture.

The second part of each of the design aids consists of *design tables*, which, in principle, can be used in a straightforward manner as an alternative to the design sheets. The tables, which are established for a selection of standard combinations of connected member sections, provide the designer with the following :

- the joint stiffnesses for design (initial and nominal);
- the design moment resistance and resistance in shear of the joint;
- an identification of the joint component which is governing the resistance moment;
- the reference lengths for the joint classification,

in addition to the material properties and the geometric parameters for the end plate, the bolts and the welds. The concept of reference length for joint classification is introduced in the Annex C to the present chapter.

The design tables can be used either to obtain the characteristics (i.e. mechanical properties of stiffness and resistance) of a given joint or to select a joint having desired characteristics.

They have been obtained using the calculation procedures given in the design sheets associated with options for the values of some parameters which generally give conservative results. However, there are some rare extreme situations where the use of the design tables alone may not be sufficient to ensure safe design. These situations are mostly related to the stress state (shear and direct stresses) of the column web panel (which is evaluated by the factors $\beta$ and $k_{wc}$) and of the column flange (factor $k_{fc}$) and which should normally be checked once the structural analysis has been completed.

The physical meanings of factors $\beta$, $k_{wc}$ and $k_{fc}$ are discussed in the next paragraphs where the recommended values for use in the design sheets are given. These recommendations have been adopted for the design tables.

The other options which have been taken when establishing the design tables, are also identified. They relate to weld and bolt sizes in end plate and cleated joints.

Finally, the joint classification reference lengths (see Annex C to the present lecture) are specified.

## 9.1 Stresses in the column web panel and flange

### 9.1.1 Factor β

The column web panel deformation is due to the actual shear force in it. This shear force ($V_{wp}$) is mostly the local shear force arising from the load-introduction forces ($F$), but, in addition, there is a contribution from the column shear forces above and below the joint (Figure 30.a). For design purposes, it is obtained by multiplying the force F by the factor $\beta$ (Figure 30.b). Strictly speaking, precise values of $\beta$, which can affect both the resistance and the stiffness of the web panel, can only be obtained by iteration. However, by using the safe values given in Table 1, there will normally be no need to resort to such an iterative procedure, which would involve re-analysing the structure with corrected $\beta$ values.

The selection of the $\beta$ values from Table 6 is based, for two-sided joint configurations, on the relative importance of the left and right moments acting in the connected beams. Usually this information is unknown in a first design step - except in particular loading and geometrical situations – what prevents the designer from selecting a correct value for $\beta$. To overcome this difficulty, the following guidelines are suggested:

*The value $\beta = 1$ can be used in a first step. In fact, the design tables have been prepared for $\beta = 1$.*

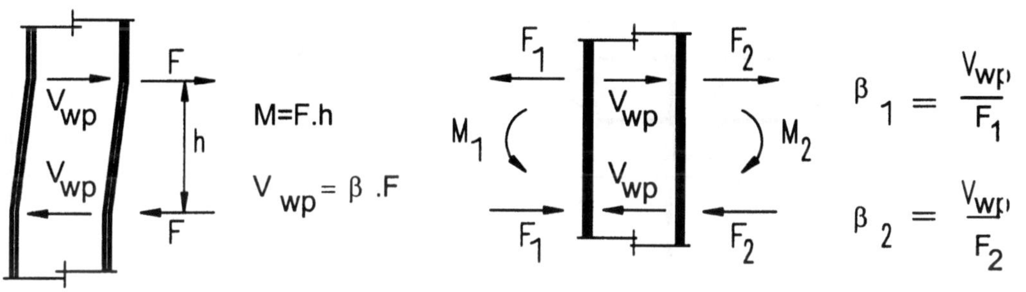


(a) Definitions (b) $\beta$ values

**Figure 30** Shear force in a column web panel

Once the global analysis with due allowance made for semi-rigidity of the joints is performed, the designer should check whether the assumption which has been made ($\beta = 1$) is satisfactory and whether it is necessary or worthwhile to update the joint design and/or the global analysis. In this process, it should be recognized that:

- *If the new $\beta$ value to be selected is equal to 0*, the results obtained from the first analysis with $\beta = 1$ are safe and conservative. They can therefore been adopted as they are, or, if preferred, the design sheets can be used to obtain more appropriate characteristics for the joint in view of a second frame analysis.
- *If the new $\beta$ value to be selected is equal to 2*, the results obtained from the first analysis with $\beta = 1$ are unsafe and the designer should perform a new analysis based on more appropriate joint characteristics obtained through the use of the design sheets with $\beta = 2$. Fortunately, such situations may be considered as exceptional cases rarely met in practice.

If, or when, a second analysis is performed, the opportunity should also be taken to make any other modifications which may appear necessary to the member and/or joint choices.

| Configuration and | Proposed value $\beta$ |
|---|---|
| 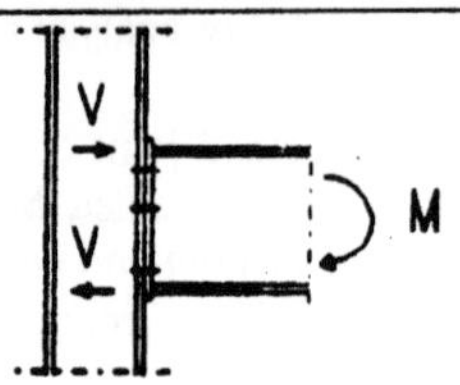  | $\beta = 1$ |
| 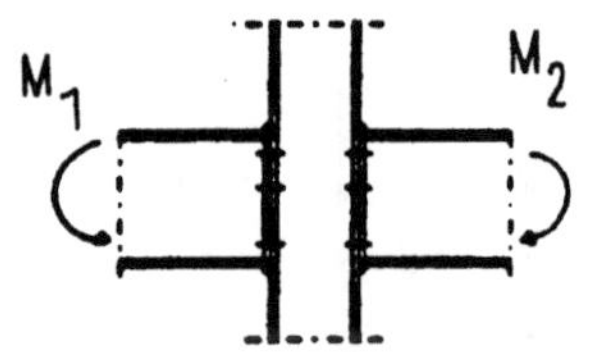  | if $\lvert M_1 \rvert \cong \lvert M_2 \rvert$<br>$\beta_1 = \beta_2 = 0$<br>if $\lvert M_1 \rvert \neq \lvert M_2 \rvert$<br>$\beta_1 = \beta_2 = 1$ |
| 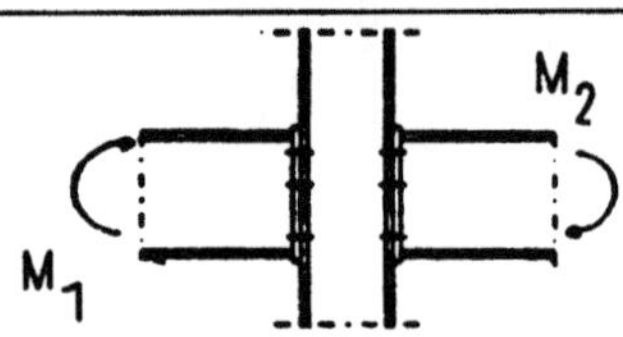  | $\beta_1 = \beta_2 = 2$<br>if $M_1$ or $M_2 \cong 0$<br>$\beta_1 = \beta_2 = 1$ |

**Table 6** Recommended $\beta$

### 9.1.2 Factor $k_{wc}$

The factor $k_{wc}$ accounts for the detrimental effect of the longitudinal web stresses (due to the normal force and the bending moment in the column) on the local design resistance of the column web in compression (local stability).

Factor $k_{wc}$ is given as follows :

$$k_{wc} = (1{,}25 - 0{,}5\frac{\sigma_{n,wc}}{f_{ywc}}) \leq 1.0 \qquad (1)$$

where $f_{ywc}$ is the yield stress of the column web and $\sigma_{n,wc}$ is the maximum longitudinal compressive (axial force and bending moment) stress in the column web at the root of the radius at a point just outside the joint (Figure 31.b). Its value is plotted in Figure 31.a.

The value of $k_{wc}$ ranges from 1,0 to 0,75 (the most unfavourable case). However, for design purposes, adopting a value of 1,0 for design will usually be safe, as in most situations: $\sigma_{n,wc} < 0{,}5\, f_{ywc}$.

The design tables have been established with $k_{wc} = 1{,}0$.

Once the global frame analysis is completed, it is essential that the designer checks that the assumption on the column stress condition is not violated . Should the value of $k_{wc}$ prove to be less than unity, the designer will have to use the design sheets to check the resistance of the joint – (by simply re-evaluating the resistance of the column web in compression). The stiffness of the joint will not be affected.

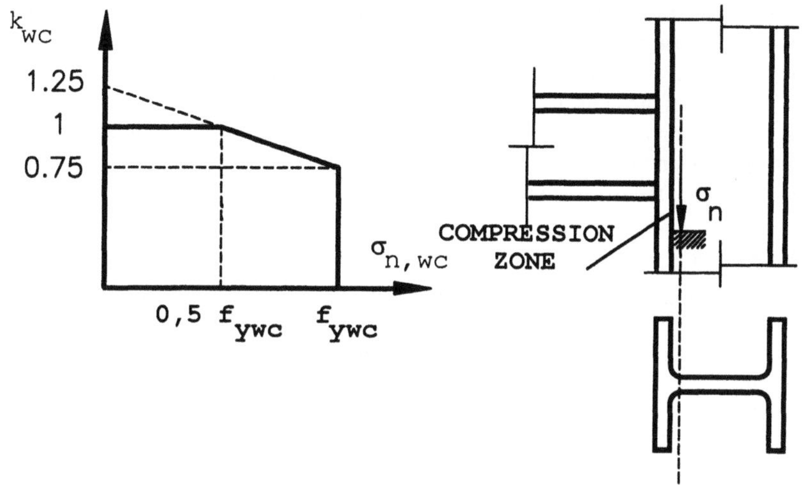


(a) Reduction factor $k_{wc}$ (b) Location of $\sigma_{n,wc}$

**Figure 31** Values of the $k_{wc}$ factor and the point where $\sigma_{n,wc}$ is calculated

## Factor $k_{fc}$

The factor $k_{fc}$ accounts for the detrimental effect of the longitudinal flange stresses (due to the normal force and the bending moment in the column) on the local resistance of the column flange in bending.

Factor $k_{fc}$ is given as follows:

$$k_{fc} = 1 \qquad \text{if } \sigma_{n,fc} \leq 180\ N/mm^2 \qquad (2.a)$$

$$k_{fc} = \frac{2 f_{yfc} - 180 - \sigma_{n,fc}}{2 f_{yfc} - 360} \qquad \text{if } \sigma_{n,fc} > 180\ N/mm^2 \qquad (2.b)$$

where $f_{yfc}$ is the yield stress of the column flange and $\sigma_{n,fc}$ is the maximum longitudinal compressive stress (axial force and bending moment) in the centroïd of the column flange. In expressions (2), $f_{yfc}$ and $\sigma_{n,fc}$ have to be expressed in N/mm$^2$.

The value of $k_{fc}$ ranges from 1,0 to 0,5 (the most unfavourable case). However, for design purposes, adopting a value of 1.0 for design will usually be safe, as in most situations: $\sigma_{n,fc} < 0{,}5\, f_{yfc}$.

The design tables have been established with $k_{fc} = 1{,}0$.

Once the global frame analysis is completed, it is essential that the designer checks that the assumption on the column stress condition is not violated . Should the value of value of $k_{fc}$ prove to be less than unity, the designer will have to use the design sheets to check the resistance of the joint (by simply revaluating the resistance of the column web in compression). The stiffness of the joint will not be affected.

## 9.2 Additional design considerations and options taken for the design tables

A number of other choices have been made when establishing the design tables, all of which have an impact on the joint design. They concern the following subjects :

- the size of the welds connecting the beam flanges and the end plates;
- the diameter of the bolts used in joints with extended end plates;
- the contribution of the web cleat to the global properties of a cleated joint;
- the slip which may occur in flange cleated joints.
- the choices for the design tables.

### 9.2.1 Weld sizes

It is assumed that double fillet welds will be used to attach the beam to the end plate in end plate joints.

The basic relations for the design of these welds are given here below for both extended and flush end plate joints in which the end plate height is greater than the beam depth. Two approaches can be used, as follows:

- Design according to the resistance of the beam .

The resistance of the fillet welds connecting the beam flanges to the end plate shall be at least equal to the plastic resistance of the flanges $N_{t.Rd}$ :

$$N_{t.Rd} = b_f t_{fb} f_{yfb} / \gamma_{M0} \tag{3}$$

where $f_{yfb}$ is the yield stress of the beam flange, $t_{fb}$ the flange thickness and $b_f$ the beam width.
Using a simplified method for weld design, one obtains:

$$2a_f b_f f_{yw,d} \geq N_{t.Rd} \tag{4}$$

with the design shear strength of the weld defined as :

$$f_{yw,d} = f_u / \sqrt{3}\beta_w \gamma_{Mw} \tag{5}$$

where $a_f$ is the weld size and $f_u$ is the design yield strength of the weaker element attached (end plate or beam flange) .

The following expression is obtained for the weld size;

$$a_f \geq \sqrt{3}\, \gamma_{Mw} \beta_w f_{yb}\, t_{fb} / 2 f_u \gamma_{M0} \tag{6.a}$$

A more economical weld design using Annex M of EC3 leads to:

$$a_f \geq \gamma_{Mw} \beta_w f_{yb}\, t_{fb} / \sqrt{2} f_u \gamma_{M0} \tag{6.b}$$

Similar expressions can be obtained for the end plate to beam web fillet weld size.

The values of the partial safety factors are: $\gamma_{Mw}$ = 1,25 and $\gamma_{M0}$ = 1,1.

The values of $\beta_w$ depend on the steel strength $f_u$ , as follows:

| | | | |
|---|---|---|---|
| $\beta_w$ | = | 0,80 | for steel S 235 |
| | = | 0,85 | for steel S 275 |
| | = | 0,90 | for steel S 355 |
| | = | 1,00 | for steel S 420 |
| | = | 1,00 | for steel S 460 |

- Design according to the resistance of the joint.

The design force in the beam flange can be conservatively estimated as:

$$F_{Rd} = M_{Rd}/(h_b - t_{fb}) \qquad (7)$$

where $M_{Rd}$ is the design moment resistance of the joint, $h_b$ and $t_{fb}$ are the beam depth and flange thickness respectively.

The design resistance of the weld shall be greater than the flange force $F_{Rd}$, multiplied by a factor $\gamma$, and using the simplified formula for weld strength one obtains:

$$2a_f b_f f_{vw,d} \geq \gamma M_{Rd} /(h_b - t_{fb}) \qquad (8)$$

The value of the factor $\gamma$ is ( EC3 revised Annex J, paragraph J.3.1.3(4)) :

$\gamma = 1{,}7$ for sway frames and

$\gamma = 1{,}4$ for non sway frames.

*However, in no case shall the weld design resistance be required to exceed the design plastic resistance of the flange, $N_{t.Rd}$, given here above.*

*It is important to note also that, as this verification does not include for the beam shear, the designer shall ensure that proper allowance has been made for this force.*

The above methods may be adapted for the design of welds in joints with flush end plate connections in which the height of the end plate is slightly less than the beam depth. The detailing of the welds for this case is shown in Figure 32.b.

For joints with small and medium size beam sections ( up to about 400mm in depth), a very simple rule, which complies with the design according to the beam resistance, is to use a weld throat size $a_f$ (Figure 32.a) of 50 % of the thickness , i.e. :

*for the double fillet welds to the beam flange,*

$$a_f \cong 0{,}5t_{fb} \qquad (9.a)$$

*for the double fillet welds to the beam web,*

$$a_w \cong 0{,}5t_{wb} \qquad (9.b)$$

*Rounded up values given by this rule have been adopted in the design tables.* An advantage of this solution is that , as the resistant section of the welds is equivalent to that of the attached beam, the verification of the beam section at the joint for combined bending, shear and axial load also ensures that the welds, chosen as above, are sufficient.

For larger joints, it may be more economical to proportion the welds to be in accordance with the joint resistance rather than with the beam resistance.

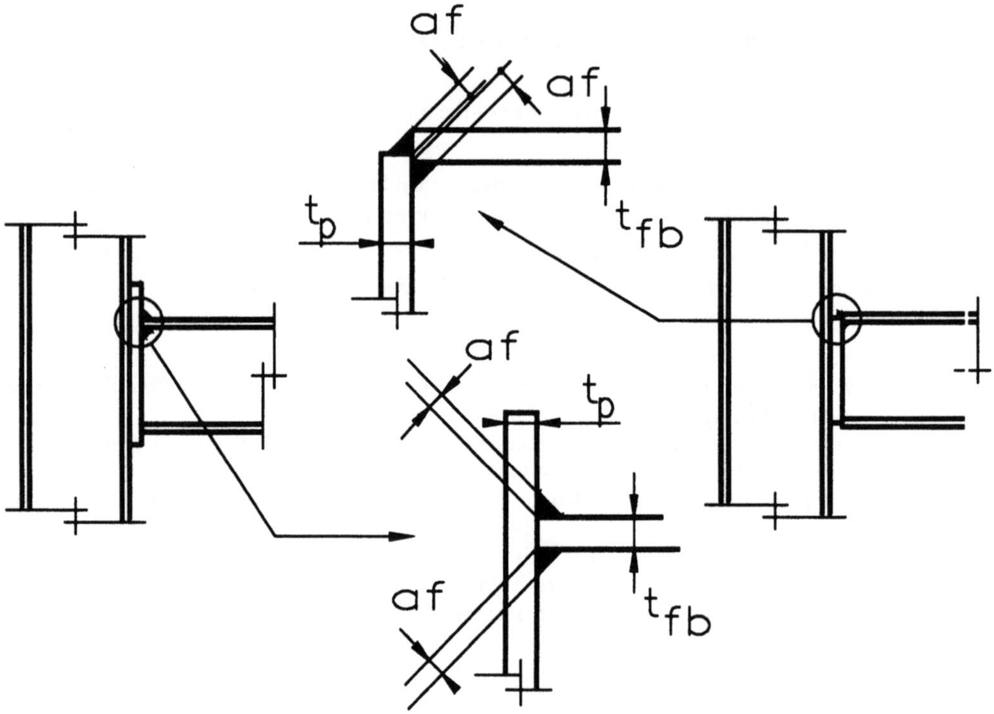


(a) Extended end plate (b) Flush end plate

**Figure 32** Beam flange-to-end plate welds

### 9.2.2 Bolt diameters

For joints with extended end plate connections, the attainment of the design moment resistance $M_{Rd}$ by full plastic redistribution of the internal bolt row forces requires a minimum ductility of the bolt-plate assemblies. In accordance with *Eurocode 3-(revised) Annex J*, the relevant requirements are :

$$\frac{d}{t_p} \geq 1{,}9\sqrt{f_{yp}/f_{ub}} \tag{10.a}$$

or

$$\frac{d}{t_{fc}} \geq 1{,}9\sqrt{f_{yfc}/f_{ub}} \tag{10.b}$$

where:

$d$ : nominal diameter of the bolt;

$f_{yp}$ : design yield strength of the end plate material;

$f_{yfc}$ : design yield strength of the column flange;
$f_{ub}$ : ultimate tensile strength of the bolt material;

$t_p$ : thickness of the end plate or flange cleat leg;

$t_{fc}$ : thickness of the column flange.

The design tables fulfil the above requirements.

### 9.2.3 Cleated joints

Joints with cleats connecting the beam flanges to the column flange, but without web cleats, are mainly considered here. As a matter of fact, web cleats can be dispensed with when the lower flange cleat connection is sufficient to transfer the beam shear force. The following conditions shall be met for that to be valid:

- the gap $g$ between the end of the beam and the face of the column shall not exceed the thickness $t_a$ of the angle cleat;
- the design shear resistance of the bolts connecting the lower cleat to the column shall be sufficient to transfer the entire shear force in the beam.

When the shear force exceeds the resistance of the lower cleat bolts, web cleats shall be provided. Some fabricators systematically add one or two web cleats for erection purposes.

When these are attached close to the beam axis and are connected by no more than two bolts (in one vertical row) to the beam web, the stiffness and the resistance of the "fully" cleated joint will not be significantly altered and their effect on the joint characteristics as a whole may then be neglected.

As the number of bolts and the height of the web cleats increases, so does the subsequent increase in the joint stiffness and resistance , which may become significant. Reference can be made to other publications on how to include their contribution.

b. Slip in the joint.

The tolerance between the bolt and the hole diameters may leave room for some slip to occur between the cleats and the members to which they are connected. The movements between the cleats and the beam flanges in particular will produce a rotation of the beam ends relative to the columns. In reality, all of this slip can be expected to take place progressively during the erection process. For this reason, it is usual to assume that it will have a negligible effect on the global response of the structure.

When no slip is to be allowed, preloaded bolts shall be used and the design shall be in accordance with the specifications for slip resistant connections of *Eurocode 3-(revised) Annex J.* This requirement may arise, for instance, when moment reversal is expected in a joint.

### 9.2.4 Choices for the design tables

The design tables have been prepared for a wide range of combinations of standard rolled sections in S235 steel. Separate tables have been prepared for joints with property classes 8.8 and 10.9 bolts. End plate thicknesses have been chosen to produce joints which have a large range of semi-rigid behaviour.

The tables can obviously be extended to cover a much wider range of joints and, in particular, those with members in higher strength steels.

The following summarises the range of cases covered, where each case is made up of a combination of elements from the relevant table:

A. Beam to column end plate joints: 364 cases for each of the following:
   Extended end plate with pc 8.8 bolts,
   Extended end plate with pc 10.9 bolts,

Flush end plate, full height, with pc 8.8 bolts,
Flush end plate, full height, with pc 10.9 bolts,
Flush end plate, short height, with pc 8.8 bolts,
Flush end plate, short height, with pc 10.9 bolts.

| ELEMENT | SECTION | SIZE |
|---|---|---|
| COLUMN | HEB | 140 TO 600 |
| BEAM | IPE | 220 TO 600 |
| PLATE | THICKNESSES | 15,20,25mm |
| BOLT | DIAMETERS | M16,M20,M24,M27 |

B. Beam to beam end plate connections: 36 cases for each of the following:
Flush end plate, full height, with pc 8.8 bolts,
Flush end plate, full height with pc 10.9 bolts,
Flush end plate, short height, with pc 8.8 bolts,
Flush end plate, short height, with pc 10.9 bolts.

| ELEMENT | SECTION | SIZE |
|---|---|---|
| BEAM | IPE | 160 TO 600 |
| PLATE | THICKNESSES | 15,20,25mm |
| BOLT | DIAMETERS | M16,M20,M24,M27 |

C. Beam to column flange cleat joints: 310 cases for each of the following:
Cleats with pc 8.8 bolts,
Cleats with pc 10.9 bolts.

| ELEMENT | SECTION | SIZE |
|---|---|---|
| COLUMN | HEB | 140 TO 600 |
| BEAM | IPE | 220 TO 600 |
| CLEAT | UNEQUAL ANGLES | 130x65x72 to 200x100x16 |
| BOLT | DIAMETERS | M16,M20,M24,M27 |

## 9.3 Shear resistance

In the case of bolted end plate joints the entire shear force in the beam shall be resisted independently by both:

- the welds between the end plate and the beam,
- and the bolts rows.

When non preloaded bolts are used, the design shear force transferred by the bolts from the end plate to the column flange should not normally exceed the sum of :

a. the total design shear resistance of the bolts in those bolts-rows that are not required to resist tension;

b. 30 % of the total design shear resistance of the bolts in those bolt-rows that are also required to resist tension.

Reference should be made to Section 1.3.3.a for more information on cleated joints.

## 9.4 Reference length for stiffness classification purposes

For typical building frames, for instance those shown in Figure 33, the length $L_b$ is defined as the span of one bay. It is assumed that the beam-to-column joints at the ends of the beam have similar joint characteristics( stiffness and resistance).

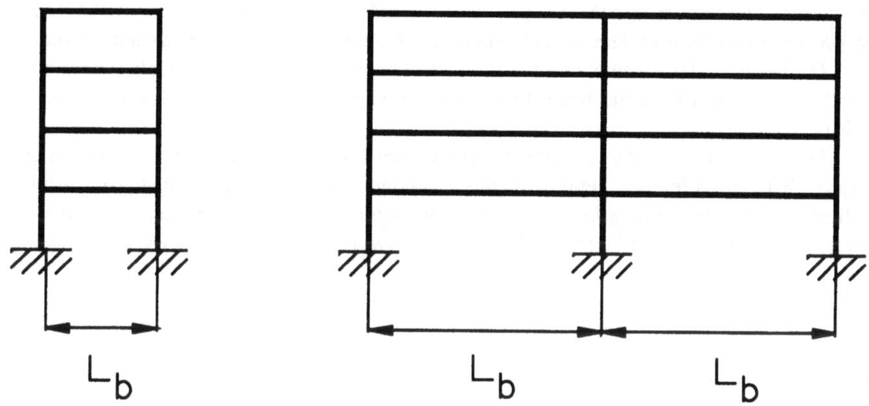


**Figure 33** Definition of $L_b$ for building frames

The beam reference lengths for joint classification, $L_{b.b}$ and $L_{b.u}$ for braced and for unbraced frames respectively, are expressed as follows, the definitions being those of the Annex C to the present chapter:

- In braced frames

$L_{b.b.1} = 8\ EI_b / S_{j.ini}$ for the boundary between rigid and semi-rigid;

$L_{b.b.2} = 0{,}5\ EI_b / S_{j.ini}$ for the boundary between semi-rigid and pinned.

- In unbraced frames

$L_{b.u.1} = 25\ EI_b / S_{j.ini}$ for the boundary between rigid and semi-rigid;

$L_{b.u.2} = 0{,}5\ EI_b / S_{j.ini}$ for the boundary between semi-rigid and pinned.

For beam splices, no distinction is made between braced and unbraced frames and a single value of the reference ($L_b$) is given :

$L_{b.1} = 25\ EI_b / S_{j.ini}$ for the boundary between rigid and semi-rigid;
$L_{b.2} = 0{,}5\ EI_b / S_{j.ini}$ for the boundary between semi-rigid and pinned.

The actual bay span $L_b$ has to be compared with these reference lengths in order to determine whether the joint has to be considered as pinned, semi-rigid or rigid in the global analysis.

When use is made of the design tables, the information provided to the designer under the column with the heading "Reference lengths" will usually be one of the following :

- *The letter R* : This means that ***the joint can be considered as rigid*** over the full range of practical lengths for which the beam in question is expected to be used. The reference lengths fall outside of this range defined as follows:

  minimum practical length: the greater of $5.h_b$ and 2,5 meters;
  maximum practical length: $40.h_b$
  where $h_b$ is the height of the beam (in meters).
- *A number followed by the letter R* : The number is the reference length in meters and the label R means that the reference length is the boundary between the semi-rigid and rigid joint classifications. ***For beam lengths greater than the reference length, the joint can be considered as rigid, while for lengths less than the reference length it shall be considered as semi-rigid.***
- *The letter S* : This means that ***the joint shall be considered as semi-rigid*** over the full range of practical lengths for which the beam in question is expected to be used. The reference lengths fall outside of this range.
- *The letter S followed by a number :* The number is the reference length in meters and the label P means that the reference length is the boundary between the pinned and semi-rigid joint classifications. ***For beam lengths greater than the reference length, the joint shall be considered as semi-rigid, while for lengths less than the reference length it can be considered as pinned.***

Example :

Consider the case of a joint with an extended end-plate connection, bolts *HR10.9*, *M16*, column *HEB240*, beam *IPE220*.

If the connection is used in a non braced frame : $L_{bu} = 2{,}7m$

If $L_b>2{,}7m$ : The connection may be considered rigid.

If $L_b<2{,}7m$ : The connection may be considered semi-rigid.

If the connection is used in a braced frame : $L_{bb} = 8{,}6m$

If $L_b>8{,}6m$ : The connection may be considered rigid.

If $L_b<8{,}6m$ : The connection may be considered semi-rigid.

# *Annex A Evaluation of the stiffness and resistance properties of the joints according to Eurocode 3-(revised) Annex J*

The component method presented in this lecture is a three step procedure for evaluating the stiffness and resistance properties of structural joints. In the first step, the list of constitutive components is established. The stiffness and/or resistance properties of the components are then derived.

The assembly of the components constitutes the third and last step of the method. As its name indicates, it consists in assembling the individual components so as to derive the mechanical properties of the whole joint. The relationship between component properties and joint properties is based on what is commonly called the "distribution of internal forces in the joint". The latter consists of determining, for a given set of external forces acting on the joint, the manner in which these forces are distributed between all the constitutive components. The force to which any given component is subjected is termed the "internal force".

This notion applies not only to structural joints but also to any cross-section in a member and a parallel between these situations is now drawn. In any cross-section of beams and columns, the distribution of internal forces in it is required so as to permit computation of its flexural rigidity or its resistance in bending, shear, torsion and/or axial compression or tension. That is why, in the following paragraphs, the word "cross-section" covers beam and column sections as much as joint sections.

The distribution of internal forces has to be carried out in a rational way, and has, from a theoretical point of view, to fulfil the following requirements :

- the internal forces given by the distribution have to be in equilibrium with the external forces acting on the cross-section;
- the compatibility of the displacements between the constitutive parts of the cross-section - namely components in the case of a joint – has to be satisfied;
- each part of the cross-section has to be able to transfer the internal force to which it is subjected;
- the maximum deformation capacity of each part of the cross-section must not be exceeded.

In the case of a beam or column H or I cross-section subjected to a bending moment, the distribution of internal forces - read stresses in this particular case - in the elastic range is generally assumed to follow the Navier rule (Figure A.1).

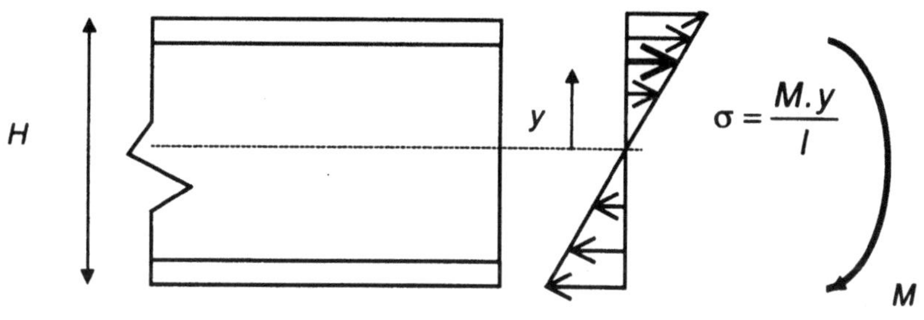


(a) Beam (b) Internal stresses (c) External force

**Figure A.1** Elastic distribution of internal forces (here stresses) in a beam profile in bending (I = second moment of area)

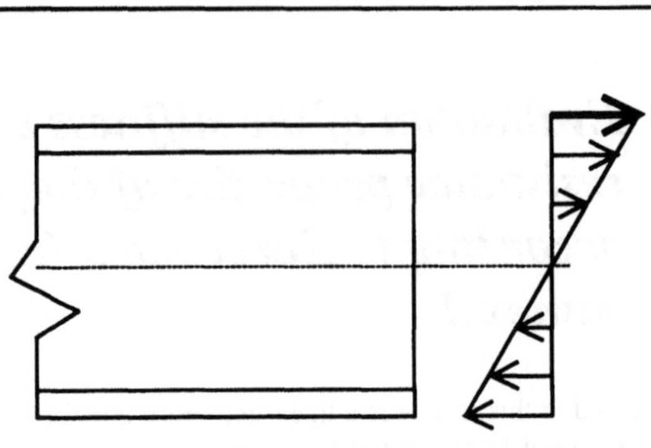


**Figure A.2** Internal forces – here stresses – corresponding to the maximum elastic design moment resistance of the cross-section

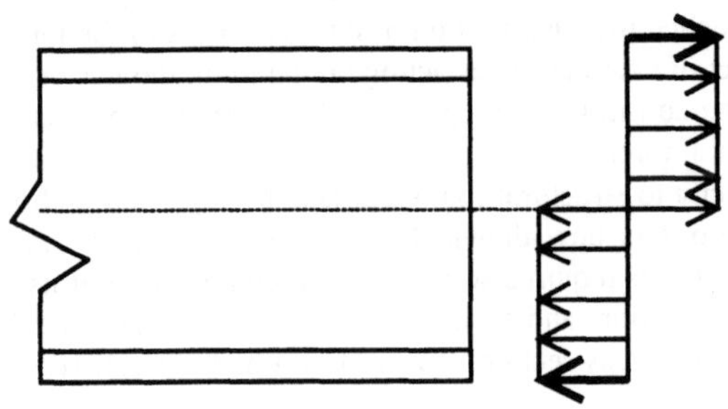


**Figure A.3** Internal forces - stresses here – corresponding to the plastic moment resistance of the cross-section

When the maximum internal stresses (at $y = \pm H/2$) reach the value of the yield stress $f_y$ of the constitutive steel divided by a partial safety factor $\gamma_M$, the maximum elastic moment resistance of the cross-section is obtained (Figure A.2). The design moment resistance is expressed as :

$$M_{Rd} = \frac{I}{H/2}\frac{f_y}{\gamma_M} = Wf_y/\gamma_M \tag{A.1}$$

where $W$ is the elastic modulus of the cross-section in bending.

To profit from the extra resistance provided by the plastification of the cross-section, reference has therefore to be made to the distribution represented at Figure A.3. The design resistance then becomes:

$$M_{Rd} = Zf_y/\gamma_M \tag{A.2}$$

where $Z$ denotes the plastic modulus of the cross-section in bending.

In these distributions, Bernoulli's assumption has been considered so as to ensure a compatibility between the elongation - or shortening - of all the constitutive fibers of the cross-section. Both elastic and plastic distributions illustrated in Figure A.2 and Figure A.3 are in equilibrium with the applied external forces and respect the plasticity criterion ($\sigma \le f_y/\gamma_M$). Therefore, they satisfy the first three of the four above-mentioned criteria which any distribution of internal forces should fulfil.

To reach the elastic or plastic moment resistance, the constitutive fibers of the section have to possess a sufficient deformation capacity so as to reach the yield stress in the case of the elastic distribution, or, in the case of the plastic distribution, to reach the yield stress and also to allow a plastic redistribution of the stresses between the adjacent fibers. This means that no premature local buckling of one of the section walls in compression and that no rupture of the material in tension must occur which would limit the moment capacity of the section. Specific criteria are provided in the codes so as to prevent the user from overestimating the resistance when such limitations apply. In these conditions, the fourth criterion on ductility is also fulfilled.

In a H or I profile, it is also common practice to replace the bending moment applied to the cross-section by a couple of forces located at the level of the centre lines of the beam flanges. The intensity of the forces is limited to the design resistance of the flanges in tension and compression, due attention being paid to the possible local buckling of the flange in compression. The web, the bending resistance of which is neglected, is usually devoted to the transfer of shear forces. Three of the four identified criteria are therefore satisfied. However the fourth condition (compatibility) is neglected. This so-called static approach is known to lead to a safe estimation of the design resistance of the section and is usually followed for sake of simplicity.

The procedure to distribute internal forces within structural joints is quite similar to that described in the foregoing paragraphs for beam and column cross-sections. In the next paragraphs, the procedure followed by *Eurocode 3-(revised) Annex J* is described; it applies to steel beam-to-column joints and beam splices where the beam(s) is (are) subject to bending moments and shear forces. For sake of simplicity and to allow for a hand calculation, two separate distribution procedures are detailed, one for the evaluation of the elastic initial stiffness and the other for the assessment of the design resistance of the joint.

The initial elastic stiffness and the design resistance are considered by *Eurocode 3* as the two main parameters characterizing the response of a joint in bending. Based on these two values, a full $M$-$\phi$ curve can then be derived as shown in Figure A.4.

Provided that the non-linear $M$-$\phi$ curve of *(revised)-Annex J* is not limited by the rotational capacity ($\phi_{Cd}$), this curve consists of three parts. Up to a level of 2/3 of the design moment resistance $M_{Rd}$ , the curve is assumed to be linear elastic and the corresponding stiffness is the so-called initial stiffness $S_{j,ini}$. Between $2/3\ M_{Rd}$ and $M_{Rd}$, the curve is non-linear. Once the moment in the joint reaches $M_{Rd}$, a yield plateau develops under further imposed rotations of the joint.

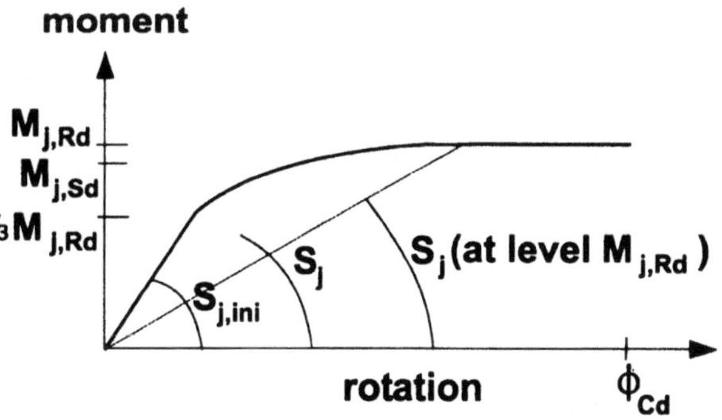


**Figure A.4** Non-linear $M$-$\phi$ curve according to Annex J

The model assumes a fixed ratio between the initial stiffness $S_{j,ini}$ and the secant stiffness at the intersection between the non-linear part and the yield plateau ($S_j$ at level $M_{Rd}$). For end-plate and welded joints, this ratio is equal to 3. For flange cleat joints, this ratio is 3,5.

The shape of the non-linear part between $2/3\ M_{Rd}$ and $M_{Rd}$ is given by the following interpolation formula :

$$S_j = \frac{S_{j,ini}}{\left(\frac{1{,}5M_{Sd}}{M_{Rd}}\right)^{\psi}} \quad \text{(A.3)}$$

where $\psi = 2{,}7$ for end-plate and welded joints and 3,1 for flange cleat joints.
In this interpolation formula, the value of $S_j$ is dependent on $M_{Sd}$.

**Stiffness assembly**

As *(revised)-Annex J* refers to the so-called component method, the rotational response of a joint is based on the mechanical properties of its different constitutive components. The advantage is that an engineer is able to calculate the mechanical properties of any joint by decomposing the joint into relevant components. Annex J gives direct guidance for end-plate, welded and flange cleat joints for this decomposition. Table A.1 shows an overview of components to be taken into account when calculating the initial stiffness for these types of joints.

| Component | Number | End-plated | Welded | Flange cleated |
|---|---|---|---|---|
| Column web panel in shear | 1 | x | x | x |
| Column web in compression | 2 | x | x | x |
| Column flange in bending | 3 | x | | x |
| Column web in tension | 4 | x | x | x |
| End-plate in bending | 5 | x | | |
| Flange cleat in bending | 6 | | | x |
| Bolts in tension | 7 | x | | x |
| Bolts in shear | 8 | | | x |
| Bolts in bearing | 9 | | | x |

**Table A-1** Overview of components for different joints

In the model, it is assumed that the deformations of the following components : a) beam flange and web in compression, b) beam web in tension and c) plate in tension or compression are included in the deformations of the beam in bending. Consequently they are not assumed to contribute to the flexibility of the joint. Haunches, when present, also fall in this category.

The initial stiffness $S_{j,ini}$ is derived from the elastic stiffnesses of the components. The elastic behaviour of each component is represented by an extensional spring. The force-deformation relationship of this spring is given by :

$$F_i = k_i . E . \Delta_i \quad \text{(A.4)}$$

where: $F_i$ is the force in the spring $i$;
$k_i$ is the stiffness coefficient of the component $i$;
$E$ is the Young modulus;
$\Delta_i$ is the deformation of the spring $i$.

The spring components in a joint are combined into a spring model. Figure A.5 shows for example the spring model for an unstiffened welded beam-to-column joint.

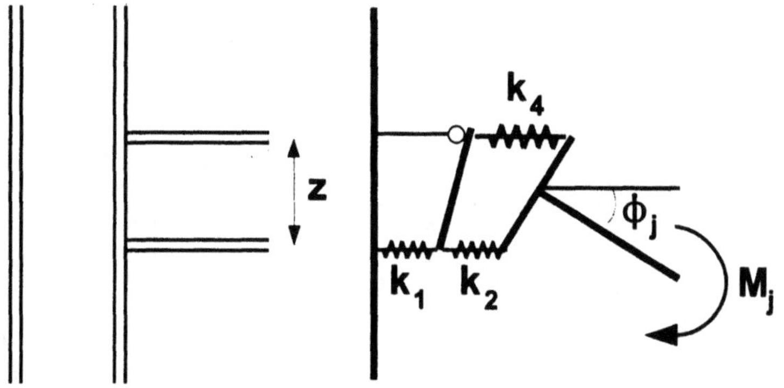


**Figure A.5** Spring model for an unstiffened welded joint

The force in each spring is equal to $F$. The moment $M$ acting on the spring model is equal to $F \cdot z$, where $z$ is distance between the centre of tension (for welded joints, located in the centre of the upper beam flange) and the centre of compression (for welded joints, located in the centre of the lower beam flange). The rotation $\phi$ in the joint is equal to $(\Delta_1+\Delta_2+\Delta_4)/z$. In other words:

$$S_{j,ini} = \frac{M}{\phi} = \frac{F_z}{\frac{\Sigma\Delta_i}{z}} = \frac{Fz^2}{\frac{F}{E}\Sigma\frac{1}{k_i}} = \frac{Ez^2}{\Sigma\frac{1}{k_i}} \tag{A.5}$$

The same formula applies for an end-plate joint with a single bolt-row in tension and for a flange cleat joint. However, the components to be taken into account are different, see Table A.1.

Figure A.6.a shows the spring model adopted for end-plate joints with two or more bolt-rows in tension. It is assumed that the bolt-row deformations for all rows are proportional to the distance to the point of compression, but that the elastic forces in each row are dependent on the stiffness of the components. Figure A.6.b shows how the deformations $k_{i,r}$ of components 3, 4, 5 and 7 are added to an effective spring per bolt-row, with an effective stiffness coefficient $k_{eff,r}$ ($r$ is the index of the row number). Figure A.6.c indicates how these effective springs per bolt-row are replaced by an equivalent spring acting at a lever arm $z$. The stiffness coefficient of this effective spring is $k_{eq}$. The effective stiffness coefficient $k_{eq}$ can directly be applied in Formula (A.5). The formulae to determine $k_{eff,r}$, $z$ and $k_{eq}$ are as follows :

$$k_{eff,r} = \frac{1}{\sum_i \frac{1}{k_{i,r}}} \tag{A.6}$$

$$z = \frac{\sum_r k_{eff,r} h_r^2}{\sum_r k_{eff,r} h_r} \tag{A.7}$$

$$k_{equ} = \frac{\sum_r h_{eff,r} h_r}{z} \tag{A.8}$$

They can be derived from the sketches of Figure A.6. The basis for these formulae is that the moment-rotation behaviour of each of the systems in Figure A.6.a to A.6.c are equal. An additional condition is that the compressive force in the lower rigid bar is equal in each of these systems.

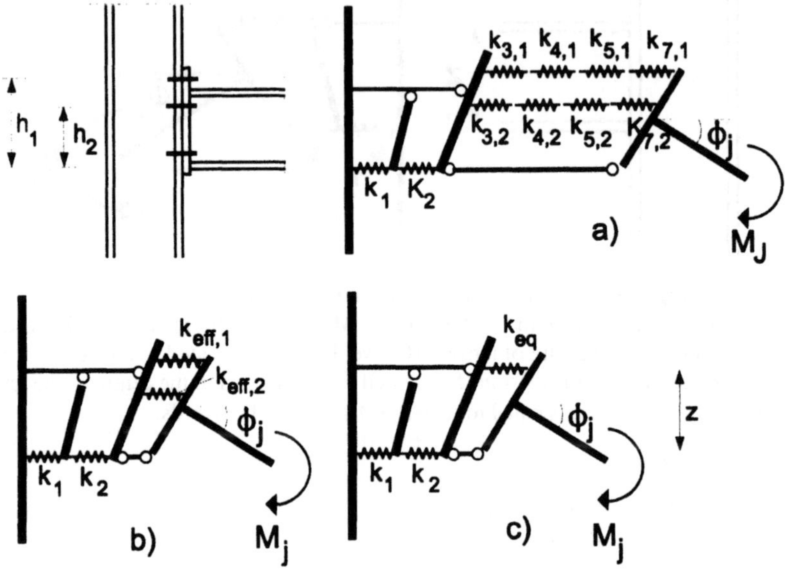


**Figure A.6** Spring model for a beam-to-column end-plated joint with more than one bolt-row in tension

In this stiffness model :

- the internal forces are in equilibrium with the bending moment;
- the compatibility of the displacements is ensured through the assumption of an infinitely rigid transverse stiffness of the beam cross-section;
- the plasticity criterion is fulfilled as long as the elastic resistance of the springs is not reached;
- no ductility requirement is likely to limit the deformation capacity of the springs in the elastic range of behaviour as far as *Eurocode 3-(revised) Annex J* components are concerned.

The solution provided by *Eurocode 3-(revised) Annex J* for the initial stiffness prediction fulfils the four main requirements which any distribution of internal forces should satisfy, from a theoretical point of view, and can therefore be considered as an " exact " one.

**Strength assembly**

The procedure for strength assembly as suggested in *Eurocode 3-(revised) Annex J* is aimed at deriving the value of the so-called design resistance of the joint. For the sake of clarity, it is not presented here in a general way but is illustrated in the particular case of beam splices with flush end-plates.

For the connection represented in Figure A.7, the distribution of internal forces is quite easy to obtain : the compressive force is transferred at the centroïd of the beam flange, while the tension force is at the level of the upper bolt-row. The resistance possibly associated with the lower bolt-row is usually neglected as it contributes in a very modest way to the transfer of bending moment in the joint (small level arm).

The design resistance of the joint $M_{Rd}$ is associated with the design resistance $F_{Rd}$ of the weakest joint component which can be one of the following : the beam and web in compression, the beam web in tension, the plate in bending or the bolts in tension. For the two last components (plate and bolts), reference is made to the concept of "idealized T-stub" introduced in *Eurocode 3-(revised) Annex J*. The bending resistance becomes :

$$M_{Rd} = F_{Rd} . z \tag{A.9}$$

where $z$ = h is the lever arm.

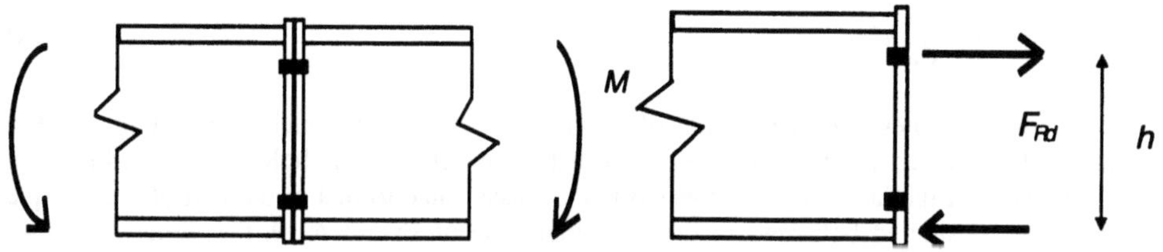


**Figure A.7** Joint with one bolt-row in tension

When more than one bolt-row has to be considered in the tension zone (Figure A.8), the distribution of internal forces is more complex.

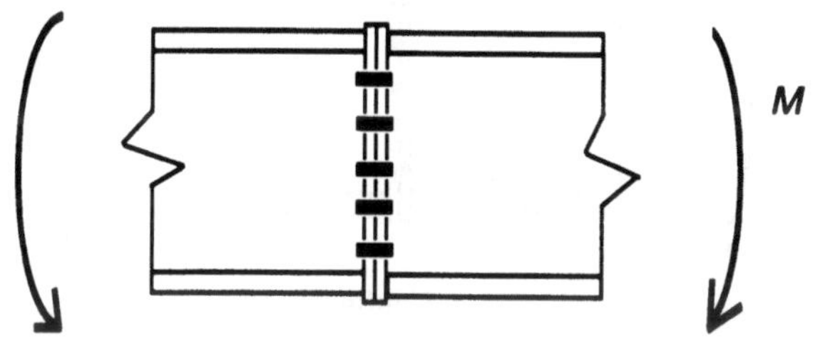


**Figure A.8 Joint with more than one bolt-row in tension**

Assume, initially, that the design of the joint leads to the adoption of a particularly thick end-plate in comparison to the bolt diameter (Figure A.9). In this case, the distribution of internal forces between the different bolt-rows is linear according to the distance from the centre of compression. The compression force $F_c$ which equilibrates the tension forces acts at the level of the centroïd of the lower beam flange. For the sake of clarity, it is shown only in Figure A.9, but not in the subsequent figures.

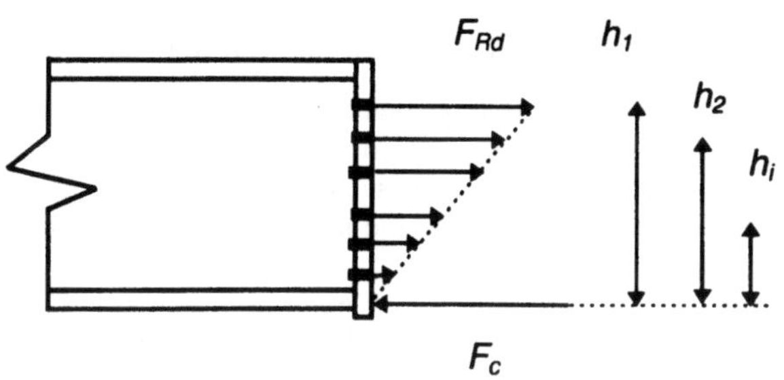


**Figure A.9** Joint with a thick end-plate

The design resistance $M_{Rd}$ of the joint is reached as soon as the bolt-row subjected to the highest stresses - in reality that which is located the farthest from the centre of compression - reaches its design resistance in tension $2B_{t.Rd}$.

Because of a limited deformation capacity of the bolts in tension no redistribution of forces is allowed to take place between bolt-rows.

It is assumed here that the design resistance of the beam flange and web in compression is sufficient to transfer the compression $F_c$ force. The tensile resistance of the beam web is also assumed not to limit the design resistance of the joint. $M_{Rd}$ is so expressed as (Figure A.9) :

$$M_{Rd} = \frac{F_{Rd}}{h_1} \sum h_i^2 \tag{A.10}$$

For thinner end-plates, the distribution of internal forces requires much more attention. When initial moment is applied to the joint, the forces distribute between the bolt-rows according to the relative stiffnesses of the latter. This stiffness is namely associated to that of the part of the end-plate adjacent to the considered bolt-row. In the particular case of Figure A.10, the upper bolt-row is characterized by a higher stiffness because of the presence of the beam flange and the web welded to the end-plate. Because of the higher stiffness, the upper bolt-row is capable of transferring a higher load than the lower bolt-rows (Figure A.10.b)

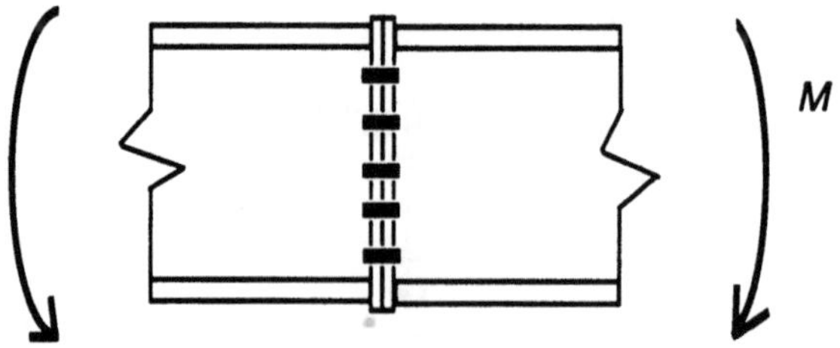


(a) Configuration

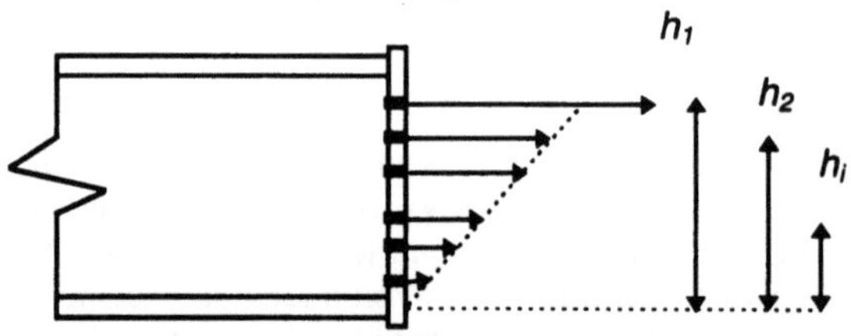


(b) Distribution of the internal forces at the beginning of the loading

**Figure A.10** Joint with a thin end-plate

In *Eurocode 3-(revised) Annex J*, it is assumed that the upper bolt-row will reach its design resistance first. This assumption is quite justified in the particular case being considered but is probably less justified for other connection types such as for end-plate connections with an extended part in the tension zone where it is usual that the second bolt-row - that just below the flange in tension - reaches its design resistance first.

The design resistance of the upper bolt-row may be associated with one of the following components : the bolts only, the end-plate only, the bolts-plate assembly or the beam web in tension. If its failure mode is ductile, a redistribution of forces between the bolt-rows can take place : as soon as the upper bolt-row reaches its design resistance, any additional bending moment applied to the joint will be carried by the lower bolt-rows (the second, then the third, etc.), each of which in their turn may reach its own design resistance.

The failure may occur in three different ways :

i. The plastic redistribution of the internal forces extends to all bolt-rows when they have sufficient deformation capacity. The redistribution is said to be "complete" and the resulting distribution of internal forces is called "plastic".

The design moment resistance $M_{Rd}$ is expressed as (Figure A.11) :

$$M_{Rd} = \sum_i F_{Rd,i} h_i \tag{A.11}$$

The plastic forces $F_{Rd,i}$ vary from one bolt-row to another according to the failure modes (bolts, plate, bolt-plate assembly, beam web, ...).

*Eurocode 3* considers that a bolt-row possesses a sufficient deformation capacity to allow a plastic redistribution of internal force to take place when :

- $F_{rd,i}$ is associated to the failure of the beam web in tension or;
- $F_{Rd,i}$ is associated to the failure of the bolt-plate assembly (including failure of the bolts alone or of the plate alone) and :

$$F_{Rd,i} \leq 1{,}9 B_{t.Rd}$$

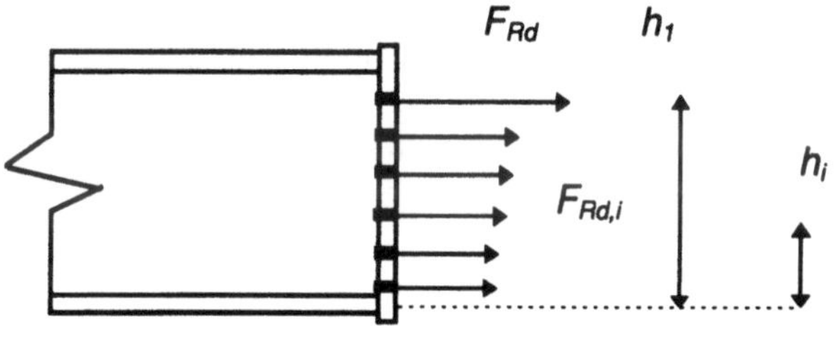


**Figure A.11** Plastic distribution of forces

ii. The plastic redistribution of forces is interrupted because of the lack of deformation capacity in the last bolt-row which has reached its design resistance ( $F_{Rd,k} > 1{,}9B_{t.Rd}$ and linked to the failure of the bolts or of the bolt-plate assembly).

In the bolt-rows located lower than bolt-row $k$, the forces are then linearly distributed according to their distance to the point of compression (Figure A.12).

The design moment resistance equals :

$$M_{Rd} = \sum_{i=1,k} F_{Rd,i} h_i + \frac{F_{Rd,k}}{h_k} \sum_{j=k+1,n} h_j^2 \tag{A.12}$$

where :

$n$ is the total number of bolt-rows;

$k$ is the number of the bolt-row, the deformation capacity of which is not sufficient.
In this case, the distribution is said "elasto-plastic".

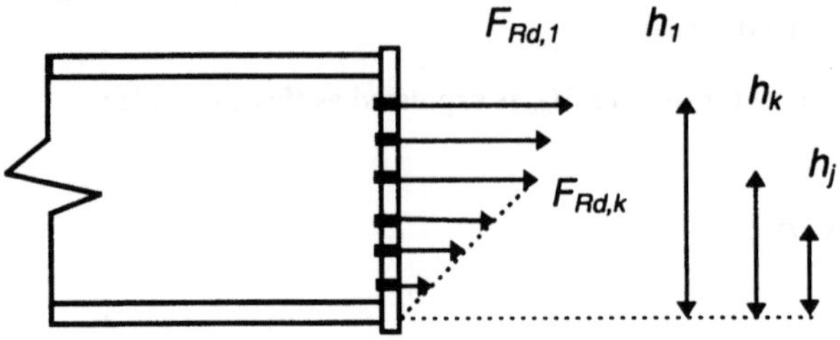


**Figure A.12** Elasto-plastic distribution of internal forces

2 The plastic or elasto-plastic distribution of internal forces is interrupted because the compression force $F_c$ attains the design resistance of the beam flange and web in compression. The moment resistance $M_{Rd}$ is evaluated with formulae similar to (A.11) and (A.12) in which, obviously, only a limited number of bolt-rows are taken into consideration. These bolts rows are such that :

$$\sum_{\ell=1,m} F_\ell = F_{c.Rd}$$

where : $m$ is the number of the last bolt-row transferring a tensile force;
$F_\ell$ is the tensile force in bolt-row number $\ell$ ;
$F_{c.Rd}$ is the design resistance of the beam flange and web in compression

The application of the above-described principles to beam-to-column joints is quite similar. The design moment resistance $M_{Rd}$ is, as for the beam splices, likely to be limited by the resistance of :

- the end-plate in bending,
- the bolts in tension,
- the beam web in tension,
- the beam flange and web in compression,

but also by that of :

- the column web in tension,
- the column flange in bending,
- the column web in compression,
- the column web panel in shear.

In *Eurocode 3-(revised) Annex J*, evaluation formulae are provided for each of these components. Annex J also presents a full example showing how to distribute the internal forces in a beam-to-column joint with a "multi-bolt-rows" end-plate connection.
This example also highlights the concept of individual and group yield mechanisms. When adjacent bolt-rows are subjected to tension forces, various yield mechanisms are likely to form in the connected plates (end-plate or column flange).

- individual mechanisms (see Figure 3-A.13.a) which develop when the distance between the bolt-rows are sufficiently large;
- group mechanisms (see Figure 3-A.13.b) including more than one adjacent bolt rows.

To each of these mechanisms are associated specific design resistances.
When distributing the internal forces, *Eurocode 3* recommends that, in a given bolt-row, the model never transfers :

- a higher load than that which can be carried when it is assumed that the considered bolt-row is the only one able to transfer tensile forces (individual resistance);
- a load such that the resistance of the whole group to which the bolt-row belongs is exceeded.

The whole procedure is rather long to apply but in Section 9 the calculation tools compatible with designer's expectations are presented.

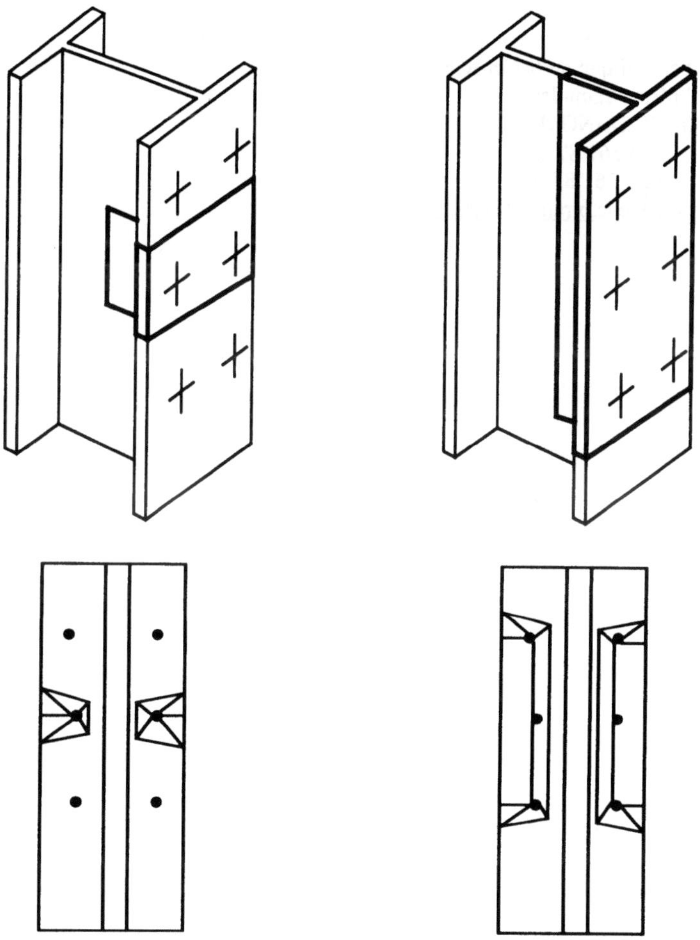

(a) Individual (b) Group mechanism

**Figure A.13** Plastic mechanisms

# *Annex B : Examples of practical design tools for joint characterisation*

In this Annex, an example of so-called *calculation procedures* and *design tables* are given. The interested reader may obtain the full set of available *procedures* and *tables* by contacting:

Mr E. PIRAPREZ
Research Centre of the Belgian Metalworking Industry (CRIF)
Parc Scientifique de la Cense Rouge
Rue du Bois Saint-Jean, 12
B-4102 SERAING (LIEGE)
Tel: ++32 4 361 87 62
Fax: ++32 4 361 87 02
E-mail: eugene.piraprez@crif.be

# *STIFFNESS AND RESISTANCE CALCULATION*

# *FOR BEAM SPLICES*

# *WITH FLUSH END-PLATES*

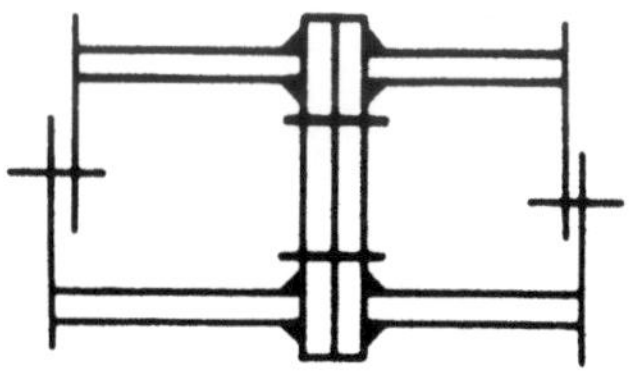

1. Calculation procedure
2. Design tables
3. Application example

# *1. CALCULATION PROCEDURE*

## *Mechanical characteristics*

| | Yield stresses | Ultimate stresses |
|---|---|---|
| Beam webs | $f_{ywb}$ | - |
| Beam flanges | $f_{yfb}$ | - |
| End-plates | $f_{yp}$ | $f_{up}$ |
| Bolts | - | $f_{ub}$ |
| | If hot-rolled profiles : $f_{ywb} = f_{yfb}$ | |

## *Geometrical characteristics*

**Joint**

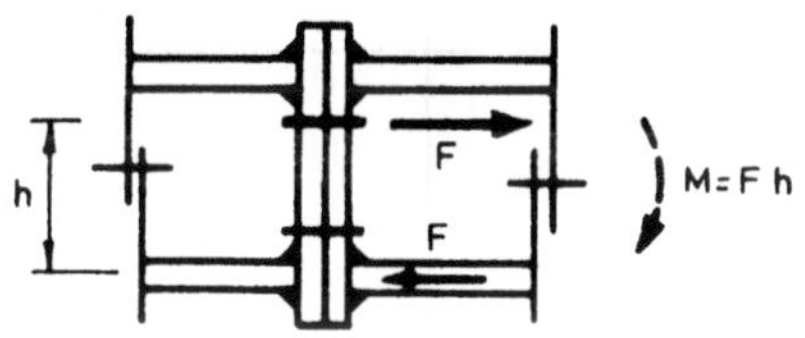


**Beams**

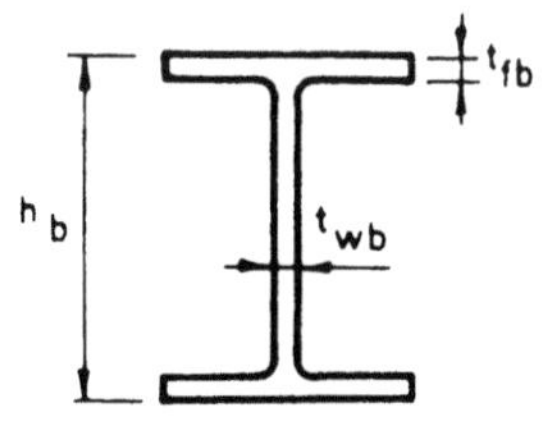


**End-plates**

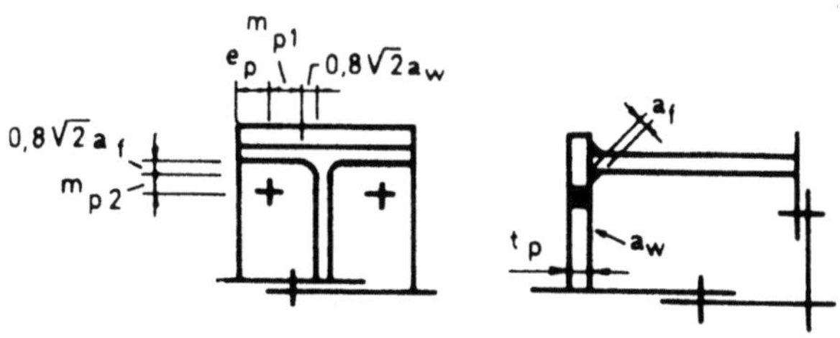


**Bolts**

$d_w$ : see figure or = $d_p$ if no washer*

$A_s$ : resistance area of the bolt

* $d_f$ is recommended in EC3 Annex J, and not $d_p$. As $d_f$ is not given in all the catalogs for bolts, $d_p$ is chosen here (safe assumption).

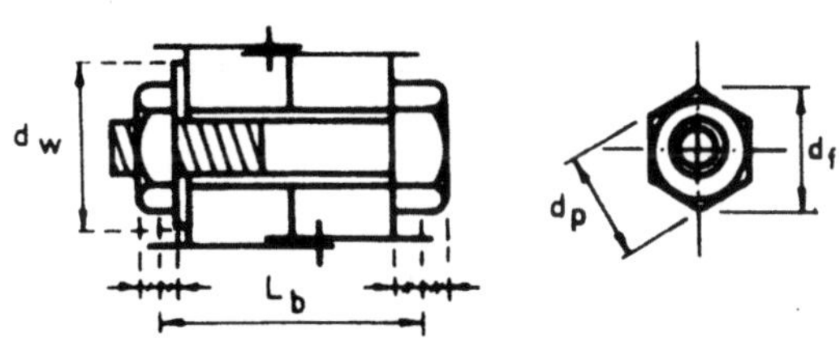

| | STIFFNESS | RESISTANCE |
|---|---|---|
| **Beam flanges in compression** | $k_3 = \infty$ | $F_{Rd,3} = M_{c.Rd} / (h_b - t_{fb})$<br>$M_{c.Rd}$ : beam design moment resistance |
| **Bolts in tension** | $k_4 = 1,6 \dfrac{A_s}{L_b}$ | $F_{Rd,4} = 2\ B_{t.Rd}$ **with** $B_{t.Rd} = F_{t.Rd}$<br>$F_{t.Rd} = \dfrac{0,9\ f_{ub}\ A_s}{\gamma_{Mb}}$ |
| **End-plates in bending** | $k_7 = \dfrac{0,85\ l_{eff,p,t}\ t_p^3}{2\ m_{p1}^3}$ | $F_{Rd,7} = \min [\ F_{ep.Rd,1}\ ;\ F_{ep.Rd,2}\ ]$<br>$F_{ep.Rd,1} = \dfrac{(8n_p - 2e_w)\ l_{eff,p,t}\ m_{pl\,p}}{2m_{p1}\ n_p - e_w(m_{p1} + n_p)}$<br>$F_{ep.Rd,2} = \dfrac{2\ l_{eff,p,t}\ m_{pl\,p} + 2\ B_{t.Rd}\ n_p}{m_{p1} + n_p}$<br>$n_p = \min [\ e_p\ ;\ 1,25 m_{p1}\ ]$<br>$m_{pl\,p} = 0,25\ t_p^2\ f_{yp} / \gamma_{Mo}$<br>$e_w = d_w / 4$ |
| | $l_{eff,p,t} = \min [\ 2\pi m_{p1}\ ;\ \alpha m_{p1}\ ]$<br>where $\alpha$ is defined in the annexed table | |
| **Beam web in tension** | $k_8 = \infty$ | $F_{Rd,8} = b_{eff,wb,t}\ t_{wb}\ f_{ywb} / \gamma_{Mo}$ |
| | $b_{eff,wb,t} = l_{eff,p,t}$ | |

| | | |
|---|---|---|
| ***JOINT*** | Initial stiffness :<br>$S_{j,ini} = E\ h^2 / \sum_{i=3,4,7,8} 1/k_i$<br>Nominal stiffness :<br>$S_j = S_{j,ini} / 3$ | $F_{Rd} = \min\ [\ F_{Rd,i}\ ]$<br>Plastic design moment resistance :<br>$M_{Rd} = F_{Rd}\ h$<br>Elastic moment resistance :<br>$\frac{2}{3} M_{Rd}$ |

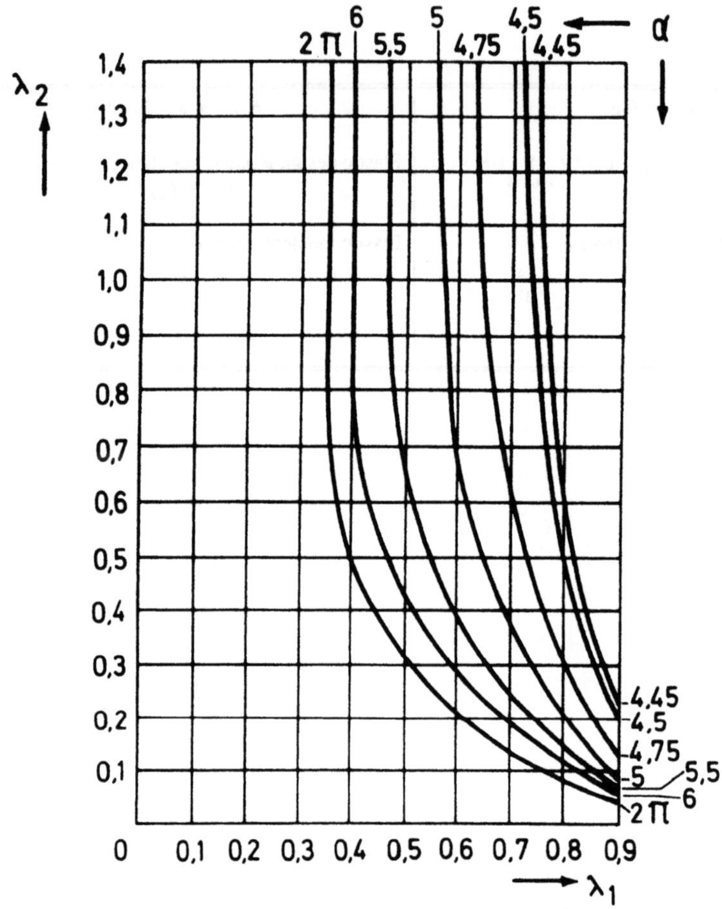


$$\lambda_1 = \frac{m_{p1}}{m_{p1} + e_p}$$

$$\lambda_2 = \frac{m_{p2}}{m_{p1} + e_p}$$

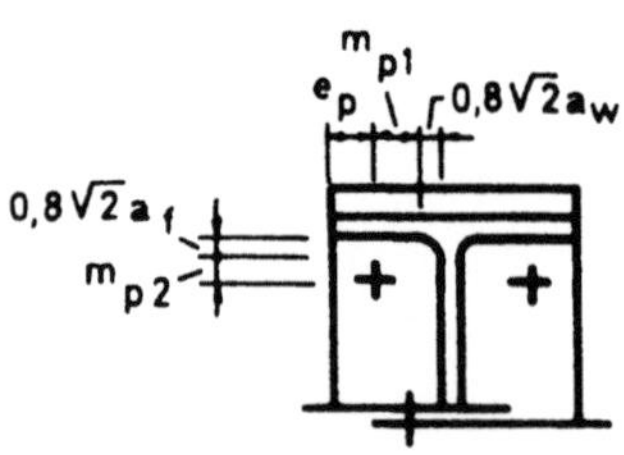


α values to evaluate the design resistance of the end-plate in tension

# 2. DESIGN TABLES

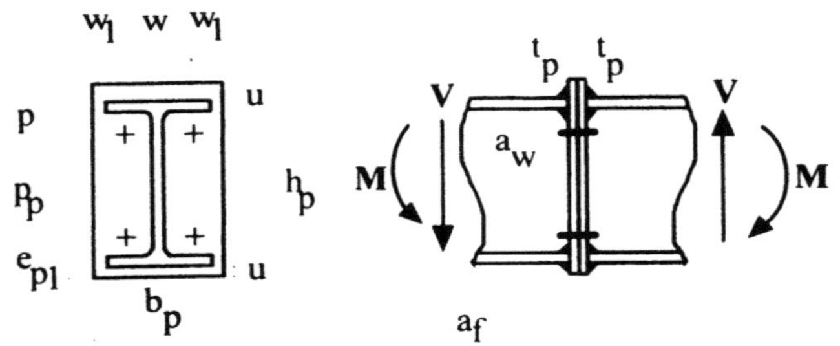


| Failure mode | Code |
|---|---|
| Beam flange in compression | BFC |
| Bolts in tension | BT |
| End plate in tension | EPT |
| Beam web in tension | BWT |

$\gamma_{M0} = 1.10$

$\gamma_{Mb} = 1.25$

| Beam | Bolts hr8.8 | End-plate: S235 (mm) | | | Connection detail (mm) | | | | | | Welds (mm) | | Rotational stiffness (kNm/rad) | | Resistance: Moment (kNm) | | Resistance: Shear (kN) | Failure mode | Reference length(m) | |
|---|---|---|---|---|---|---|---|---|---|---|---|---|---|---|---|---|---|---|---|---|
| S235 | | $t_p$ | $b_p$ | $h_p$ | p | $p_p$ | $e_{p1}$ | w | $w_1$ | u | $a_w$ | $a_f$ | $S_{j,ini}$ | $S_{j,ini}/2$ | $M_{Rd}$ | $2/3M_{Rd}$ | $V_{Rd}$ | Code | $L_{bb}$ | $L_{bu}$ |
| IPE220 | M16 | 15 | 140 | 240 | 60 | 120 | 60 | 90 | 25 | 10 | 3 | 5 | 15433 | 5144 | 24.1 | 16.1 | 157 | EPT | 3.0-R | S |
| IPE240 | M16 | 15 | 140 | 260 | 60 | 140 | 60 | 90 | 25 | 10 | 4 | 5 | 20098 | 6699 | 27.2 | 18.1 | 157 | EPT | 3.3-R | S |
| IPE270 | M16 | 15 | 154 | 290 | 65 | 160 | 65 | 90 | 32 | 10 | 4 | 6 | 26826 | 8942 | 32.4 | 21.6 | 157 | EPT | 3.6-R | S |
| | M20 | 20 | 154 | 290 | 65 | 160 | 65 | 90 | 32 | 10 | 4 | 6 | 42892 | 14297 | 53.8 | 35.9 | 245 | EPT | R | 7.1-R |
| IPE300 | M16 | 15 | 170 | 320 | 65 | 190 | 65 | 90 | 40 | 10 | 4 | 6 | 36564 | 12188 | 38.9 | 25.9 | 157 | EPT | 3.8-R | 12.0-R |
| | M20 | 20 | 170 | 320 | 65 | 190 | 65 | 90 | 40 | 10 | 4 | 6 | 57607 | 19202 | 64.3 | 42.8 | 245 | EPT | R | 7.6-R |
| IPE330 | M16 | 15 | 180 | 350 | 65 | 220 | 65 | 90 | 45 | 10 | 4 | 6 | 47398 | 15799 | 44.8 | 29.9 | 157 | EPT | 4.2-R | 13.0-R |
| | M20 | 20 | 180 | 350 | 65 | 220 | 65 | 90 | 45 | 10 | 4 | 6 | 74007 | 24669 | 73.8 | 49.2 | 245 | EPT | 2.7-R | 8.3-R |
| | M24 | 20 | 180 | 350 | 75 | 200 | 75 | 110 | 35 | 10 | 4 | 6 | 62600 | 20867 | 78.5 | 52.3 | 352 | EPT | 3.2-R | 9.9-R |
| IPE360 | M16 | 15 | 210 | 400 | 75 | 250 | 75 | 90 | 60 | 20 | 5 | 7 | 60854 | 20285 | 50.1 | 33.4 | 157 | EPT | 4.5-R | 14.0-R |
| | M20 | 20 | 210 | 400 | 75 | 250 | 75 | 90 | 60 | 20 | 5 | 7 | 93626 | 31209 | 82.5 | 55.0 | 245 | EPT | 2.9-R | 9.1-R |
| | M24 | 20 | 210 | 400 | 85 | 230 | 85 | 110 | 50 | 20 | 5 | 7 | 85645 | 28548 | 96.8 | 64.5 | 352 | EPT | 3.2-R | 10.0-R |
| IPE400 | M16 | 15 | 220 | 440 | 75 | 290 | 75 | 90 | 65 | 20 | 5 | 7 | 78661 | 26220 | 56.7 | 37.8 | 157 | EPT | 4.9-R | 15.4-R |
| | M20 | 20 | 220 | 440 | 75 | 290 | 75 | 90 | 65 | 20 | 5 | 7 | 120698 | 40233 | 93.4 | 62.3 | 245 | EPT | 3.2-R | 10.1-R |
| | M24 | 20 | 220 | 440 | 85 | 270 | 85 | 110 | 55 | 20 | 5 | 7 | 113428 | 37809 | 112.7 | 75.1 | 352 | EPT | 3.4-R | 10.7-R |
| | M27 | 25 | 220 | 440 | 95 | 250 | 95 | 130 | 45 | 20 | 5 | 7 | 118284 | 39428 | 139.3 | 92.9 | 458 | EPT | 3.3-R | 10.3-R |
| IPE450 | M16 | 15 | 230 | 490 | 75 | 340 | 75 | 90 | 70 | 20 | 5 | 8 | 104399 | 34800 | 65.0 | 43.3 | 157 | EPT | 5.4-R | 17.0-R |
| | M20 | 20 | 230 | 490 | 75 | 340 | 75 | 90 | 70 | 20 | 5 | 8 | 159614 | 53205 | 107.1 | 71.4 | 245 | EPT | 3.6-R | 11.1-R |
| | M24 | 20 | 230 | 490 | 85 | 320 | 85 | 110 | 60 | 20 | 5 | 8 | 152172 | 50724 | 130.4 | 86.9 | 352 | EPT | 3.7-R | 11.6-R |
| | M27 | 25 | 230 | 490 | 95 | 300 | 95 | 130 | 50 | 20 | 5 | 8 | 162176 | 54059 | 165.8 | 110.5 | 458 | EPT | 3.5-R | 10.9-R |
| IPE500 | M16 | 15 | 240 | 540 | 80 | 380 | 80 | 100 | 70 | 20 | 6 | 9 | 118633 | 39544 | 72.4 | 48.3 | 157 | EPT | 6.8-R | S |

| | | End-plate: S235 (mm) | | | Connection detail (mm) | | | | | | Welds (mm) | | Rotational stiffness (kNm/rad) | | Resistance | | | Failure mode | Reference length(m) | |
|---|---|---|---|---|---|---|---|---|---|---|---|---|---|---|---|---|---|---|---|---|
| | | | | | | | | | | | | | | | Moment (kNm) | | Shear (kN) | | | |
| | | $t_p$ | $b_p$ | $h_p$ | p | $p_p$ | $e_{p1}$ | w | $w_1$ | u | $a_w$ | $a_f$ | $S_{j,ini}$ | $S_{j,ini}/2$ | $M_{Rd}$ | $2/3M_{Rd}$ | $V_{Rd}$ | Code | $L_{bb}$ | $L_{bu}$ |
| | M20 | 20 | 240 | 540 | 80 | 380 | 80 | 100 | 70 | 20 | 6 | 9 | 187003 | 62334 | 119.3 | 79.5 | 245 | EPT | 4.3-R | 13.5-R |
| | M24 | 20 | 240 | 540 | 90 | 360 | 90 | 120 | 60 | 20 | 6 | 9 | 176462 | 58821 | 145.7 | 97.1 | 352 | EPT | 4.6-R | 14.3-R |
| | M27 | 25 | 240 | 540 | 100 | 340 | 100 | 140 | 50 | 20 | 6 | 9 | 190544 | 63515 | 183.3 | 122.2 | 458 | EPT | 4.2-R | 13.3-R |
| IPE550 | M16 | 15 | 250 | 590 | 80 | 430 | 80 | 100 | 75 | 20 | 6 | 9 | 149023 | 49674 | 80.7 | 53.8 | 157 | EPT | 7.6-R | S |
| | M20 | 20 | 250 | 590 | 80 | 430 | 80 | 100 | 75 | 20 | 6 | 9 | 233999 | 78000 | 132.9 | 88.6 | 245 | EPT | 4.8-R | 15.1-R |
| | M24 | 20 | 250 | 590 | 90 | 410 | 90 | 120 | 65 | 20 | 6 | 9 | 222310 | 74103 | 162.7 | 108.5 | 352 | EPT | 5.1-R | 15.9-R |
| | M27 | 25 | 250 | 590 | 100 | 390 | 100 | 140 | 55 | 20 | 6 | 9 | 245405 | 81802 | 210.7 | 140.5 | 458 | EPT | 4.6-R | 14.4-R |
| IPE600 | M16 | 15 | 260 | 640 | 80 | 480 | 80 | 100 | 80 | 20 | 7 | 10 | 188438 | 62813 | 88.9 | 59.3 | 157 | EPT | 8.2-R | S |
| | M20 | 20 | 260 | 640 | 80 | 480 | 80 | 100 | 80 | 20 | 7 | 10 | 291826 | 97275 | 146.5 | 97.7 | 245 | EPT | 5.3-R | 16.6-R |
| | M24 | 20 | 260 | 640 | 90 | 460 | 90 | 120 | 70 | 20 | 7 | 10 | 280299 | 93433 | 179.7 | 119.8 | 352 | EPT | 5.5-R | 17.2-R |
| | M27 | 25 | 260 | 640 | 100 | 440 | 100 | 140 | 60 | 20 | 7 | 10 | 314622 | 104874 | 239.6 | 159.7 | 458 | EPT | 4.9-R | 15.4-R |

# Annex C : Joint stiffness classification "Reference length" concept

Generally speaking, joint stiffness classification is a means to obtain an indication of the importance of the joint rotational response on the overall structural behaviour and, in particular, of its effect on the forces acting at the joints. More specifically, it provides a means of indicating whether an assumption that a given joint is "pinned" or "rigid" for the global frame analysis is justified or not. When used in this manner, the joint classification check can be an alternative to modelling the joints in the structure. When a chosen joint is found to be classified as "semi-rigid", while the analysis assumed it to be "rigid", one has the choice between two procedures of corrective action. The first would be to modify the joint design so that it now becomes sufficiently rigid to allow it to be classified as "rigid", and the second (and more precise approach) would be to re-analyse the structure, this time including for the modelling of the joint behaviour in the structural model.

In order to know if a joint can be classified as "rigid", "semi-rigid" or "pinned", the concept of beam "reference length" is introduced, which is explained in Figure C.1 for a beam with end joints.

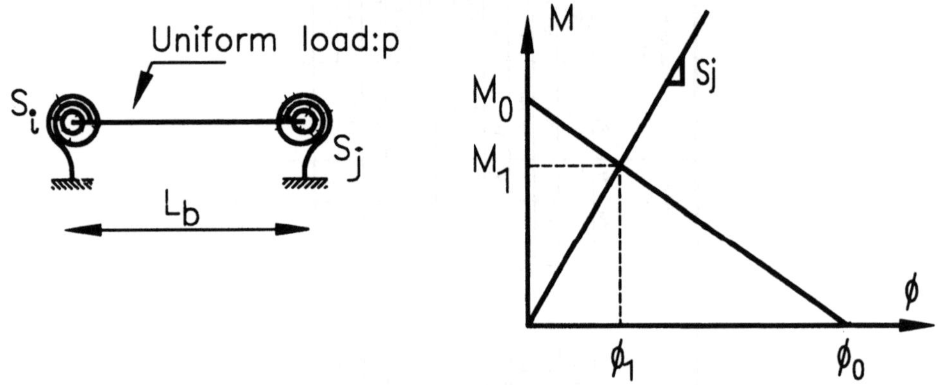


$M_o = pL_b^2 / 12 (fully fixed)$ $\phi_o = pL_b^3 / 24EI_b (perfectly pinned)$

(a) Different end joints (b) Identical end joints

**Figure C.1** Restrained beam

The general equilibrium or non sway "slope deflection" equations for the beam i j (Figure C.1.a), where clockwise moments and rotations are defined as positive, are as follows :

$$\overline{M}_{ij} = M_{ij} + \frac{2EI_b}{L_b}(2\phi_i + \phi_j) \tag{C.1.a}$$

$$\overline{M}_{ji} = M_{ji} + \frac{2EI_b}{L_b}(2\phi_j + \phi_i) \tag{C.1.b}$$

where

| | |
|---|---|
| $\overline{M}_{ij}, \overline{M}_{ji}$ : | actual end moment at nodes *i* and *j* respectively |
| $M_{ij}, M_{ji}$ : | fully fixed end moments (same beam, same loading) |
| $\phi_i, \phi_j$ : | actual beam end rotations. |
| $E$ : | modulus of elasticity for steel |
| $I_b$, $L_b$ : | second moment of area and beam span respectively. |

For identical joint conditions at the beam ends (Figure C.1.b), the following holds:

$$Mij = -Mji = M_0 \tag{C.2.a}$$

$$\overline{M}_{ij} = -\overline{M}_{ji} = M_1 \tag{C.2.b}$$

$$\phi_i = -\phi_j = \phi_1 \tag{C.2.c}$$

$$S_{j,i} = S_{j,j} = S_j \tag{C.2.d}$$

$$Fj,i = -Fj,j = Fj \tag{C.2.e}$$

$$M_{j,i} = -M_{j,j} = M_j \tag{C.2.f}$$

Equation (C.1) then becomes :

$$M_1 = M_0 + \frac{2EI_b}{L_b}\phi_1 \tag{C.3}$$

The equilibrium equation of the spring, the stiffness of which is $S_j$, gives:

$$M_j = S_j F_j \tag{C.4}$$

Compatibility of the rotations and equilibrium of the moments at the spring to beam interface requires that:

$$F_j = F_1 \tag{C.5.a}$$

$$M_1 + S_j\phi_1 = 0 \tag{C.5.b}$$

obtained for the left hand end of a uniformly loaded beam:

$$M_1 = -\frac{pL_b^2}{12} \cdot \frac{S_j}{S_j + \frac{2EI_b}{L_b}} \tag{C.6}$$

Equation (C.5) demonstrates the influence of the joint stiffness on the end moment. For a joint having a very high stiffness compared to that of the beam, the beam behaves as almost fixed ended. When the joint stiffness is very small compared to that of the beam, the beam behaves as almost pin ended. How the beam behaves essentially depends on the relative stiffnesses of the joint and the beam.

The stiffness of a given beam section varies only with its length, It can be deduced that, for a beam section with properties $I_b$ and $L_b$ and having end joints of stiffness $S_j$ :

- when $L_b$ is relatively large (very flexible beam), the joint tends to behave like a fully fixed joint ( $M_1 \Rightarrow -pL_b^2/12$ ),
- when $L_b$ is relatively short (very stiff beam), the joint tends to behave like a pinned joint ( $M_1 \Rightarrow 0$).

Keeping these observations in mind, two characteristic lengths, $L_{b1}$ and $L_{b2}$ , may be defined:

1) $L_{b1}$ such that :

- if $L_b \geq L_{b1}$ the joint may be considered as rigid,

- if $L_b < L_{b1}$ the joint shall to be considered as semi-rigid;

2) $L_{b2}$ such that :

- if $L_b > L_{b2}$ the joint shall to be considered as semi rigid,

- if $L_b \leq L_{b2}$ the joint may be considered as pinned,

where $L_{b2} < L_{b1}$

Taking the lengths in increasing order , the length $L_{b2}$ represents the boundary between the "pinned" and "semi-rigid" classifications, and the length $L_{b1}$ represents the boundary between the "semi-rigid" and "rigid" classifications. Their values depend on the joint stiffness and on the given beam section properties. It is quite important to stress that, in accordance with *Eurocode 3-(revised)Annex J*, the joint stiffness to be considered for stiffness classification is the elastic initial one.

# PART III

# PROCEDURES TO VERIFY ROTATION CAPACITY

**U. Kuhlmann and F. Kühnemund**
**University of Stuttgart, Stuttgart, Germany**

**Abstract:** This document compares the definition and the verification of rotation capacity of steel joints and members and suggests a coherent definition. It describes the verification procedure of rotation capacity for members and joints and points out how deemed-to-satisfy criteria can be derived from scientific investigations. Finally special attention is given to component tests and theoretical studies aiming at the determination of their load deformation behaviour. The dominant influence of the component "column web in compression" is given special consideration when the rotation capacity of a joint is derived.

## 1 Rotation Capacity of Members

### 1.1 Introduction

Rising prices in structural engineering demand efficient calculation procedures for structural systems with high exploitation of the cross sections.

Steel and composite structural systems offer possibilities to satisfy these demands. One possibility is to use the material's plastic sources. Statically indeterminate systems bear the ability of moment redistribution. The loadbearing capacity of a system based on plastic design is remarkably higher than that of a elastic based calculation.

The ultimate load of a structural system according to plastic hinge theory depends on the development of a kinematic hinge mechanism. This mechanism presupposes a certain rotation capacity of the hinges that are first to form. To allow full redistribution of bending moments in a structure the first hinges must be able to rotate without losing the plastic bearing capacity of the sections. According to Eurocode 3 (1992) sufficient rotation capacity for plastic design is guaranteed by maximum b/t ratios for the flanges and the webs of I-profiles. Chapter 1 illustrates a procedure for defining such critical b/t ratios by comparison of available with required rotation capacity values.

In the following a short example is given to illustrate the background. Figure 1 shows a continuous beam with two spans and a uniformly distributed load. The corresponding load-deformation curve is given in Figure 2.

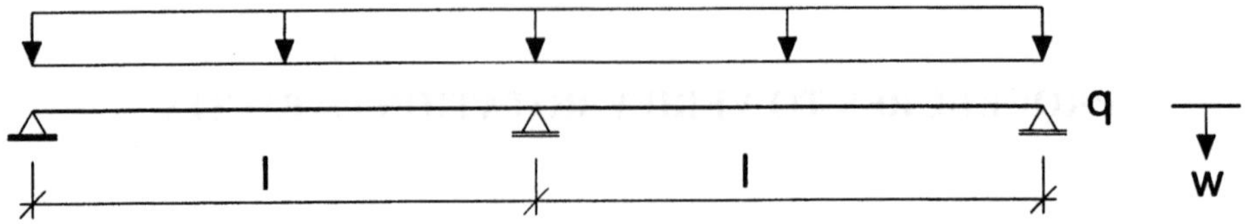


**Figure 1.** Continuous beam with uniformly distributed load

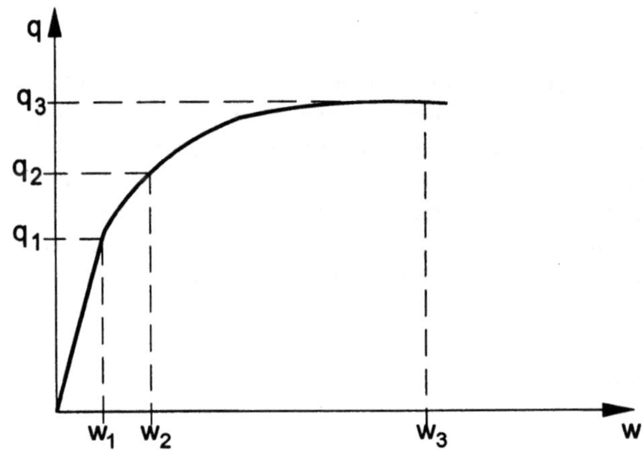


**Figure 2.** Load-deformation curve

Under load $q_1$ plastic stresses in the compression flange of the support cross section are reached. Increasing the load the plastic parts in the beam expand with decreasing flexural rigidity. As a consequence the deformations w of the span grow clearly.

At load step $q_2$ a plastic hinge develops at the support section. A fully plastic stress distribution is assumed at the support. The belonging bending moment corresponds to the maximum plastic resistance of the support section.

Following the load path the system is now considered as two single span beams. It is assumed that the plastic hinge is able to rotate without that the resistance of the cross section decreases. With rising loads the plasticity also expands in the span sections until the midspan moment reaches its ultimate value, too. Large deformations in the span and large rotations of the hinge at the support section go along with this stage. Finally another two plastic hinges have developed in the spans. A kinematic failure mechanism occurs.

The short example shows that when using plastic design methods indeterminate systems dispose additional bearing capacities. The rotation of the support hinge after passing the load $q_2$ is the required rotation capacity $\phi_{req}$ for the plastic moment redistribution. The available rotation capacity $\phi_{avail}$ of the member must be at least of the same size. Consequently the following equation has to be fulfilled for plastic calculation:

$$\Phi_{avail} \geq \Phi_{req} \tag{1}$$

In the following background information is given how to check sufficient rotation capacity of members.

## 1.2 Available Rotation Capacity of Members

Figure 3 shows a typical moment-rotation curve for members (see Kuhlmann, 1986, 1989). According to strain hardening of the material the curve exceeds the plastic moment of the member. But the bearing capacity is limited. Already before reaching the maximum moment value first local buckles occur mostly at the beam flanges. After reaching a maximum value the curve decreases slowly and steadily without abrupt break down but accompanied by extensive plastic deformations.

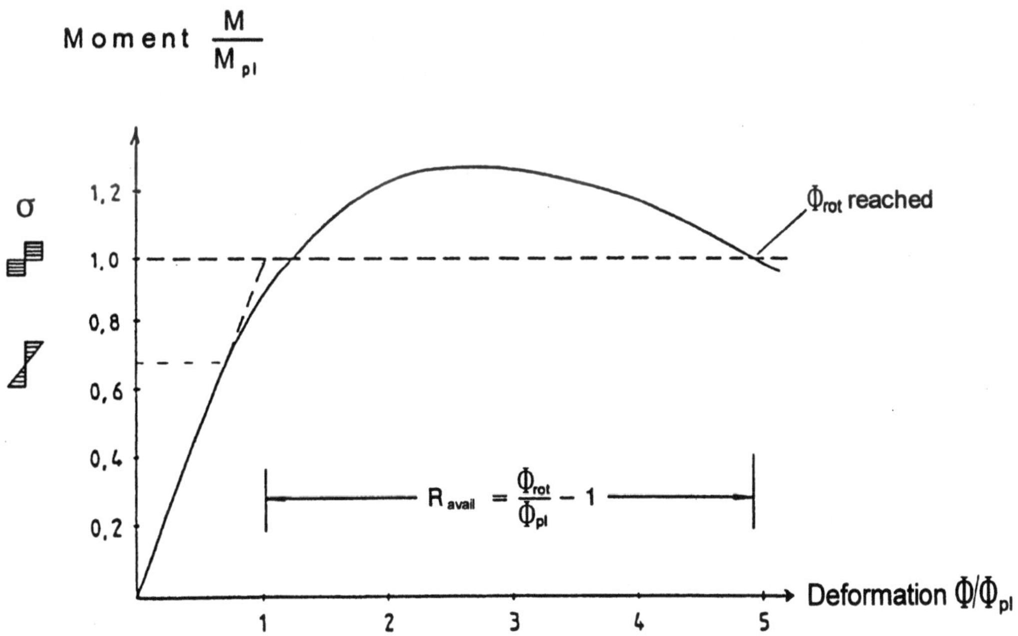


**Figure 3.** Moment-rotation curve for members

In general rotation capacity is the ability of a plastic hinge to keep the full plastic moment resistance while rotating. The example given above illustrates that definition. The first hinge at the support section must be able to rotate without decreasing moment, in order that plastic hinges in the span can develop.

The available rotation capacity of a member is accordingly defined as the difference between the rotation belonging to the first point of intersection between the plastic plateau and the M-$\phi$-curve and the rotation belonging to the point where the graph reaches the plastic plateau again.

$$\Phi_{avail} = \Phi_{rot} - \Phi_{pl} \tag{2}$$

Also a dimensionless definition for the rotation capacity is possible:

$$R_{avail} = \frac{\Phi_{rot}}{\Phi_{pl}} - 1 \tag{3}$$

To gain complete M-$\phi$-curves for members tests were executed and calculation procedures were developed and verified by these tests. By tests and calculations several parameters were varied. Figure 4 summarises the main results of the parametric analysis. The graph shows the dependence of the rotation capacity R on the b/t ratio of the flange. It can be seen that for increasing slenderness values the calculated R values decrease.

The rotation capacity has been calculated for sections with two different webs: Web A corresponds to webs of small rolled sections, e.g. an HEA 200; whereas web C represents a very slender one. Especially for medium b/t values, the restraint by the web seems to be of importance. In contrast, the span to flange width ratios L/b, varied here from 7 to 14, produce different values of rotation capacity in the whole region under consideration. From this graph, for a given section and structural system, the available rotation capacity can be obtained.

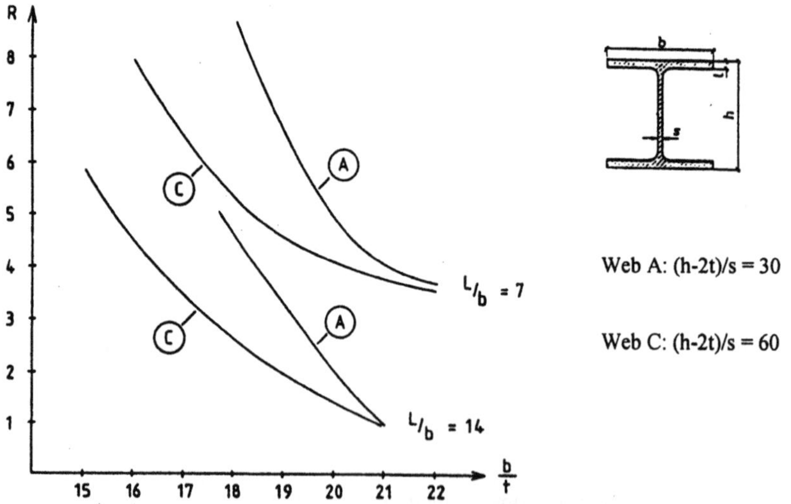


**Figure 4.** Interrelation between slenderness values and rotation capacity of beams (see Kuhlmann, 1986)

The calculation procedures named above consider the non-linear (elastic/plastic/strain hardening) behaviour of the material and the linear and non-linear buckling of the cross section.

## 1.3 Required Rotation Capacity

The evaluation of the available rotation capacity is based on a simply supported beam. However the redistribution of bending moments can be observed only in indeterminate structures. Therefore, to evaluate the required rotation capacity, indeterminate structures have to be studied.

The required rotation capacity of a plastic hinge is derived from the structural analysis of systems. It is defined as the difference between the rotation $\phi_{pl}$ when a plastic hinge develops and the rotation $\phi_{Fl}$ of the considered hinge when the final hinge of the system has developed.

$$\Phi_{req} = \Phi_{Fl} - \Phi_{pl} \tag{4}$$

$$R_{req} = \frac{\Phi_{Fl}}{\Phi_{pl}} - 1 \tag{5}$$

The required rotation capacity depends on the structural system. Figure 5 illustrates the definition for a three span beam. This structural system loaded by a point load in midspan is known as "paradox of limit analysis". The collapse load derived by simple plastic analysis is independent of the outer span length, although for very long outer spans the restraint at the inner supports may be too weak to develop the plastic moment. This structural system therefore forms a critical system for continuous beams in general.

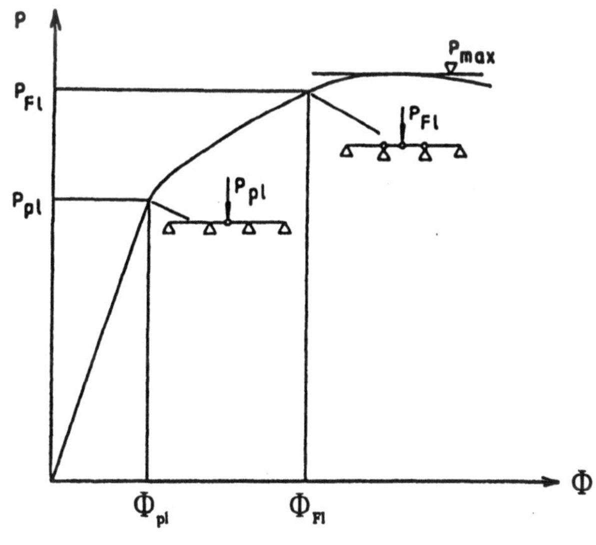


**Figure 5.** Definition of required rotation capacity for a three span beam

Whereas $P_{pl}$ belongs to the load where according to the elastic-plastic hinge theory the first hinge forms, $P_{Fl}$ represents the ultimate load belonging to the failure mechanism according to the elastic-plastic hinge theory neglecting the effects of strain-hardening and plastic zones. The maximum load $P_{max}$ exceeds the theoretical collapse $P_{Fl}$ of the simple plastic analysis based on the hinge mechanism. As a consequence, this example can be regarded as appropriate to plastic design.

The required rotation has to be compared to the available rotation calculated or tested for simple beams loaded with a single load in the middle. As a basis for this comparison one span beams are cut from the whole structural system at the points of inflection of the moment diagram. Figure 6 shows that procedure for a three span beam.

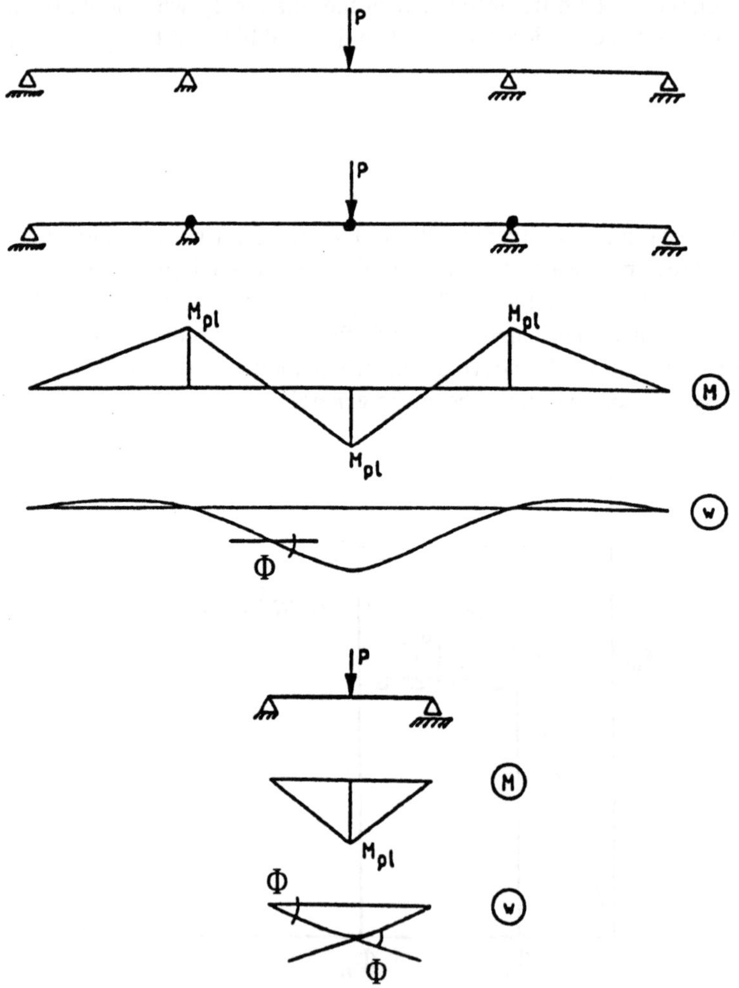


**Figure 6.** Derivation of required rotation capacity values with substitute one span beams

In dependence of the structural system or the kind of loading the values of the required rotation $R_{req}$ differ. To get reference values for certain systems (e.g. continuous beams, frames) and by means of them permissible dimensions for those systems and kind of loading, upper boundaries for required rotations for unfavourable systems and loading conditions are to be found. A disadvantageous system for rotation capacity is a three span beam with a single load in the midspan as explained before. Investigations for required rotation capacity values are illustrated in the following based on this system.

For a HE-200A the required rotation capacity is graphed in Figure 7 in dependence of the span ratio and the ratio between the span of the middle field and the width b of the cross section.

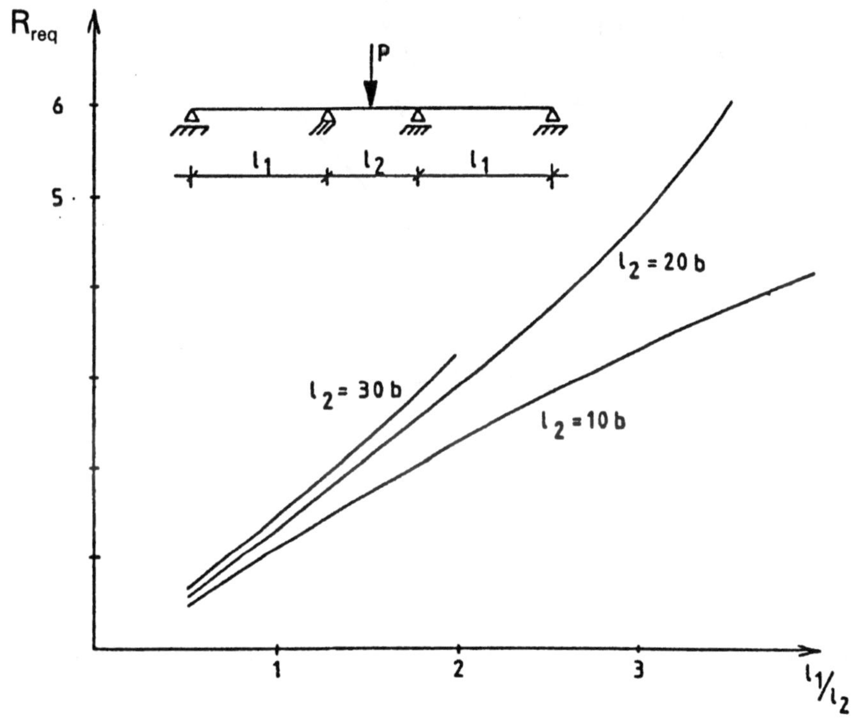


**Figure 7.** Required rotation capacity for HE-200A profile (see Kuhlmann, 1986)

Figure 8 illustrates for the same profile and a middle span length of $l_2 = 20 \cdot b$ the required rotation capacity for two loading combinations: a single load in the middle span with and without a uniformly distributed load. It shows the behaviour of the required rotation capacity with increasing outer span length ($l_1$) compared with the inner length ($l_2$).

For an extreme value of $l_1/l_2$ of 3, the required rotation capacity reaches a value of about 5. This is the case for a beam without any distributed load.

Considering an additional small distributed load, caused for example by weight, the behaviour changes completely. Instead of a continuous increase, the required rotation capacity starts decreasing when $l_1/l_2$ reaches 2.

The increase of the outer span length does not only diminish the restraint at the inner supports but also raises the loading on the outer spans. The latter finally causes the reduction of the required rotation capacity. For the loads under consideration the required rotation capacity does not exceed a value of 2.

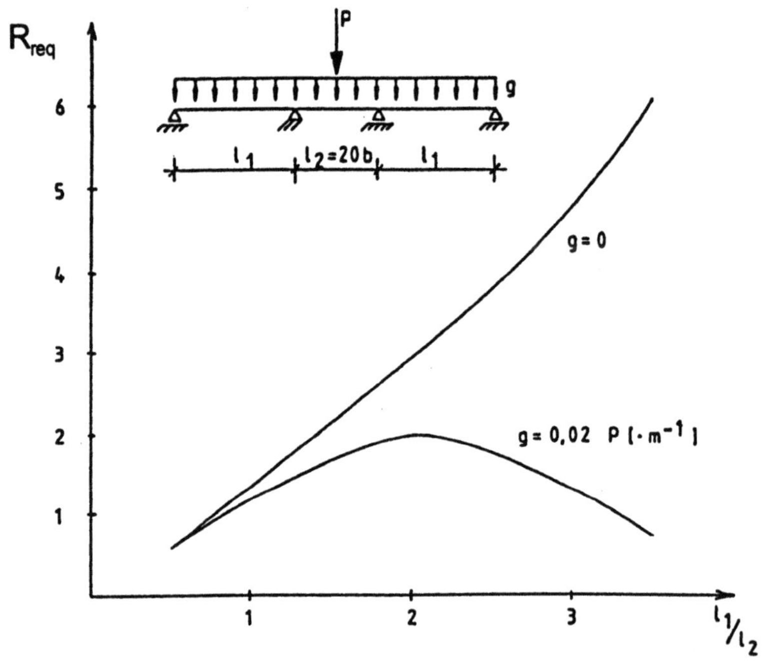


**Figure 8.** Influence of a uniformly distributed load on required rotation capacity (see Kuhlmann, 1986)

### 1.4 Procedure for the Check of Sufficient Rotation Capacity

**Check of sufficient rotation capacity by slenderness values.** The application of plastic hinge theory requires that the following equation is fulfilled:

$$\Phi_{avail} \geq \Phi_{req} \tag{1}$$

This procedure is inadequate for practical design. To derive practical design rules it is essential to simplify the assessment. The comparison of available and required rotation capacity is replaced by the check of profile slenderness values. The derivation of these slenderness values is explained in the following.

Investigations for several structural systems lead to boundary values for required rotation capacity. As one can draw from the example of chapter 1.3 the boundary value for a three span continuous beam with uniformly distributed load and a single load in the midspan is $R_{req} = 2$ (see Figure 8). Slenderness values which satisfy the condition of $R_{req} = 2$ are found e.g. by means of Figure 4. The available rotation capacity must be at least $R_{avail} = 2$. The corresponding slenderness value is given by the graph as shown in Figure 9.

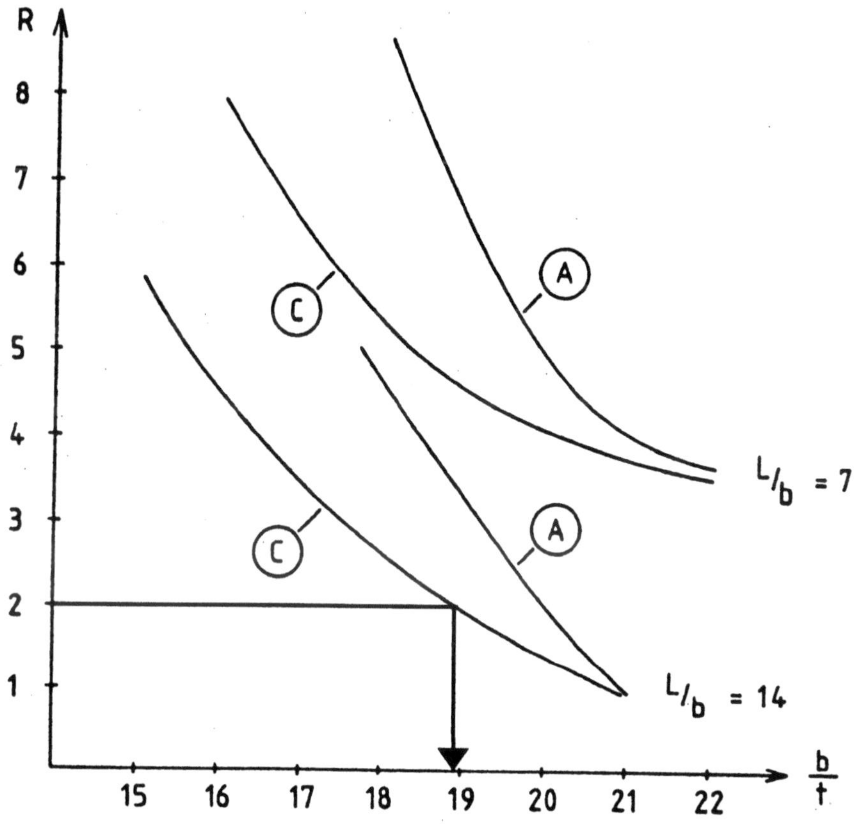


**Figure 9.** Interrelation between rotation capacity, flange slenderness and span length (see Kuhlmann, 1986)

**Regulation according to standards.** Present European standards (Eurocode 3, 1992, and Eurocode 4, 1992) refer to this method of verification. From thorough scientific investigations boundary values are derived due to flange and web slenderness and the stress distribution in the cross section.

The rotation capacity of members is limited among others by local buckling. In dependence of the slenderness values of its parts Eurocode 3 (1992) classifies the sections (see also Figure 10) to allow for different kinds of design methods:

*Class 1 cross sections are those which can form a plastic hinge with the rotation capacity required for plastic analysis.*

*Class 2 cross sections are those which can develop their plastic moment resistance, but have limited rotation capacity.*

*Class 3 cross sections are those in which the calculated stress in the extreme compression fibre of the steel member can reach its yield strength, but local buckling is liable to prevent development of the plastic moment resistance.*

*Class 4 cross sections are those in which it is necessary to make explicit allowances for the effects of local buckling when determining their moment resistance or stress resistance.*

In case the slenderness values according to class 1 (see, e.g., Eurocode 3, table 5.3.1) are satisfied no further check of rotation capacity is necessary.

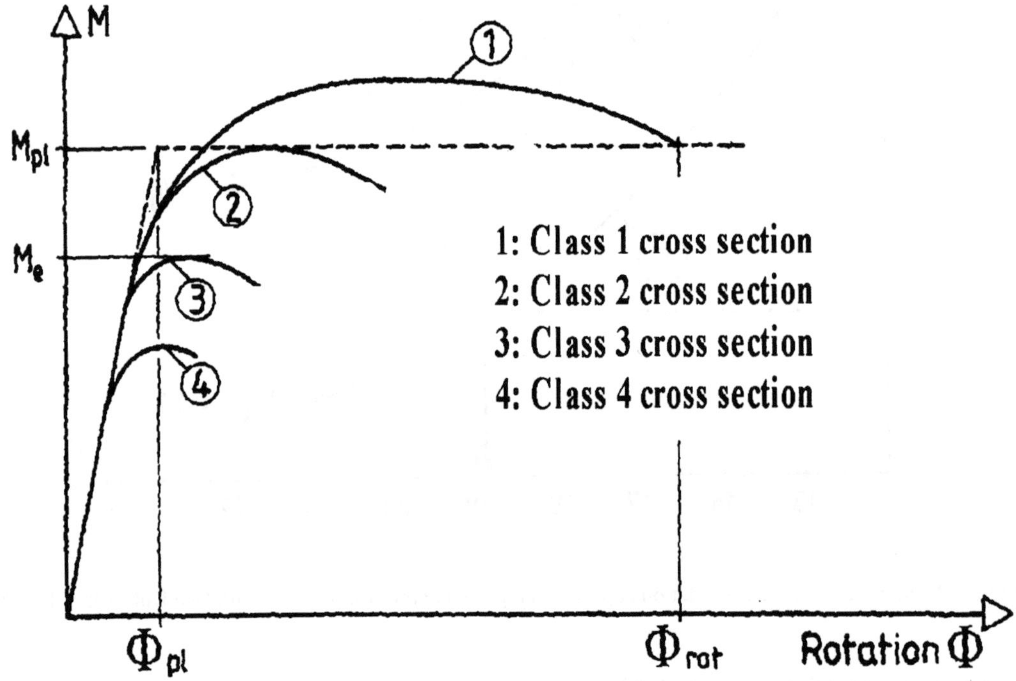


**Figure 10.** Rotation capacity in dependence of cross section classes

## 2 Rotation Capacity of Steel Joints

### 2.1 Definition of Rotation Capacity

**Definition of rotation capacity for members.** Rotation capacity characterises the ability of a plastified section to rotate while maintaining its design moment resistance (see also chapter 1). Figure 11 shows the available rotation capacity of a member as defined in Eurocode 3 (1992).

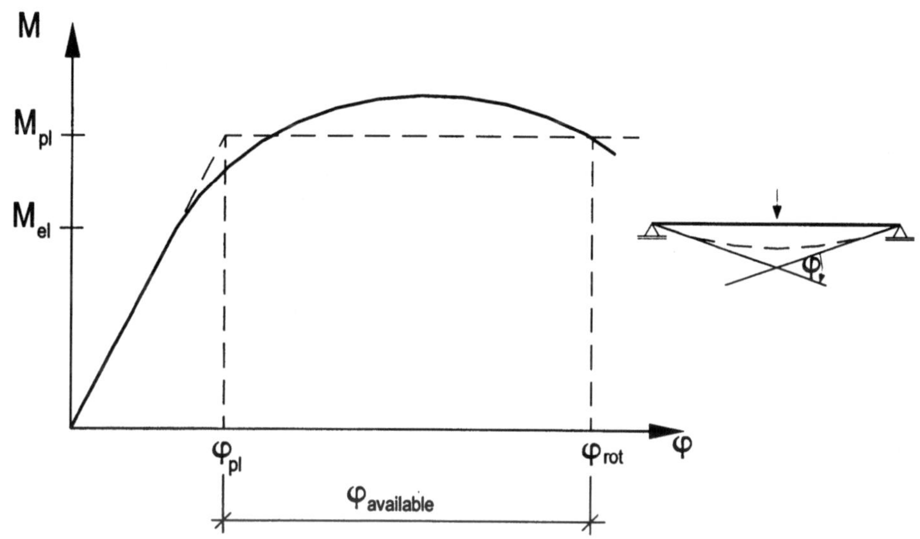


**Figure 11.** Definition of rotation capacity for a member

Due to strain-hardening effects the real moment exceeds the ideal plastic moment $M_{pl}$. With increasing rotations the moment graph passes a maximum value and decreases until it reaches the level of the ideal plastic moment $M_{pl}$ again. The available rotation capacity of a member is defined as the rotation from the point $M_{pl}$ is reached first (under assumption of ideal-elastic/ideal-plastic behaviour) to the point the real M-$\phi$-curve reaches the level of $M_{pl}$ again.

**Definition of rotation capacity for steel joints according to Eurocode 3, revised Annex J (1997).** The definition of available rotation capacity for steel joints differs from the definition for members:

Eurocode 3 (1992, chap. 6.9.5):
*„The design rotation capacity $\phi_{Cd}$ of a beam to column connection shall be taken as the rotation achieved at the maximum design moment resistance of the connection."*

The revised Annex J (1997) of Eurocode 3 shows a more elaborated moment-rotation characteristic for a joint, see Figure 12. Following this new definition the rotation $\phi_{Cd}$ exceeds the value $\phi_{Xd}$, the rotation when the maximum design moment resistance is first achieved.

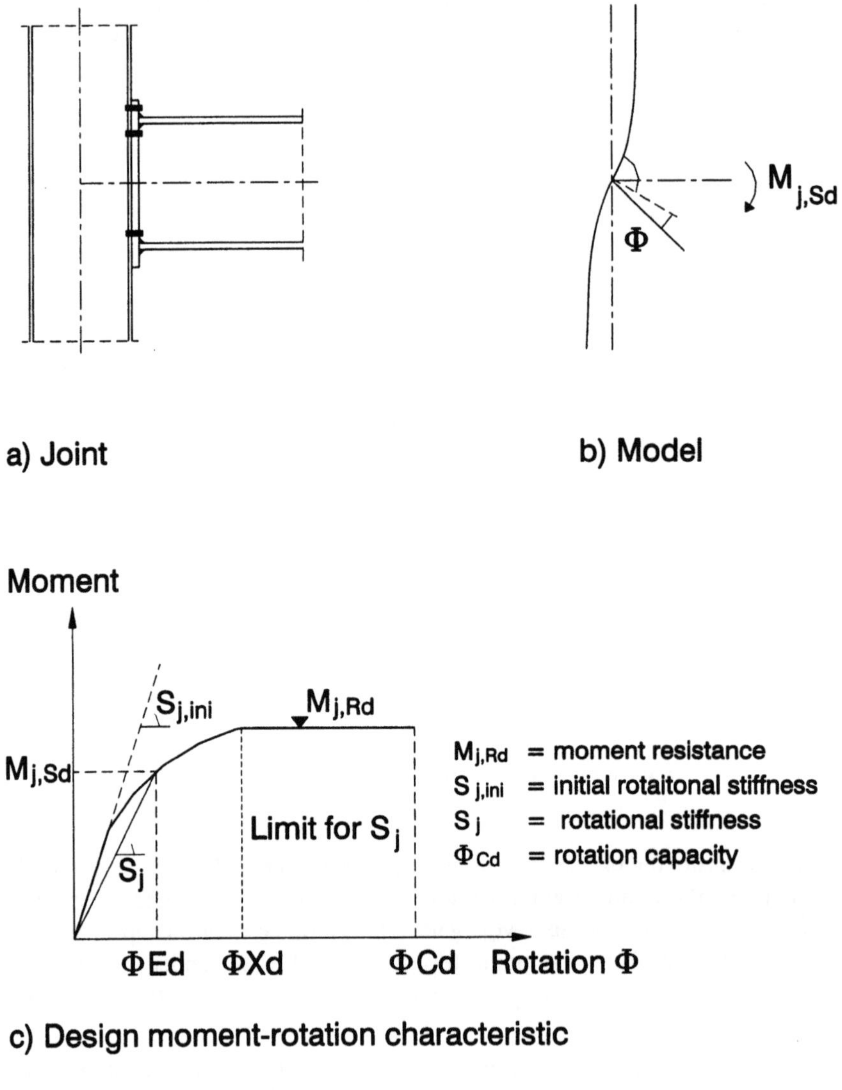


**Figure 12.** Moment-rotation characteristic for a joint according to Eurocode 3, revised Annex J (1997)

The main differences of this definition of the joint rotation capacity in comparison to the definition of the member rotation capacity (see Figure 11) is:

- the available rotation capacity of a joint includes elastic rotations.

As both definitions of rotation capacity, the one for members as well as the one for joints, are finally used to clarify whether plastic design can be accomplished to a structure or not, they should be related to a common basis.

**A coherent definition for members and joints.** The total rotation of a joint is composed of three different contributions, see Figure 13:

$\phi_{el}$ = the purely elastic rotation reached at the level of the elastic design moment resistance

$\phi_{tr}$ = the transient rotation between the level of the elastic design moment resistance and the maximum design moment resistance

$\phi_{pl}$ = the plastic rotation after reaching the maximum design moment resistance to the point the graph reaches this level of moment resistance again or the length of the plastic plateau

Only the plastic rotation $\phi_{pl}$ is decisive for the moment redistribution in a system, if one of the first hinges is formed within a joint. Only this value is comparable to the member rotation which forms the basis of the available rotation capacity.

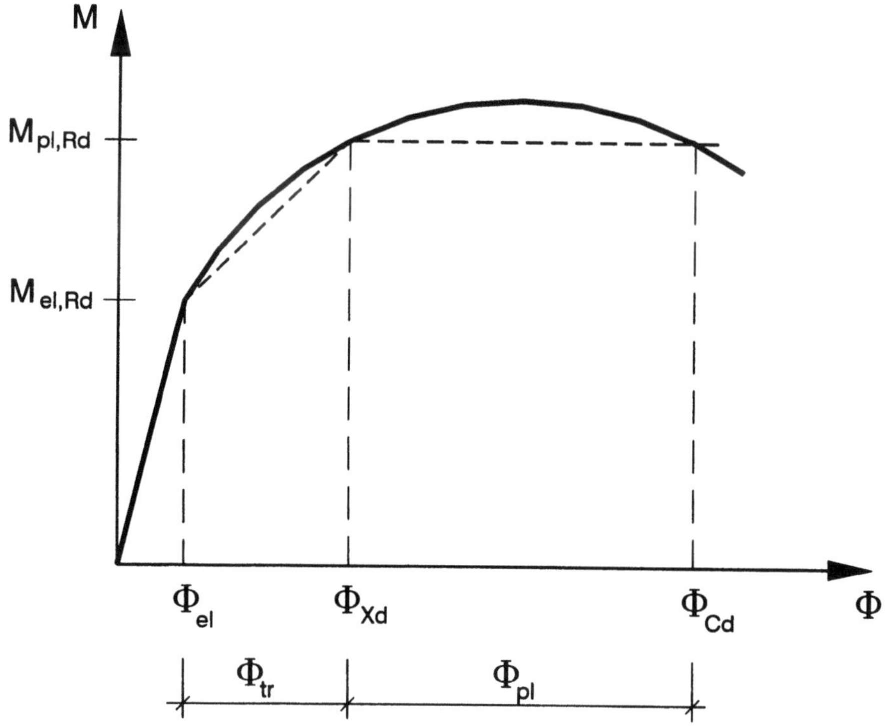


**Figure 13.** Definition of rotation capacity for a joint in correspondence with the member definition

## 2.2 Determination of the Joint Rotation Capacity From the Component's Deformation Capacity

**Deformation capacity of components.** The overall behaviour of a joint is characterised by the behaviour of its single components. Consequently also the rotation capacity of a joint is bound by the deformation capacity of its single components. Therefore before regarding the available rotation capacity of a joint, the available deformation capacity of its components has to be determined. Only as a second step the rotation capacity is derived from the components' deformation capacity. Concerning their load-deformation behaviour components can be distinguished in several classes due to their mode of failure:

1. Components with limited ductility, e. g.:
   - column web in compression
   - beam flange and web in compression

   After passing a maximum load level the load-deformation curves of these components decrease due to buckling effects. At a certain point the load-deformation curves reach the level of the characteristic load $F_{Rc}$ again.
   → The characteristic available deformation capacity of these components may be defined as the deformation $w_c$ belonging to this point.

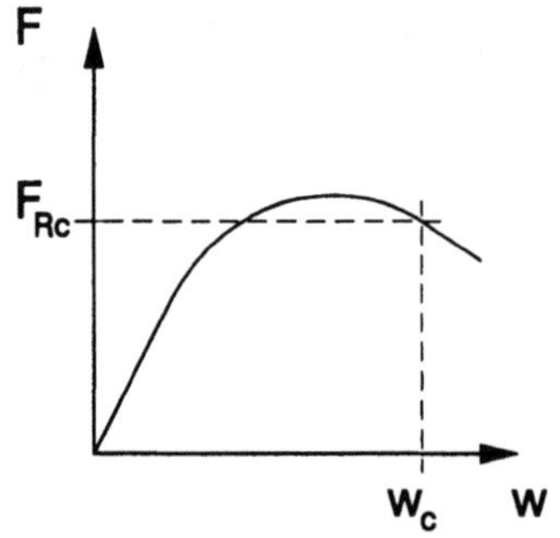


**Figure 14.** Components with limited ductility

2. Components with high ductility, e. g.:
   - column web in shear
   - column flange in bending
   - end-plate in bending
   - flange-cleat in bending

   These components change mostly by membrane effects from an initial carrying mode into a second carrying mode, which allows increasing deformations with increasing load. The deformation capacity is very high or nearly unlimited.

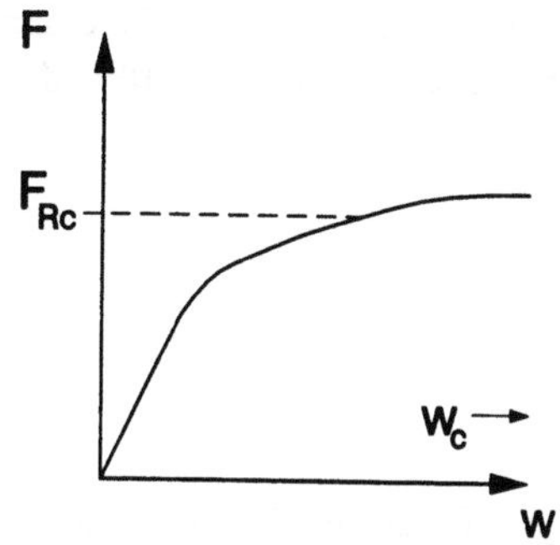


**Figure 15.** Components with high ductility

3. Components with brittle failure, e. g.:
   - bolts in tension

   These components allow only little deformations with increasing load, then sudden failure occurs.
   → A characteristic deformation capacity may be defined as the deformation belonging to the point of failure.

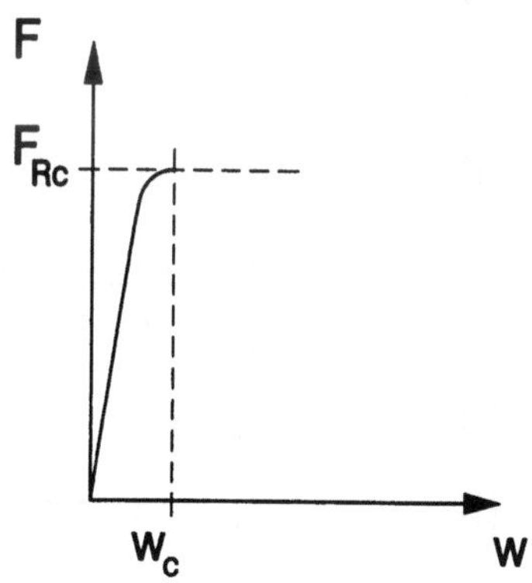


**Figure 16.** Components with brittle failure

**Interplay of the component's deformation capacity for the joint rotation capacity.** When calculating the rotation capacity of a joint from the deformation capacity of its components special features resulting from the interplay between the different joint components have to be considered.

As the joint is manufactured of more than one product, the components may show significantly different behaviour compared to the behaviour theoretically adopted.

Various effects, as
- strain hardening
- material overstrength
- high post-limit stiffness
- imperfections
- residual stresses, etc.

may cause a different sequence of failure for the various components than assumed theoretically. Under the worst circumstances ductile components may still behave elastically while components with non-ductile or even brittle behaviour are overloaded and fail prematurely.

As an example a joint bolted by an end-plate to the column is regarded. Only the following components are taken into account:
- bolts in tension ($B_t$)
- end-plate in bending ($F_{ep}$)
- column web in compression ($F_{c,wc}$)

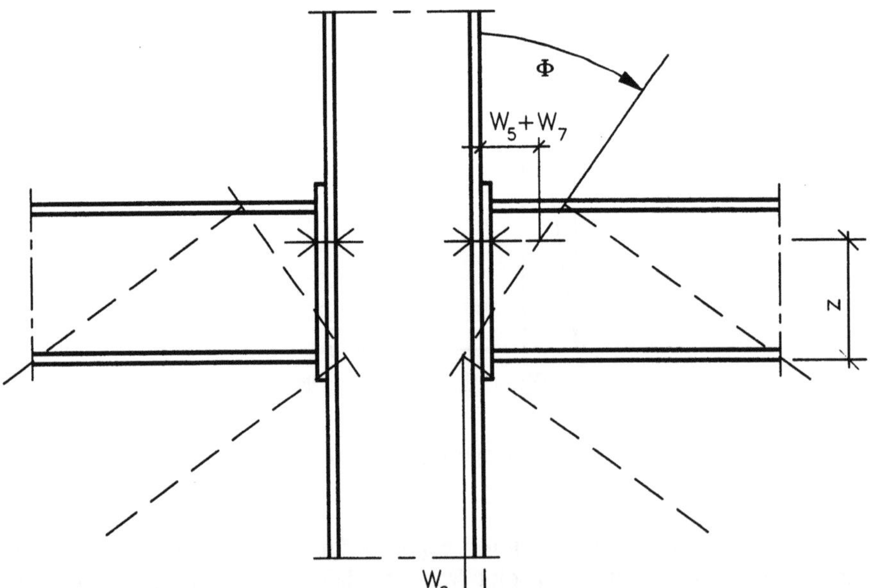


**Figure 17.** Joint with bolted end plate

Case 1: max $F_{ep}$ < max $F_{c,wc}$ < max $B_t$

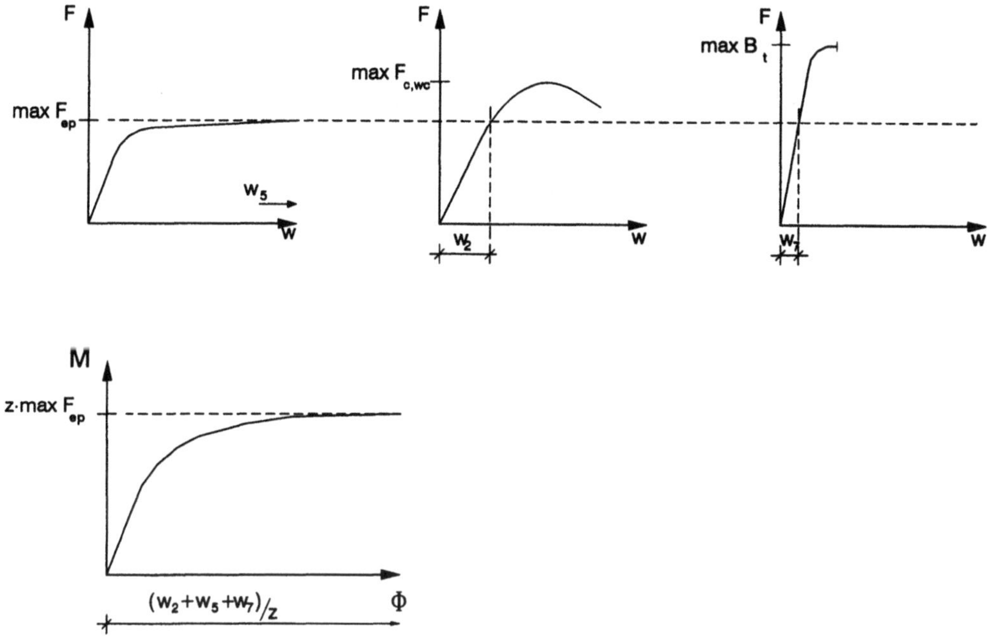


**Figure 18.** max $F_{ep}$ < max $F_{c,wc}$ < max $B_t$

In case 1 the component „end-plate in bending“ has the lowest ultimate load. The behaviour of this component provides high or nearly unlimited available rotation capacity of the joint. The other components contribute to the rotation capacity of the joint in the degree they are deformed when the ultimate load of the component „column flange in bending“ has been reached.

Case 2: max $F_{c,wc}$ < max $F_{ep}$ < max $B_t$

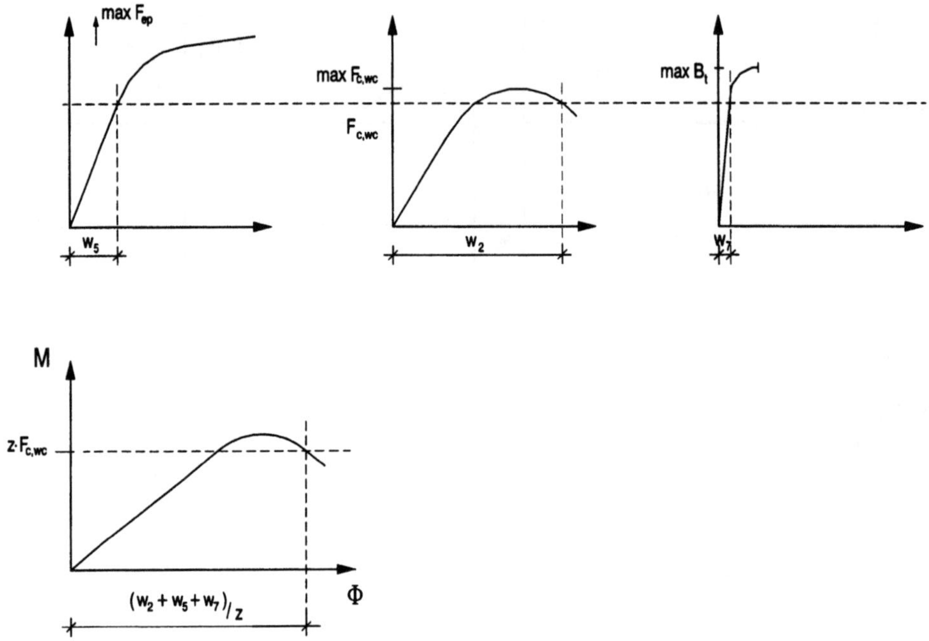


**Figure 19.** max $F_{c,wc}$ < max $F_{ep}$ < max $B_t$

Due to strain-hardening, high post-limit stiffness, overstrength, etc. the ultimate load of one single component (e. g., max $F_{ep}$ of the component „end-plate in bending") may exceed its design value, see Figure 19. At the same time the ultimate load of another component (e. g., max $F_{c,wc}$ of the component „column web in compression") may be reduced unexpectedly, for example due to high imperfections, residual stresses, material understrength, etc.

These differences between real and design component behaviour may lead to a changed failure sequence for the joint. As shown in case 2 the overall behaviour of the joint is dominated by the low ultimate load of the component „column web in compression". The limited deformation capacity of this component finally leads to a limited overall rotation capacity of the joint.

Case 3: max $B_t$ < max $F_{c,wc}$ < max $F_{ep}$

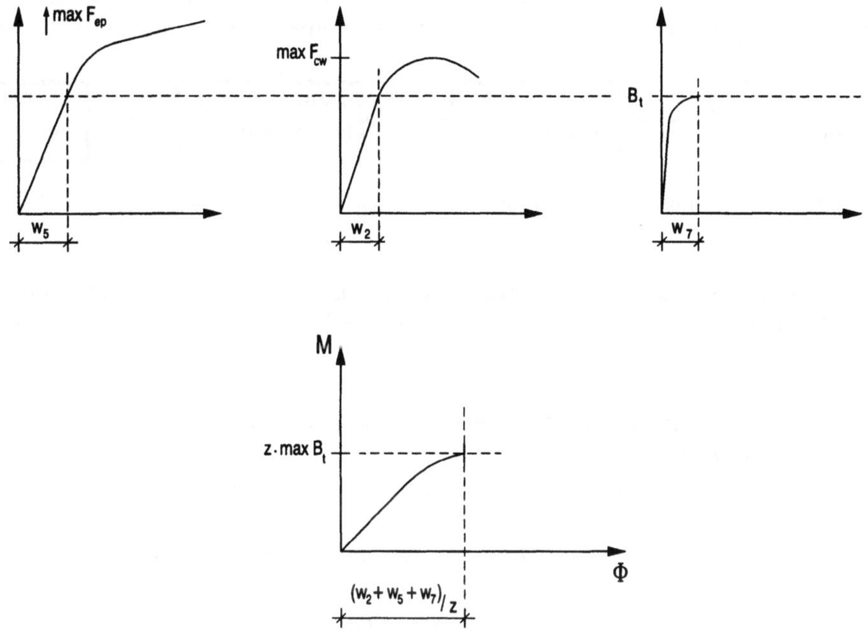


**Figure 20.** max $B_t$ < max $F_{c,wc}$ < max $F_{ep}$

In case 3 due to over-/understrength effects mentioned above neither the component „end-plate in bending", nor the component „column web in compression", but the component „bolts in tension" fails. This leads to the most critical case: brittle failure of the joint.

These three cases show how the deformation capacity of the components and their interplay influence the rotation capacity of a joint. The importance of over-/understrength effects for the available rotation capacity of joints is clearly demonstrated. Therefore for deriving rotation capacity values of joints the component behaviour has to be investigated under consideration of possible over-/understrength effects of the components which of course should also be measured in tests.

For practical design it seems reasonable to consider only the deformation capacity of the dominant component with the lowest ultimate load and to neglect the elastic or plastic deformations of the other components (e. g., in case 2 only the deformation capacity of the component „column web in compression" should be considered). Nevertheless especially for a brittle failure as in case 3 a margin of safety concerning the deformation capacity has to be defined.

### 2.3 Verification of Rotation Capacity

**Verification of sufficient rotation capacity of members.** For the verification of sufficient rotation capacity of members a simplified procedure is given in Eurocode 3 (1992), see Figure 21.

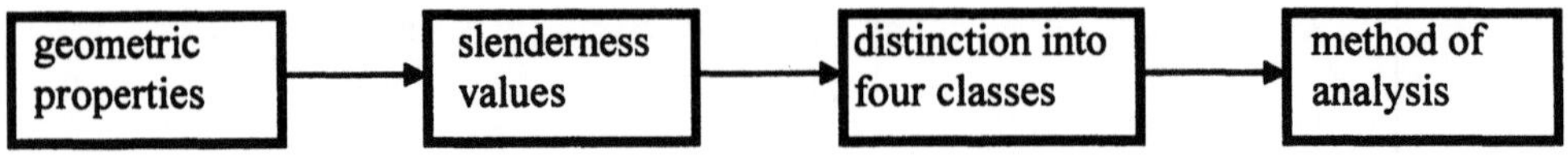


**Figure 21.** Simplified procedure for the verification of sufficient rotation capacity of members

Due to their web and flange slenderness sections can be subdivided into four classes, that allow different methods of analysis. An explicit verification of sufficient rotation capacity is not necessary.

The four classes are defined as follows (see also chapter 1):
Eurocode 3 (1992, chap. 5.3.2):

*Class 1 cross sections are those which can form a plastic hinge with the rotation capacity required for plastic analysis.*

*Class 2 cross sections are those which can develop their plastic moment resistance, but have limited rotation capacity.*

*Class 3 cross sections are those in which the calculated stress in the extreme compression fibre of the steel member can reach its yield strength, but local buckling is liable to prevent development of the plastic moment resistance.*

*Class 4 cross sections are those in which it is necessary to make explicit allowances for the effects of local buckling when determining their moment resistance or stress resistance.*

These deemed-to-satisfy criteria, by which four classes of sections and the corresponding analytical procedures are identified, are the results of a thorough scientific investigation which started by testing beams, followed by sophisticated calculations.

The simplified procedure for the verification of sufficient rotation capacity of members with *class 1* sections according to the scheme in Figure 21 was derived from scientific investigations comparing **available** and **required** rotation capacity. Sufficient rotation capacity is provided if the available rotation capacity of a member is higher than the required rotation capacity of this member, see equation (1) in chapter 1.

**Available** rotation capacity is gained from experiments or calculations based on experimental values of members separated from the structure, whereas **required** rotation capacity of a member results from the rotations under a certain load in a certain structure. For the comparison between available and required rotation capacity it is necessary, that both values are derived by similar means: If available rotation capacity is defined according to realistic moment-rotation curves gained from tests, then required rotation capacity must be defined by similar means, i. e., required rotation capacity must be gained from system investigations under consideration of the real system behaviour.

The following diagram shows the path followed by the derivation of simplified rules for the verification of sufficient rotation capacity of members:

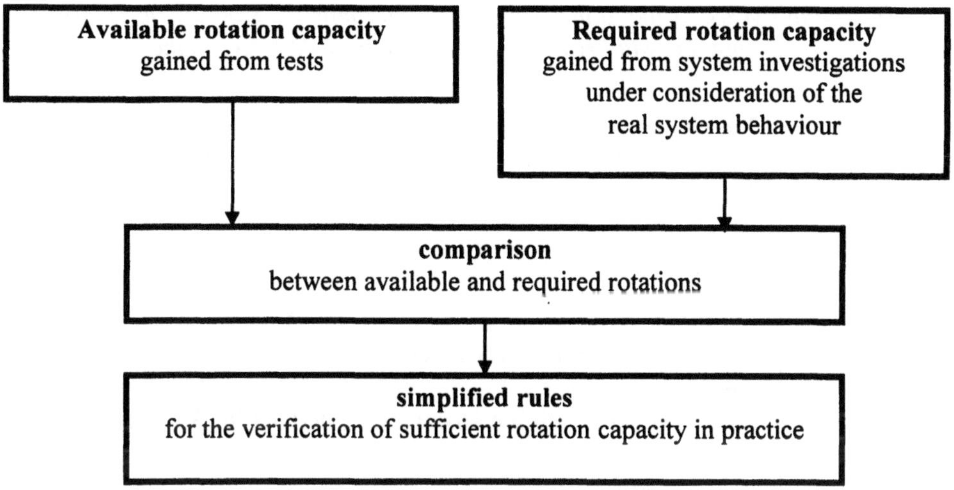


**Figure 22.** Derivation of simplified rules for the verification of sufficient rotation capacity of members

**Verification of sufficient rotation capacity of steel joints.** Usually the resistance of semi-rigid steel joints is less than that of the connected members, but as a matter of fact the highest moments often occur at the joints and not in the beams or columns. Therefore in systems with semi-rigid connections plastic hinges will most probably form at the joints and not in the adjacent members. To allow rigid-plastic analysis the rotation capacity of the joints has to be checked:

Eurocode 3, revised Annex J (1997, chap. J.5):

> „*When plastic global analysis is used, partial-strength joints should have sufficient rotation capacity...*"

At present Eurocode 3, revised Annex J (1997), chap. J.5 gives only some basic principles for the verification of sufficient rotation capacity for a limited number of beam-to-column joints with a specific modus of failure. No general method of verification exists.

As there should be a common basis for the definition for the rotation capacity of members and joints there should also be a common method for the verification of rotation capacity of members and joints.

A first step towards a common method of verification of rotation capacity is a classification of joints analogous to that of members. Jaspart (1996, 1999) refers to this point and gives definitions for joint classes:

*Class 1 joints: able to reach $M_{R,d}$ and with a sufficiently good rotation capacity to allow plastic design of the frame.*

*Class 2 joints: able to reach $M_{R,d}$ but with a reduced plastic rotation capacity. A plastic verification of the section is anyway allowed.*

*Class 3 joints: where brittle failure (or instability) limits the moment resistance and does not allow full redistribution of the internal forces in the joint.*

As for members this classification and the corresponding deemed-to-satisfy criteria must be based on thorough scientific investigations. During this procedure a very strict distinction has to be made between the

- test level and the
- model level.

Concerning the definition of resistance Jaspart (1996/1997) points out the importance of a clear distinction between test and model level of resistance. The same is true for the definition of rotation capacity values.

At the **test level** real actual moment-rotation curves are derived for different types of joints. All non-linearities like e.g.

- non-linearities due to the material
- non-linearities due to the joint behaviour itself
- non-linearities due to the plastic zones developing in the adjacent members or
- non-linearities due to second order effects in the columns

are included in this evaluation by testing. Of course analytical procedures which follow the real testing behaviour and include all the different non-linearities don't need any modelling, classification or check of rotation capacity.

In contrast modelling, classification and check of rotation capacity are necessary at the **model level.** But the procedure is strongly dependent of the level of sophistication. In a first step the actual moment-rotation curve is transformed in a more or less linear design curve with design values for the maximum design moment and a plastic plateau for the region where the real moment exceeds the maximum design moment. On this level a verification of the rotation capacity is needed. Required rotation values derived from the system must be compared to available rotation values derived from joint design curves. This way of verification follows the load path each time a real system has to be verified. This verification includes the rotation control of the joint for each load step considered. Simplification for the adopted joint behaviour as well as for the joint stiffness within the system calculation are possible, see e.g. the Innsbruck joint model. Experiences with this way of verification have been gathered (see Tschemmernegg and Huber, 1987).

As the system calculation according to plastic hinge theory only considers the rotation of a hinge after it has formed i.e. after reaching the maximum design moment resistance only the plastic rotation value of the joint design curve $\phi_{pl}$ is of interest, see Figure 13.

At a final stage of the development of a design procedure the direct comparison of the required and the available rotations should be replaced by deemed-to-satisfy criteria. These must be derived by systematical investigations for a range of typical systems and joints. System investigations for various system and load situations then lead to a set of boundary values for required joint rotations. These values are compared to the available joint rotations to define limits which joints have to fulfil to be appropriate for a certain joint class i.e.

- class 1 joints — which allow rigid-plastic analysis or
- class 2 joints — which just allow an elastic-plastic design or usage as „final hinge“ or
- class 3 joints — which even limit the moment resistance to values less than the section value $M_{b,pl}$.

For the comparison between available and required rotations it is necessary, that both values are derived by similar means: If the available rotation capacity is defined according to realistic moment-rotation curves gained from tests, then required rotation capacity must be defined by similar means, i. e., required rotation capacity must be gained from system investigations under consideration of the real system behaviour.

Therefore a deep knowledge of the joint behaviour and its influence on the structure's response is necessary. Deriving values for available and required rotation capacities of semi-rigid joints thus requires the consideration of several non-linearities that influence the rotation-behaviour of a joint in a system.

These non-linearities are e.g. (see Figure 23, 24 and 25)

- non-linearities due to the joint behaviour itself
- non-linearities due to the plastic zones developing in the adjacent members
- non-linearities due to second order effects in the columns

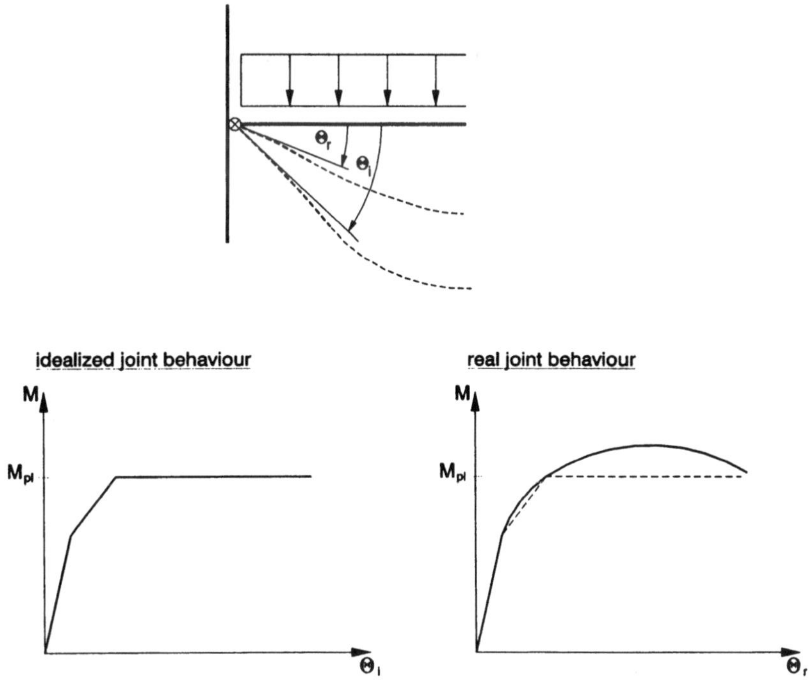


**Figure 23.** Non-linearity of the joint behaviour

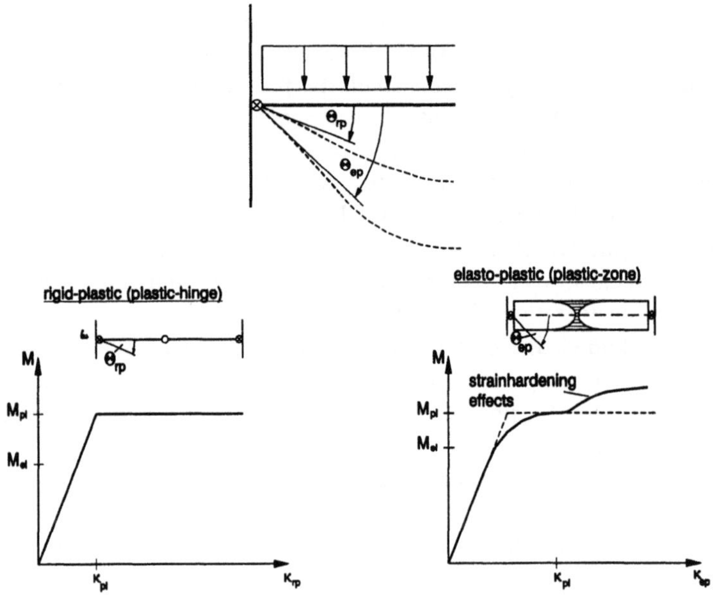


**Figure 24.** Non-linearity of the plastic zones developing in the connected members

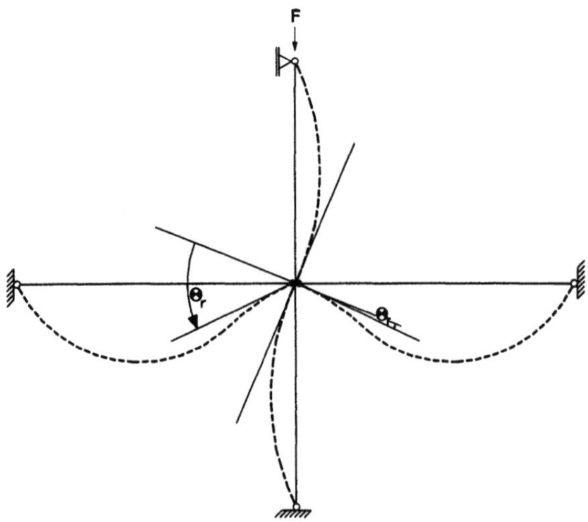


**Figure 25.** Non-linearity from second order effects in the columns

**Interrelation of system and joint behaviour with the verification procedure.** As stated in chapter 2.2 joints should be classified according rotation capacity values derived from their components' deformation capacity. Wald et al. (1997) e.g. suggest a procedure how to predict rotation values on the basis of the component method. As pointed out before the various components show different modes of failure. One must consider this behaviour when checking rotation capacity and deriving deemed-to-satisfy criteria for joint classes (Kuhlmann, 1997).

System behaviour:

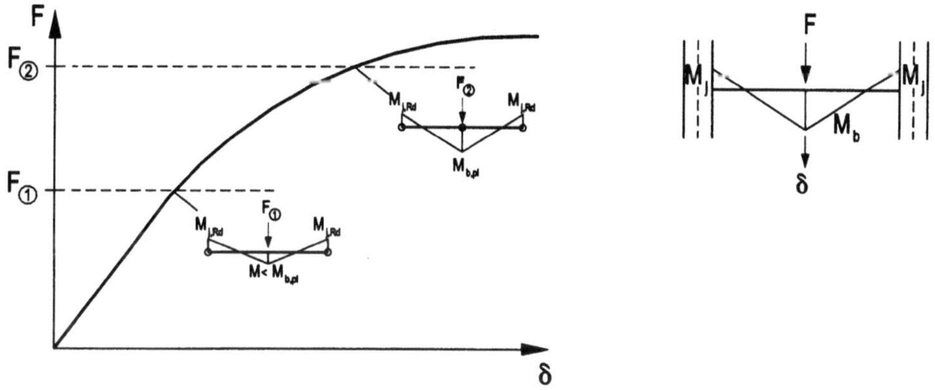


Joint behaviour:

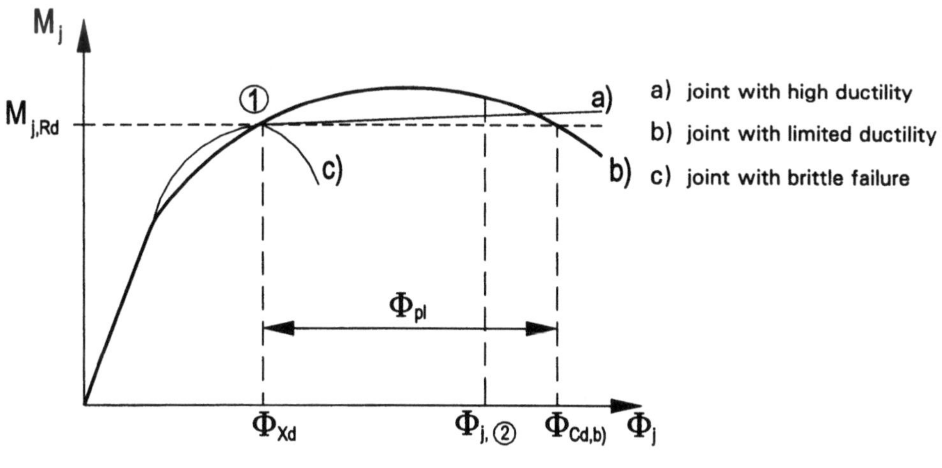


**Figure 26.** Load deformation curve for the system behaviour and corresponding moment-rotation curves for the joints

| | | |
|---|---|---|
| Case a) | $\Phi_{pl} \rightarrow \infty$ | no limitation |
| Case b) | $\Phi_{j,②} - \Phi_{Xd} \stackrel{!}{<} \Phi_{pl}$<br>$F \leq F_{②}$ | full mechanism possible |
| Case c) | $\Phi_{j,①} \stackrel{!}{<} \Phi_{Xd}$<br>$F \leq F_{①}$ | conservative, safe side |

**Figure 27.** Verification cases

For the system in Figure 26 the load reaches the value of $F_{①}$ when the joint moment resistance of $M_{j,Rd}$ has just been reached, but the maximum sagging moment in the span has not yet adopted the value of $M_{b,pl}$. Caused by the semi-rigid behaviour, the joint rotations have already reached the value of $\phi_{Xd}$. When now increasing the load F, the behaviour depends on the characteristics of the joint:

Case a) is a very ductile joint with nearly unlimited rotation ability, the actual resistance even exceeds the design joint resistance $M_{j,Rd}$. In this case the load F surpasses the system load $F_{②}$ which refers to the ideal hinge mechanism based on the design value of the moment $M_{j,Rd}$. Caused by non-linearities like strain-hardening and the post-limit behaviour of the joint the local moment exceeds $M_{j,Rd}$ so that also the real moment distribution may be different to that adopted by simple plastic hinge analysis.

Case b) represents a joint with a limited ductility. The strength slightly exceeds the design moment resistance $M_{j,Rd}$, but after a certain rotation value $\phi_{pl}$ the resistance becomes less than $M_{j,Rd}$. For the system it is decisive whether the joint rotation $\phi_{j,②}$ belonging to the full hinge mechanism load $F_{②}$ lies within this region or not. According to Figure 26 the value of the difference of ($\phi_{j,②}$ - $\phi_{Xd}$) is less than $\phi_{pl}$. Thus the assumption of the full hinge mechanism is admissible.

Case c) is a joint with a brittle behaviour. The maximum strength $M_{j,Rd}$ can just be reached, but no further rotation is possible. As a consequence the full hinge system load of $F_{②}$ cannot be reached either. A consideration of this system according to plastic hinge analysis is excluded.

**Safety considerations for the verification of rotation capacity.** As stated above this verification procedure may be adopted for a single system and joint configuration as well as for a range of systems and joints to define deemed-to-satisfy criteria.

As all this happens on the **model level,** careful consideration has to be given to the safety margin between characteristic and design values of moment resistance as well as of rotation capacity. Not only the strength of a component has to be divided by a safety factor $\gamma_m$, but also the deformation capacity of a component should be reduced to a design value.

It seems reasonable to derive design values for the rotation capacity by dividing the characteristic deformation values of their components by an appropriate safety factor $\gamma_{rot}$, see Figure 28.

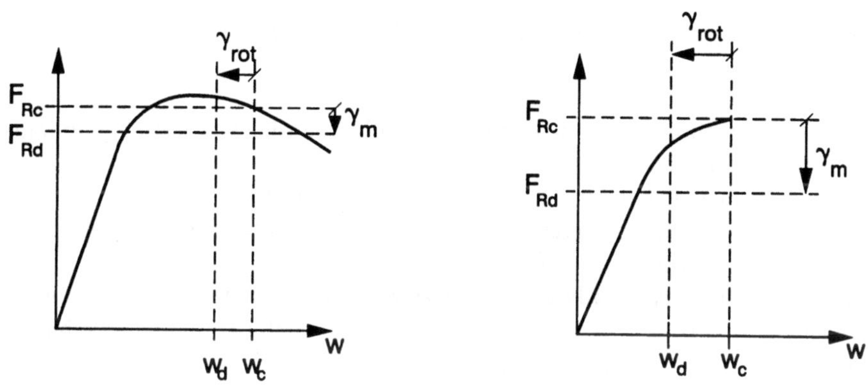


**Figure 28.** Suggestion for a safety factor for components with limited ductility and with brittle failure

For components with brittle failure the safety factor $\gamma_{rot}$ should be chosen higher than the safety factor $\gamma_{rot}$ for components with limited ductility.

### 2.4 Conclusions

To allow plastic analysis for systems with semi-rigid joints sufficient rotation capacity of the joints has to be checked. The available rotation capacity for a certain type of joint has to be compared with the required rotation capacity for this joint derived from system investigations. The available rotation capacity of a joint can be determined on the basis of the deformation capacity of its components. The various component ductility decisively influences the ductile overall behaviour of the joint. A strong distinction between test and model level is essential. Safety margins ought to be defined for deformation values.

For the determination of available rotation capacity of steel joints several models can be used (see Klein, 1985, Tschemmernegg et al., 1997, Sibai, 1991, Jaspart, 1991, Vandegans, 1995). These models were derived by different ways of research, but they were all checked against test results and therefore provide safe and accurate predictions of available rotation capacity.

For the derivation of values for required rotation capacities also different paths may be followed (see Vandegans and Jaspart, 1996, Kattner, 1999, Boender 1995, Boender et al., 1996, Xiao, 1996, Li et al., 1995, Nethercot et al., 1995, Ahmed and Kirby, 1997) . The main differences lie in the consideration of the several non-linear effects influencing the required rotations.

The level of comparison between available and required rotation capacity should be brought in one line, i. e. values for available rotation capacities derived under consideration of the real joint behaviour should be compared with values of required rotation capacities derived as well under consideration of the real system behaviour.

## 3 Research Work on the Rotation Capacity of Steel Joints

### 3.1 Introduction

Among the various components two types especially contribute to a satisfying value of the available rotation capacity:

- any kinds of plate in bending like end-plates or column flanges
- and components under compression suffering from stability phenomena like column webs or beam flanges in the compression zones.

Whereas the first category usually shows a nearly unlimited ductility, see e.g. Figure 15, the latter is characterised by a decreasing branch of the load-deformation curve which thus also limits the overall ductility of the joints, see e.g. Figure 19. To solve the question of overall rotation capacity of a joint therefore a thorough investigation of the behaviour of the components under compression is essential. That is the reason why the component "column web in compression" is given a closer look in the following.

To characterise a joint component's load-deformation behaviour different possibilities exist:

1. By testing:
   Load-deformation curves gained by tests include all kind of realistic material and system influences as strain-hardening, material overstrength, post-limit stiffness, imperfections or residual stresses. But of course the number of variable parameters which can be looked at is limited because of the important expenses for each test specimen.
2. By numerical simulations:
   By help of Finite-Element studies the parameter field of the tests can be extended; test results are used to calibrate the numerical models to achieve an optimum approximation of the real behaviour.
3. By mechanical models:
   With the help of such models an analytical description of the load-deformation behaviour is possible. This is particularly important because the formulas describing the load-deformation behaviour of the single components form the basis for the modelling of the overall rotation behaviour of the joint.

In the following for the component "column web in compression" own test results and corresponding theoretical studies are compared. The results of these investigations show the decisive influence of an axial force in the column on the deformation capacity of this component.

### 3.2 Tests on the Component "Column Web in Compression"

**Introduction.** In 1997 a first test series of six tests (Series I: Specimen A1 - A3 and B1 - B3) were performed at the University of Stuttgart (see Kuhlmann and Fürch, 1997). The aim was

- to investigate not only strength and stiffness of the component "column web in compression", but also the deformation capacity especially in the post-critical region and
- to investigate the influence of axial forces on the load-deformation behaviour of the component "column web in compression" which is essential for the behaviour of the columns in typical semi-rigid joints.

In 1999 the first test series of 1997 was supplemented by a second test series of 10 tests (Series II: Specimen A4 - A8 and B4 - B8) (see Kuhlmann and Kühnemund, 1999, and Kuhlmann et al., 2000). Again the axial force in the column was the main parameter. Table 1 gives an overview on all 16 tests. To have a reference value without the influence of an axial force also in each series reference tests without axial force in the column (A3, B1 and A8, B8), one for each profile type HE 240A and HE 240B, were executed.

**Table 1.** Overview on all 16 tests for the component "column web in compression"

| Number | Profile | Ratio of $N/N_{pl}$ |
|---|---|---|
| A1 | HEA 240 | 0,67 |
| A2 | HEA 240 | 0,60 |
| A3 | HEA 240 | 0,00 |
| A4 | HEA 240 | 0,11 |
| A5 | HEA 240 | 0,22 |
| A6 | HEA 240 | 0,33 |
| A7 | HEA 240 | 0,47 |
| A8 | HEA 240 | 0,00 |
| B1 | HEB 240 | 0,00 |
| B2 | HEB 240 | 0,46 |
| B3 | HEB 240 | 0,65 |
| B4 | HEB 240 | 0,11 |
| B5 | HEB 240 | 0,24 |
| B6 | HEB 240 | 0,36 |
| B7 | HEB 240 | 0,58 |
| B8 | HEB 240 | 0,00 |

**Test set up and test procedure.** Figure 29 illustrates the scheme of the test set up. The real joint is turned around by 90°. The axial force of the column is applied as constant horizontal force, whereas the compression force F belonging to the beam moment M at the real joint is introduced as vertical force from both sides of the section.

a) Real joint

b) Scheme of test set up

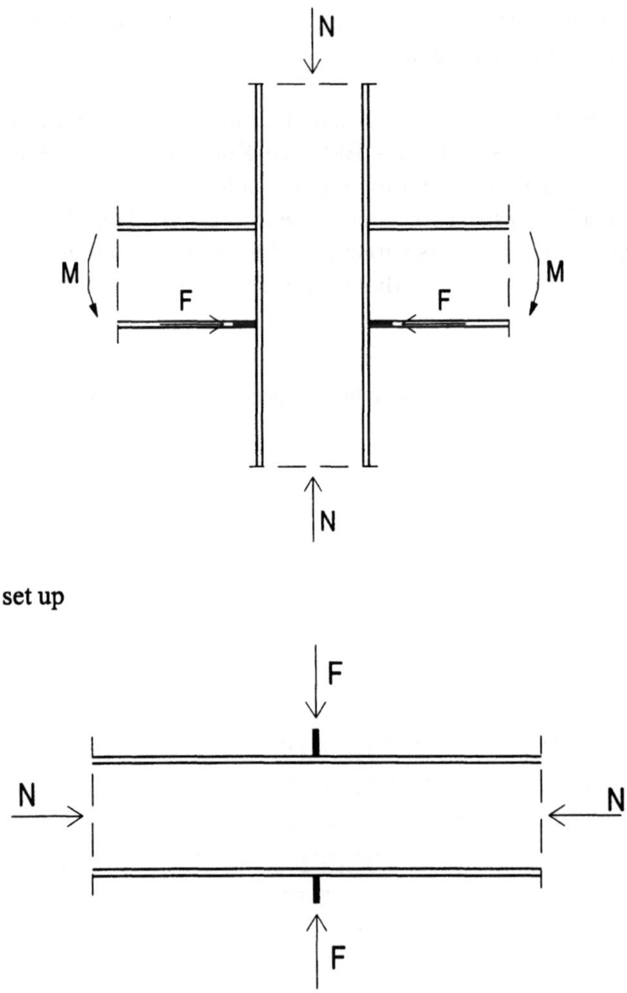


**Figure 29.** Real joint and scheme of test set up of the component test

The test set up is shown in Figure 30. Whereas the axial force of the column was applied by an horizontal hydraulic jack built within an horizontal frame of end-plates and 4 rods, the compression force for the web was applied by the vertical testing machine. Careful measurements were taken as well of the geometry and material of the tests specimen as of the deformations during the tests. All test specimen were provided with a measuring grid (see Figure 31) before executing the tests. It formed the basis for the measuring of the geometrical data like web thickness or flange width and supported the visualisation of the buckling web. All tests were controlled by deformation in order to follow the decreasing part of the curves. Furthermore about ten times the tests were stopped for relaxation.

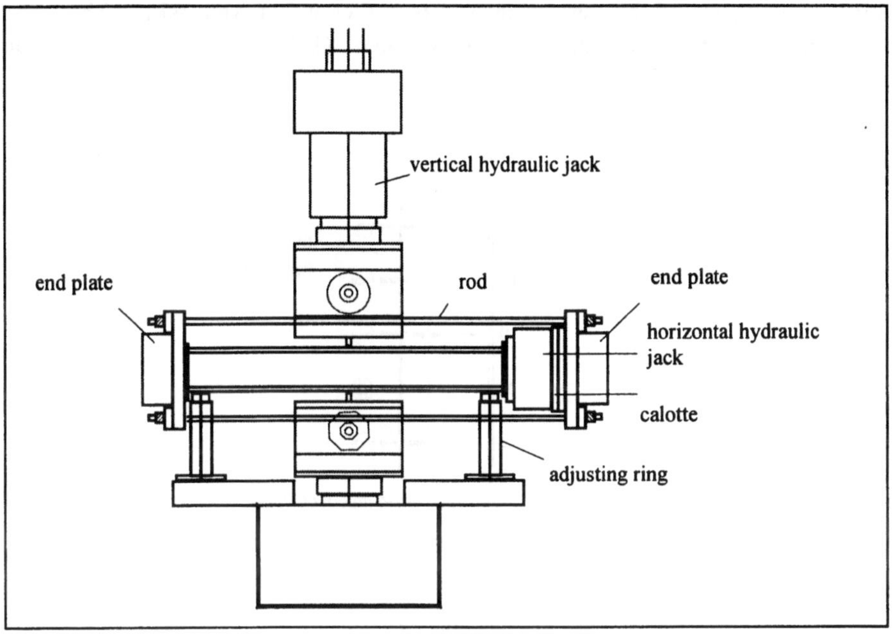


**Figure 30.** Test set up

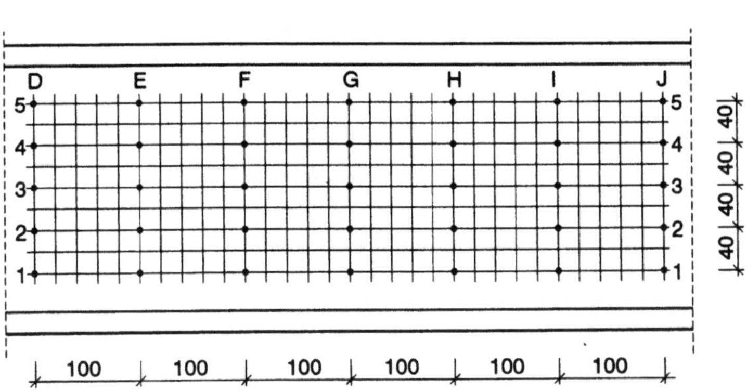


**Figure 31.** Measuring grid of the test specimen

**Measurements.** Details of the measurements are given in Figure 32 and 33. The main measured quantities are

- the compression force F (1)
- the corresponding deformations w (2) and (10) respectively.

These data form the basis for the load-deformation curves F-w as given e.g. in Figure 36. The axial force in the column was applied by an special calibrated hydraulic jack (3). To get information about the growing buckle in the column web a measuring system with five gauges was developed, (5) – (9). At every stop for relaxation the horizontal deformations of distinguished points of the measuring grid (see Figure 31) were taken. Three strain gauges fixed near and in the axis of the compression force , (11) – (13) covered the spreading of the compression force within the column web.

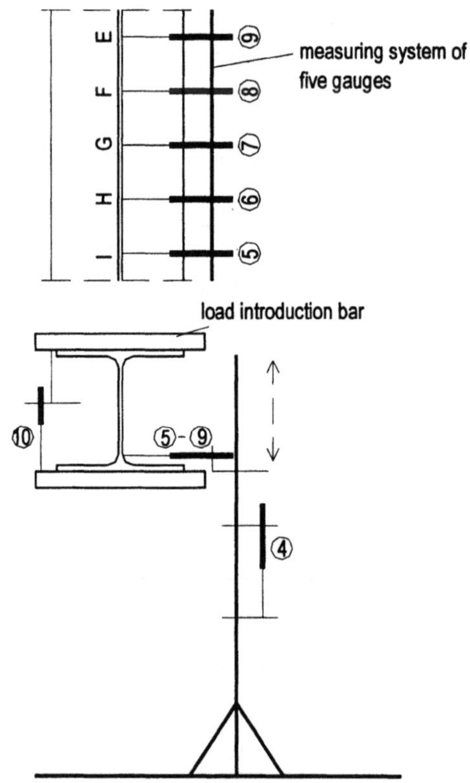


**Figure 32.** Measuring system for the buckle in the column web

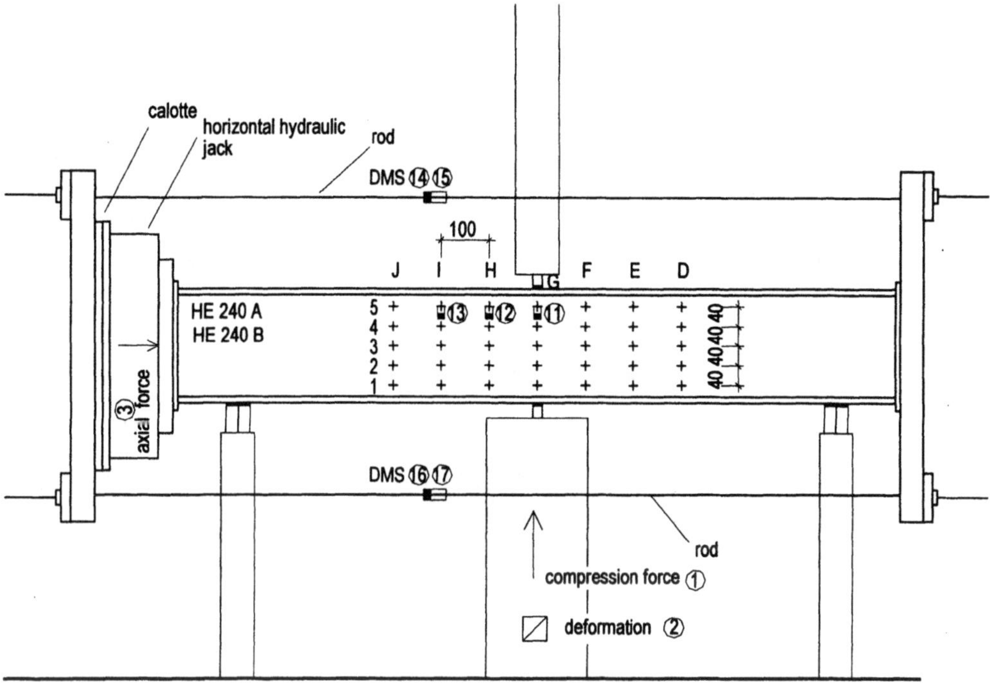


**Figure 33.** Measurements

**Measurement of geometry.** Before executing the tests the geometry of each test specimen was measured in detail. The measured quantities are the flange width b and thickness $t_f$, the web height d and web thickness $t_w$, the height h of the profile and the length L of the test specimen (see Figure 34). All measured quantities were taken at several distinguished points; for distinction indices are used.

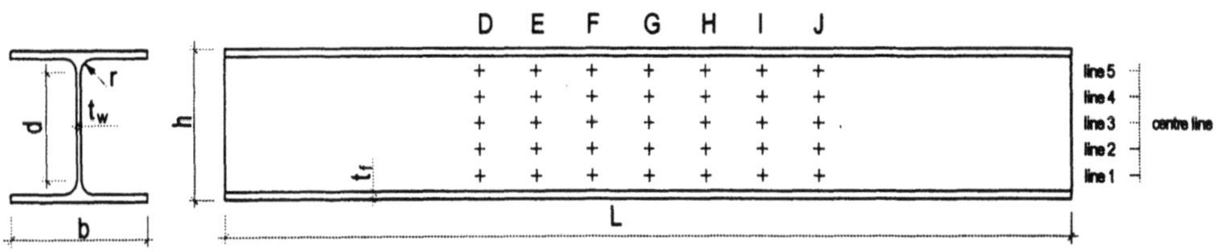


**Figure 34.** Measured quantities of geometry

For same quantities average values are calculated, which form the basis for determining the area values of the real cross section. The plastic axial force $N_{pl}$ is finally calculated under consideration of these average values.

**Measurement of material.** In the same way as special view was given to the geometry also the real material properties were determined. For both test series deformation controlled tension tests were executed to gain corresponding stress-strain curves. As the behaviour of the tension test specimen differ in dependence of the various cross sectional parts as flanges or web, several tension tests were executed for each profile type (see Table 2).

**Table 2.** Tension test specimen

| Test series | Cross section part | Orientation of tension test specimen |
|---|---|---|
| Series I<br>Profile HEA 240 and<br>Profile HEB 240 | Top flange | Longitudinal direction |
| | Bottom flange | Longitudinal direction |
| | Web | Longitudinal direction |
| | Web | Transverse direction |
| Series II<br>Profile HEB 240 and<br>Profile HEB 240 | Top flange | Longitudinal direction |
| | Bottom flange | Longitudinal direction |
| | Web | Longitudinal direction |
| | Web | Transverse direction |

### 3.3 Test Results

**Geometry.** Table 3 compares the measured values of the cross section areas with those from the standards. Whereas the measured values of test specimen A1 - A3 are bigger than the nominal values the test specimen of the second series (A4 - A8) dispose of smaller cross section areas than theoretically given. Also for the HEB profiles of the first test series the real values are smaller than the values from the code. There is almost no difference between the theoretical and real cross section areas of test specimen B4 - B8.

**Table 3.** Cross section areas of all 16 test specimen (test series I in bold letters)

| Test specimen | Nominal values [cm²] | Measured values [cm²] | Difference [%] (Nominal values = 100%) |
|---|---|---|---|
| **A1** | **76,84** | **78,78** | **1,03** |
| **A2** | **76,84** | **78,70** | **1,02** |
| **A3** | **76,84** | **78,53** | **1,02** |
| A4 | 76,84 | 74,58 | 0,97 |
| A5 | 76,84 | 74,87 | 0,97 |
| A6 | 76,84 | 74,35 | 0,97 |
| A7 | 76,84 | 75,19 | 0,98 |
| A8 | 76,84 | 74,44 | 0,97 |
| **B1** | **105,99** | **102,60** | **0,97** |
| **B2** | **105,99** | **102,43** | **0,97** |
| **B3** | **105,99** | **102,33** | **0,97** |
| B4 | 105,99 | 105,66 | 0,99 |
| B5 | 105,99 | 106,00 | 1,00 |
| B6 | 105,99 | 105,80 | 0,99 |
| B7 | 105,99 | 105,70 | 0,99 |
| B8 | 105,99 | 106,23 | 1,01 |

**Material.** As an example Figure 35 contains the stress-strain relationship for test specimen HEA 240, series II. It is clear from the diagram, that the measured yield strength is higher than the theoretical values from the codes. Table 4 refers to this point and presents for all tension tests the comparison between nominal and real values.

**Table 4.** Yield strength of tension test specimen

| Tension test specimen | Cross section part | Measured values [kN/cm²] | Nominal values [kN/cm²] |
|---|---|---|---|
| Test series I HEA 240 | Flange | 28.85 | 23,5 |
| | Web | 28,97 | |
| Test series I HEB 240 | Flange | 24,64 | |
| | Web | 29,07 | |
| Test series II HEA 240 | Flange | 26,32 | |
| | Web | 27,63 | |
| Test series II HEB 240 | Flange | 26,13 | |
| | Web | 28,04 | |

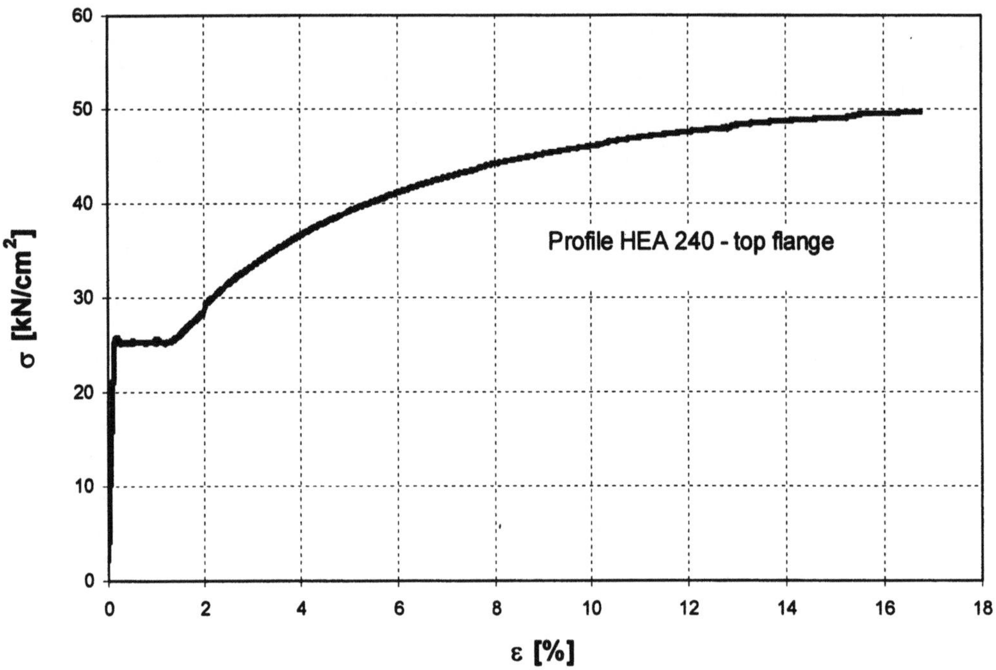


**Figure 35.** Stress-strain curve for profiles HEA 240, test series II, top flange

**Load-deformation behaviour.** As main results the applied compression force F is referred to the measured deformation w of the web in direction of the force F. The load-deformation curves clearly show the typical very ductile behaviour of the component "column web in compression". In Figure 36 the test results of three HEA 240 profiles of the first test series of 1997 are compared to each other. Whereas test specimen A3 had no axial force, A1 and A2 were simultaneously loaded by axial forces of 60% and 67% of $N_{pl}$. The diagram shows clearly the strong influence of the axial force for the load-deformation behaviour of the web.

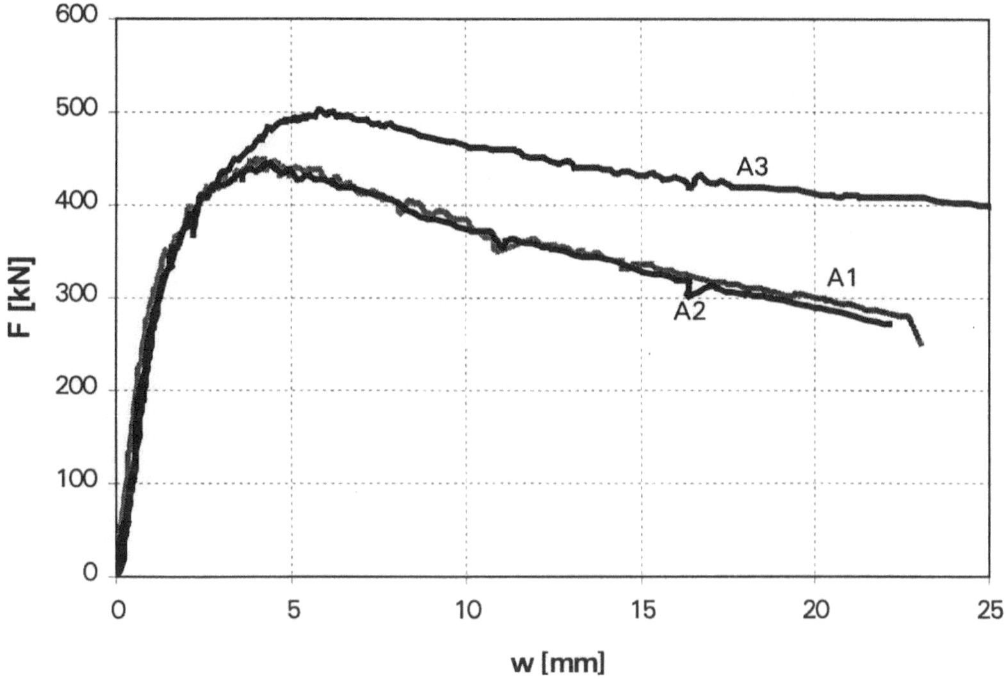


**Figure 36.** Typical load-deformation curves

Tests series A and B differ from each other by the choice of the profile. Thus the behaviour of the slender profile HEA 240 always showed a smaller resistance and deformation capacity than the corresponding test specimen of series B made of HEB 240, see Figure 37.

These comparisons point out as most important factors of influence
- the geometry of the profile and the
- ratio of axial force $N/N_{pl}$.

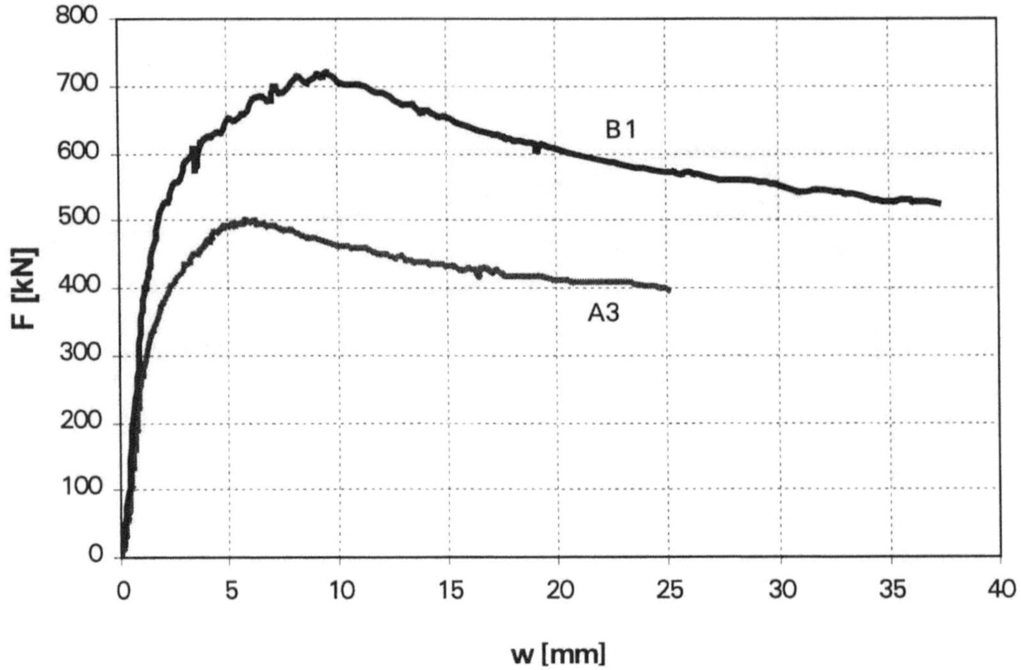


**Figure 37.** Comparison of the load-deformation behaviour of HEA 240 profiles and HEB 240 profiles

As the test series of 1997 only covered three specimen for each profile type no overall statement concerning the deformation ability of the component "column web in compression" in dependence of the axial force in the column was possible. For this reason a second test series was necessary with the same profiles but different ratios of $N/N_{pl}$. The test set up was the same as for the first series; also the schedule of measurement was quite similar to gain comparable test results. Though same profiles for both series were used, the resistance of the two series differed (see Table 7). Consequently reference tests without axial forces (specimen A8 and B8) were necessary also for the second series. In Figures 38 and 39 the results of the second test series are given. Again the strong influence of the axial force on the maximum strength and the decreasing character of the load-deformation behaviour beyond the curve maximum is obvious.

The main characteristic values of such load-deformation curves - stiffness, strength and deformation capacity - are discussed in the following and complemented by corresponding theoretical models in chapter 3.4.

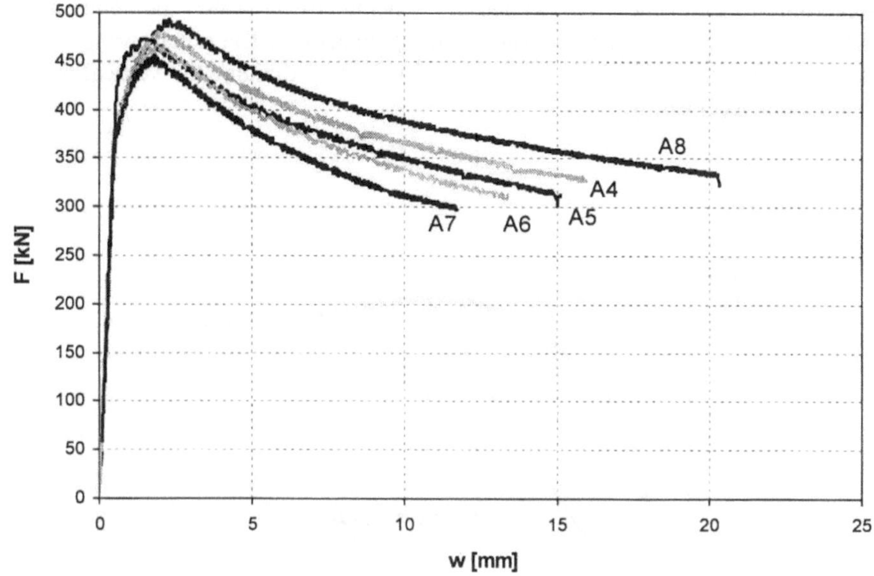


**Figure 38.** Load-deformation curves for test series II (1999); profiles HEA 240

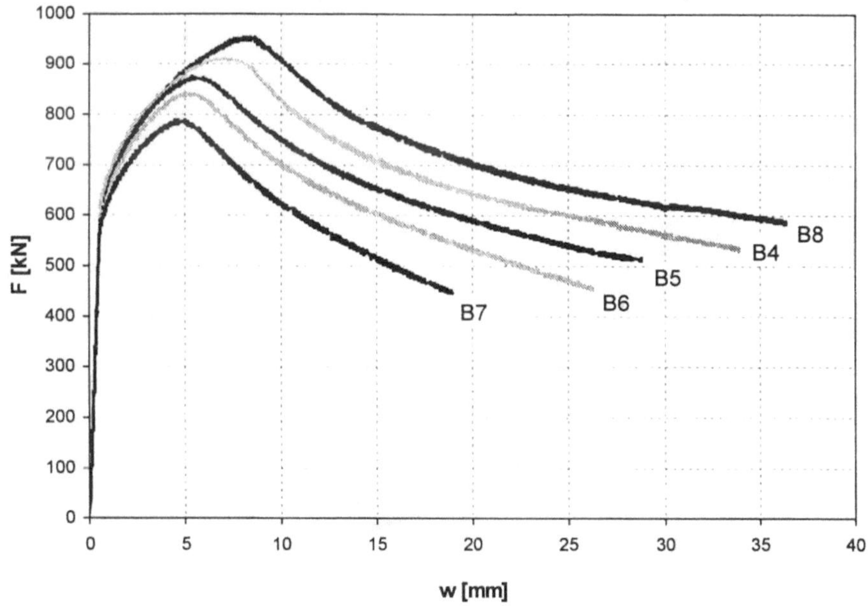


**Figure 39.** Load-deformation curves for test series II (1999); profiles HEB 240

**Stiffness.** According to Eurocode 3 (1992) and to Tschemmernegg (1982) the component load deformation behaviour is interpreted as non-linear spring behaviour. The actual spring stiffness C is derived, as shown in Figure 40, from the measured values by dividing the vertical force F by the corresponding displacement w.

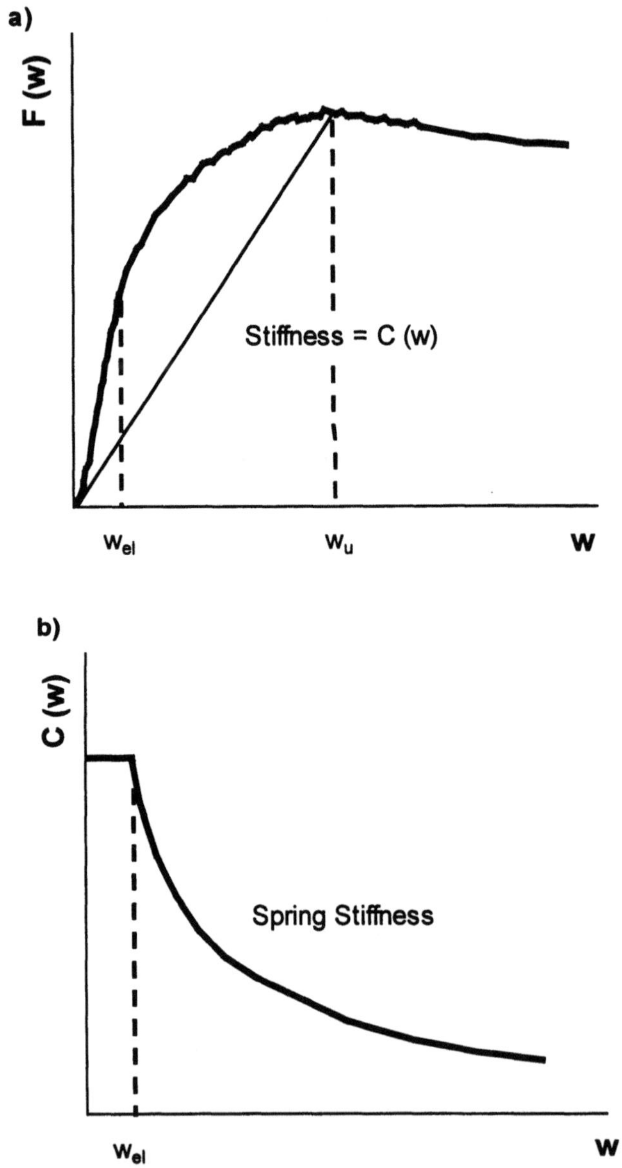


**Figure 40.** Derivation of spring stiffness

Diagrams as in Figure 41 depicts the behaviour of the spring stiffness with increasing deformation w (see Reichert, 1998).

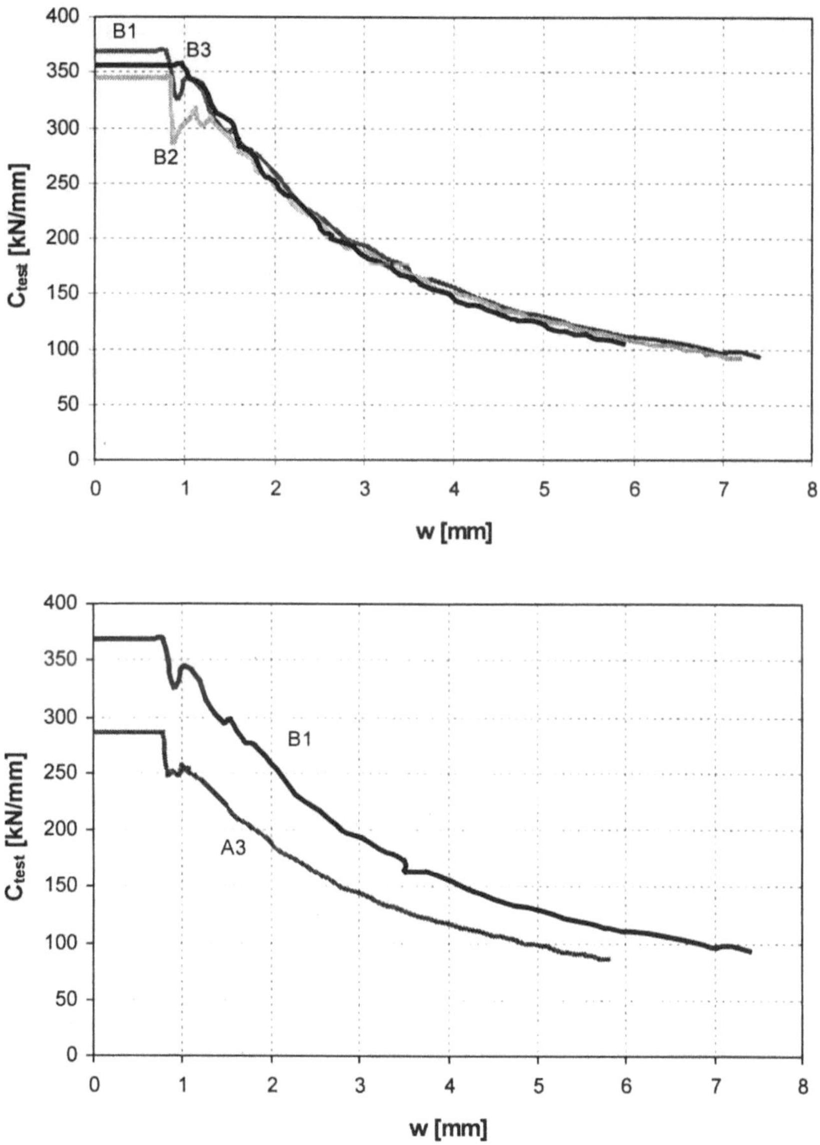


**Figure 41.** Measured spring stiffness versus displacement

Obviously the spring stiffness behaviour does not differ much within one series i.e. for one type of profile with different axial forces. But it is strongly dependent of the chosen profile, as shown in the lower diagram.

Figure 42 shows for test specimen A2 as a typical example a comparison between the load-displacement curve of Eurocode 3, revised Annex J (1997), the non-linear and the simplified linear one, and the measured test curve.

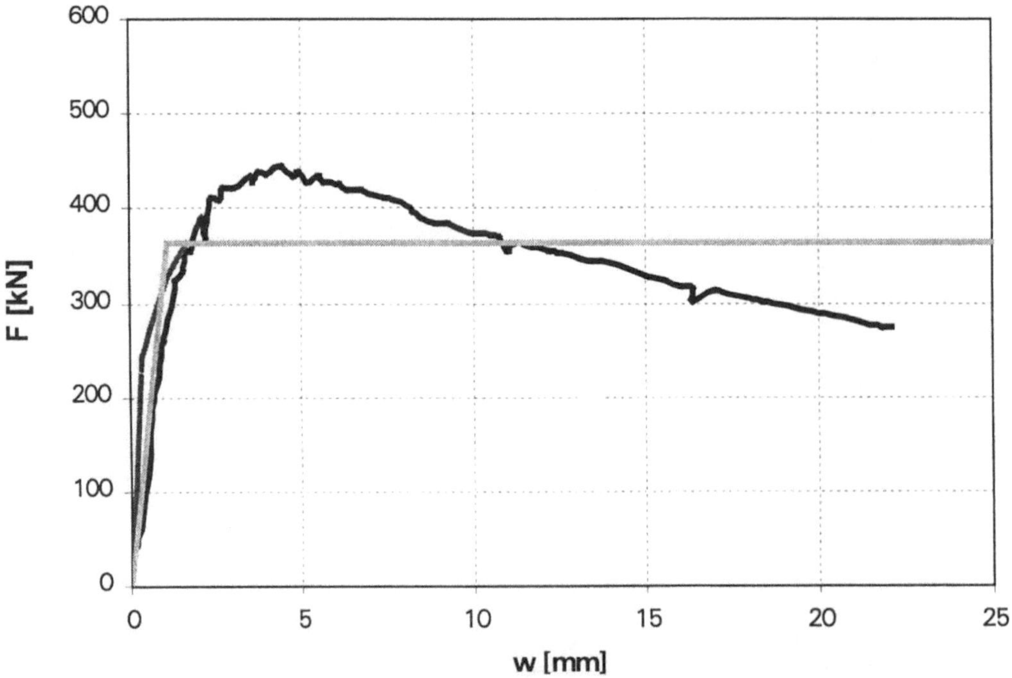


**Figure 42.** Comparison of curves according to Eurocode 3, revised Annex J (1997) and test curve of test A2

It can be seen that the measured initial stiffness of the load - deformation curve is lower than the values theoretically adopted by Eurocode 3, revised Annex J (1997), see chapter 3.4. Table 5 compares these initial spring stiffness values for series I, where $S_{j,ini}$ represents the theoretical value of the non – linear Annex J model, whereas $C_{j,test}$ belongs to the measured stiffness.

**Table 5.** Comparison between Eurocode 3, revised Annex J (1997) values of stiffness and measured test values for series I

| Test | $C_{j,test}$ [kN/cm] | $S_{j,ini}$ [kN/cm] | Difference [%] |
|---|---|---|---|
| A1 | 6938 | 13684 | 49,3 |
| A2 | 5225 | 13654 | 61,7 |
| A3 | 6224 | 13597 | 54,2 |
| B1 | 7245 | 17836 | 59,4 |
| B2 | 6850 | 17820 | 61,5 |
| B3 | 7531 | 17858 | 57,8 |

The high differences are relatively constant for both profile types. There is also no clear tendency in dependence of the ratio of axial forces. Some of the additional deformations may result of the load introduction which are not taken into account by the modelling.

Huber (1999) suggested a procedure how to evaluate test curves. One aspect is that the initial gradient of the curve does not reveal the real behaviour because of imperfections of the test set up. To exclude imperfections of the test set up an unloading and reloading of the linear branch is recommended. Test series I does not follow such a procedure of unloading and reloading in contrast to the second series. The results of the second series indicate a better agreement of test results and theoretical model though the difference is still quite remarkable (see Table 6).

**Table 6.** Comparison between Eurocode 3, revised Annex J (1997) values of stiffness and measured test values for series II

| Test | $C_{j,test}$ [kN/cm] | $S_{j,ini}$ [kN/cm] | Difference [%] |
|---|---|---|---|
| A4 | 8782 | 12603 | 30,3 |
| A5 | 7159 | 12639 | 43,4 |
| A6 | 6708 | 12651 | 47,0 |
| A7 | 7430 | 12942 | 42,5 |
| A8 | 7575 | 12805 | 40,8 |
| B4 | 11398 | 20944 | 45,6 |
| B5 | 9943 | 21086 | 52,8 |
| B6 | 9198 | 20980 | 56,1 |
| B7 | 10162 | 21019 | 51,6 |
| B8 | 10416 | 21221 | 50,9 |

**Strength.** The measured strength F obviously exceeds the plastic resistance. Table 7 shows a comparison of the maximum load $F_{u,test}$ with the plastic resistance $F_{pl}$ according to Eurocode 3, revised Annex J (1997), see chapter 3.4. The final column of Table 7 contains the differences in percentages. The differences surely result partly of the negligence of the strain hardening in the code model. As an additional effect also the contribution of the flange in addition to the carrying capacity of the web may cause a higher strength of the test specimens.

**Table 7.** Comparison between Eurocode 3, revised Annex J (1997) values of strength with measured test values

| Test | $F_{u,test}$ [kN] | $F_{pl}$ [kN] | Difference [%] |
|---|---|---|---|
| A1 | 464,64 | 423,41 | 8,8 |
| A2 | 453,29 | 421,26 | 7,1 |
| A3 | 531,51 | 418,60 | 21,2 |
| A4 | 480,55 | 377,01 | 21,5 |
| A5 | 473,08 | 377,93 | 20,1 |
| A6 | 467,06 | 380,74 | 18,5 |
| A7 | 455,17 | 393,63 | 13,5 |
| A8 | 492,47 | 384,40 | 21,9 |
| B1 | 755,21 | 599,77 | 20,6 |
| B2 | 678,09 | 598,88 | 11,7 |
| B3 | 629,12 | 578,02 | 8,1 |
| B4 | 909,90 | 630,71 | 30,7 |
| B5 | 873,71 | 634,68 | 27,4 |
| B6 | 842,05 | 631,69 | 25,0 |
| B7 | 788,06 | 632,96 | 19,7 |
| B8 | 953,20 | 639,03 | 33,0 |

Obviously the axial force in the profile has a strong influence on the ultimate load of the test specimen. Figure 43 compares the ultimate loads of the tests for the HEA 240 profiles in dependence of the ratio $N/N_{pl}$. $N_{pl}$ is calculated according to the measured values.

The strength values of test specimen A3 and A8 differ very much from each other though same nominal profiles (HEA 240) without an axial force were used. The different areas of cross section and the different yield strengths of these profiles (see Table 3 and 4) may cause these deviation.

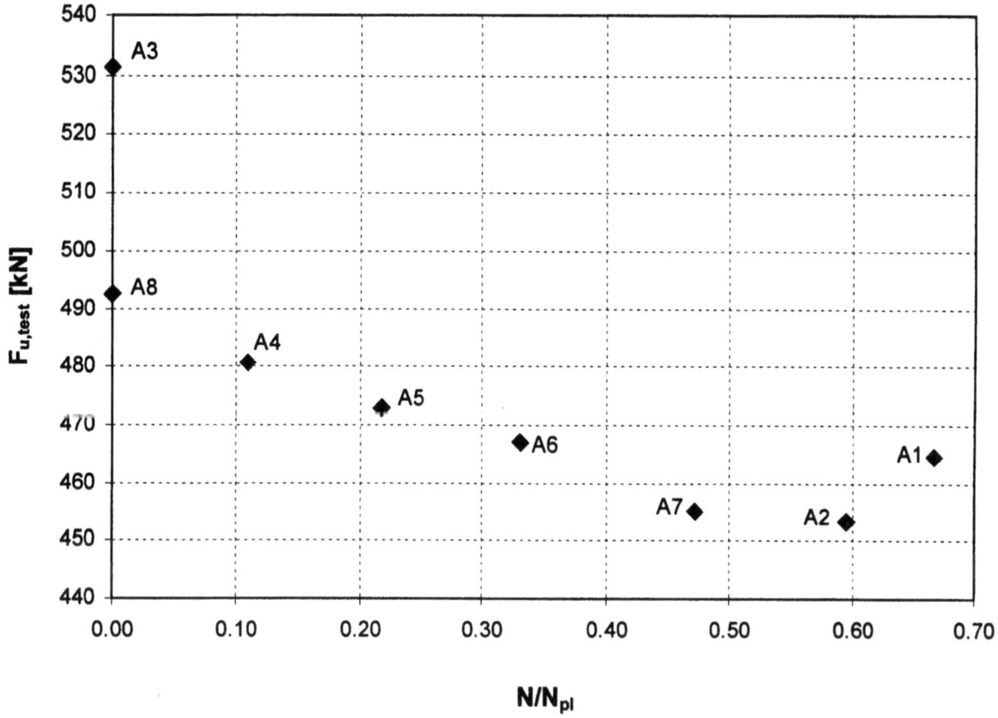


**Figure 43.** Ultimate loads of the HEA 240 profiles in dependence of ratio $N/N_{pl}$

**Deformation capacity.** The main objective of these tests was to evaluate the deformation capacity of the component "column web in compression" at the level of the plastic resistance. In contrast to former tests this time also the behaviour of the specimen after the vertical force F had passed its maximum was considered. To reach the characteristic plastic resistance of the component $F_{pl}$ a deformation $w_{pl}$ is needed. Only the additional deformation ($w_{max} - w_{pl}$) may be used for a sufficient rotation capacity for the redistribution of internal forces within the system without reducing the resistance of the component under its plastic value. So the deformation capacity is defined as distance ($w_{max} - w_{pl}$). And as can be seen by the test curve in Figure 44, the contribution of the descending part of the load-deformation curve $w_{decr}$ is of greater importance than the increasing part $w_{incr}$.

Table 8 contains deformation capacity values of all executed tests. Also the contribution of the increasing and decreasing parts are given. The last column of Table 8 contains the ratio $w_{decr}/w_{avail}$. Percentage values of more than 50 % support the importance of the decreasing part for the available deformation capacity of the component "column web in compression".

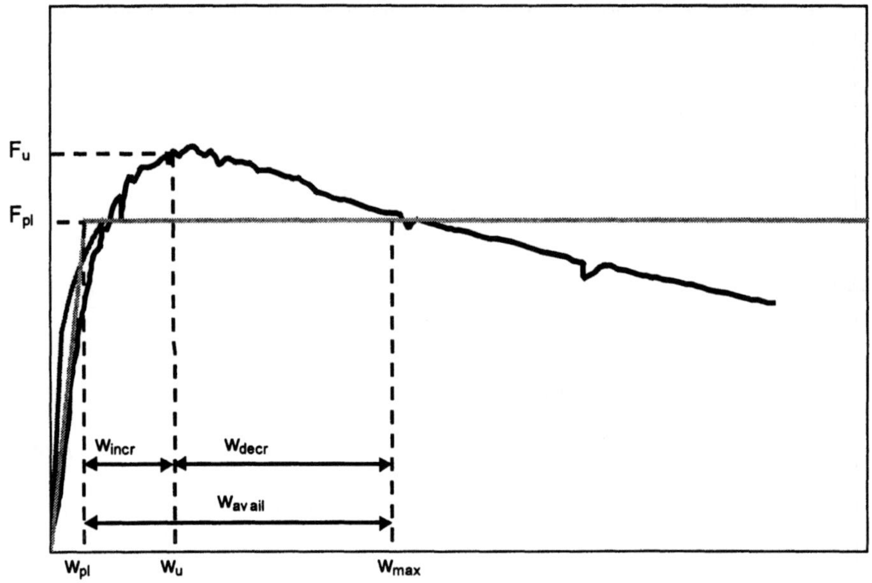


**Figure 44.** Definition of deformation capacity

**Table 8.** Deformation capacity values of the tests

| Test specimen | $N/N_{pl}$ | $w_{pl}$ [mm] | $w_u$ [mm] | $w_{avail}$ [mm] | $w_{incr}$ [mm] | $w_{decr}$ [mm] | $w_{decr}/w_{avail}$ [%] |
|---|---|---|---|---|---|---|---|
| A1 | 0,67 | 1,24 | 2,12 | 2,42 | 0,88 | 1,54 | 64 |
| A2 | 0,60 | 1,33 | 2,21 | 2,14 | 0,88 | 1,26 | 59 |
| A3 | 0,00 | 1,07 | 2,83 | 11,85 | 1,76 | 10,09 | 85 |
| A4 | 0,11 | 0,51 | 1,94 | 7,99 | 1,43 | 6,56 | 82 |
| A5 | 0,22 | 0,49 | 1,36 | 6,79 | 0,87 | 5,92 | 87 |
| A6 | 0,33 | 0,66 | 1,86 | 5,63 | 1,20 | 4,43 | 79 |
| A7 | 0,47 | 0,74 | 1,84 | 3,57 | 1,10 | 2,47 | 69 |
| A8 | 0,00 | 0,62 | 2,31 | 10,05 | 1,69 | 8,36 | 83 |
| B1 | 0,00 | 1,40 | 4,73 | 11,46 | 3,33 | 8,13 | 71 |
| B2 | 0,46 | 1,64 | 3,28 | 4,81 | 1,64 | 3,17 | 66 |
| B3 | 0,65 | 1,61 | 2,94 | 3,05 | 1,33 | 1,72 | 56 |
| B4 | 0,11 | 0,67 | 7,01 | 21,03 | 6,34 | 14,69 | 70 |
| B5 | 0,24 | 0,89 | 5,35 | 15,52 | 4,46 | 11,06 | 71 |
| B6 | 0,36 | 0,93 | 5,05 | 12,61 | 4,12 | 8,49 | 67 |
| B7 | 0,58 | 1,01 | 4,95 | 8,79 | 3,94 | 4,85 | 55 |
| B8 | 0,00 | 0,88 | 8,30 | 26,34 | 7,42 | 18,92 | 72 |

For each of the profiles HEA 240 and HEB 240 the values of deformation capacity of the tests with axial forces are divided by the deformation values of the reference test specimen without axial force. As shown in Figure 45 for the profiles HEA 240 and in Figure 46 for the profiles HEB the deformation capacity decreases nearly linearly with increasing axial force.

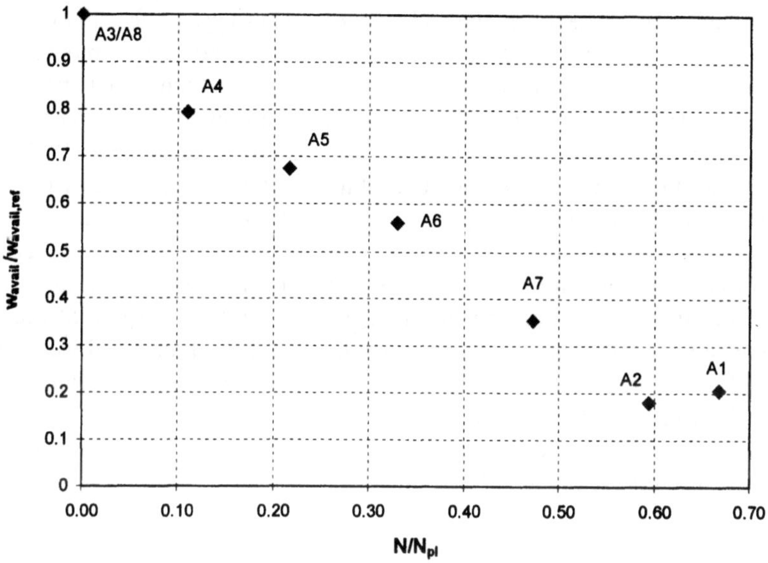


**Figure 45.** Ratio of deformation capacity in dependence of the axial force (HEA 240)

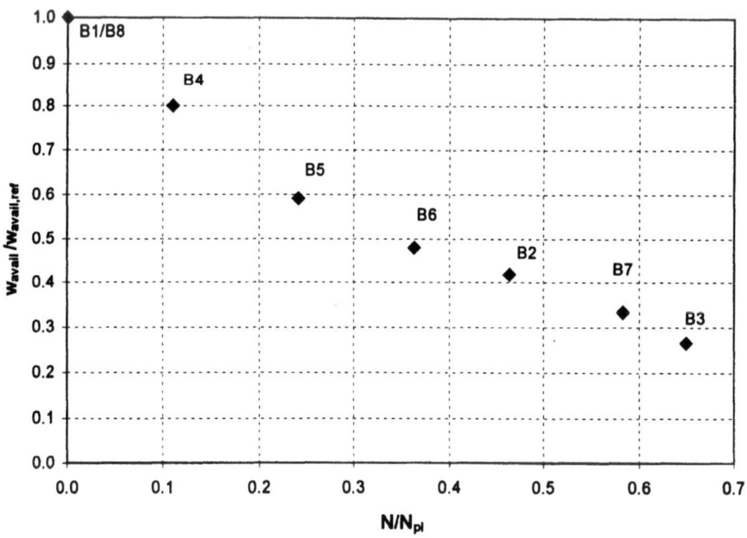


**Figure 46.** Ratio of deformation capacity in dependence of the axial force (HEB 240)

### 3.4 Theoretical Models

**Introduction.** As mentioned above by means of mechanical models an analytical description of the load deformation behaviour is possible. So far few models are given in literature for characteristic points of the load-deformation curve of the component "column web in compression". Studies of Tschemmernegg and Huber (1987) suggest a formula for the elastic $F_{el}$ and the plastic force $F_{pl}$. Eurocode 3, revised Annex J (1997) contains a model for the initial stiffness in the elastic part and provides a similar procedure to get the secant stiffness belonging to the plastic part. Also a model for the plastic strength $F_{pl,N}$ under consideration of an axial force in the column, which is similar to the formula of Tschemmernegg and Huber (1987), is available in the code. But as shown before the measured ultimate strengths of the tests $F_{u,test}$ differ very much from the theoretical values of $F_{pl}$. For this reason with reference to investigations of Ungermann (1990) a new formula for the calculation of the ultimate strength is suggested.

Especially the decreasing part of the load-deformation curve is of great importance for the deformation capacity of the component itself as well as for the interplay of this component with other joint components. Therefore a yield line model for the flange and web behaviour has been developed under consideration of the axial force in the profile (see Kühnemund, 2001). The real load-deformation curve is now approximated by a sequence of different models. All named models will be illustrated in the following. As a result a characteristic model curve similar to a polygonal course is found (see Figure 47).

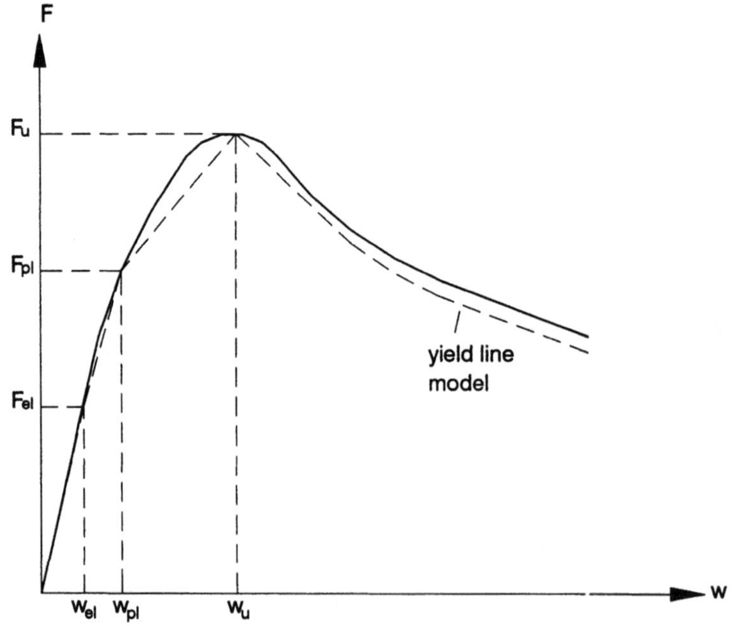


**Figure 47.** Model curve

**Stiffness model.** According to Eurocode 3, revised Annex J (1997), chapter J.4.1 "*the rotational stiffness of a joint should be determined from the flexibilities of its basic components, each represented by its elastic stiffness coefficient $k_i$*". The rotational stiffness $S_j$ (see also Figure 12) may be obtained from formula (6), which is taken from the code:

$$S_j = \frac{E \cdot z^2}{\mu \cdot \sum_i k_i} \tag{6}$$

where:

- $k_i$ is the stiffness coefficient for a basic joint component i
- z is the lever arm
- $\mu$ is the stiffness ratio $S_{j,ini}/S_j$
- $S_{j,ini}$ is the initial rotational stiffness of the joint, given by formula (6) with $\mu$ = 1,0

The stiffness ratio $\mu$ depends on the amount of the acting forces (for components) or moments (for joints) respectively (see equation (8)):

$$\text{if } F_{j,Sd} \le \frac{2}{3} \cdot F_{j,Rd} : \quad \mu = 1 \tag{7}$$

$$\text{if } \frac{2}{3} \cdot F_{j,Rd} < F_{j,Sd} \le F_{j,Rd} : \quad \mu = \left(1{,}5 \cdot \frac{F_{j,Sd}}{F_{j,Rd}}\right)^{\psi} \tag{8}$$

where:

- $F_{j,Sd}$ is the applied force on the component
- $F_{j,Rd}$ is the plastic resistance
- $\psi$ is a coefficient to consider the type of the connection

The stiffness coefficient $k_i$ for the component "column web in compression" should be obtained from the following expression:

$$k_i = \frac{0{,}7 \cdot b_{eff,c,wc} \cdot t_{wc}}{d_c} \tag{9}$$

where:

- $b_{eff,c,wc}$ is the effective width of the column web in compression
- $d_c$ is the clear depth of the column web
- $t_{wc}$ is the thickness of the column web

With the presented formulas from Eurocode 3, revised Annex J (1997) as well the initial (elastic) stiffness of the component "column web in compression" as the (secant) stiffness belonging to the plastic resistance $F_{pl}$ can be determined. A comparison between the theoretical values for the initial stiffness according to the given model and the measured values is shown in chapter 3.3.

On the basis of models for the stiffness and the formulas for the elastic and plastic resistance (see below) the characteristic deformations $w_{el}$ and $w_{pl}$ can be determined:

$$w_{el} = \frac{F_{el}}{S_{j,ini}} \tag{10}$$

$$w_{pl} = \frac{F_{pl}}{S_j} \tag{11}$$

**Theoretical models for $F_{el}$ and $F_{pl}$.** The models are based on the idea of load spreading (see Figure 48) of the compression force F in the fillet radius and column flange with different angles. An effective width $b_{eff}$ is received in the web for the calculation of the elastic and plastic force. This effective width is multiplied with the yield strength and the thickness of the column web to get the elastic and plastic resistance (see formula (12)). Reduction factors ω and ρ cover effects like plate buckling of the web and shear in the web panel.

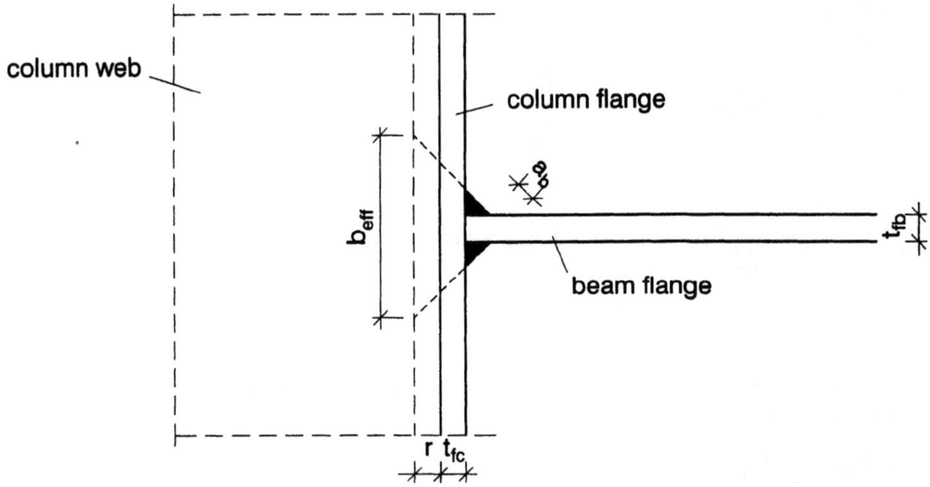


**Figure 48.** Models for the elastic and plastic resistance $F_{el}$ and $F_{pl}$

$$F_{c.wc.Rd} = \frac{\omega \cdot \rho \cdot b_{eff,c,wc} \cdot t_{wc} \cdot f_{y,wc}}{\gamma_M} \tag{12}$$

where:

- ρ is the reduction factor for plate buckling
- ω is a reduction factor to allow for the possible effects of shear in the column web
- $b_{eff,c,wc}$ is the effective width of the belonging web to the component "column web in compression"

$$b_{eff,c,wc} = t_{fb} + 2 \cdot \sqrt{2} \cdot a_b + n \cdot (t_{fc} + r) \tag{13}$$

where:

- $t_{fb}$ is the thickness of the beam flange
- $a_b$ is the throat thickness of the welded seam
- $t_{fc}$ is the thickness of the column flange
- $r$ is the fillet radius

Tschemmernegg and Huber (1987) as well as Eurocode 3, revised Annex J (1997) set the factor n equal to 5 for the plastic resistance of the component. For the elastic resistance Tschemmernegg and Huber (1987) suggest to set n equal to 2,5 because of the smaller load spreading on that level.

**Ultimate load model.** A better modelling of the ultimate load values of the tests was reached by modifying a model which is founded on an orthotropic plate buckling mechanism for the web (see Ungermann, 1990). Exchanging the yield strength in the corresponding formula by ultimate strength values, using a calibration factor of 0,87 for HEA profiles and 0,88 for HEB profiles and multiplying the formula by a term taking the axial force into account leads to an equation (see equation (14) for HEA 240 profiles), which results in good agreement between the measured ultimate load values $F_{u,test}$ and the theoretical values $F_{u,model}$ for all executed test (see Figures 49 and 50).

$$F_{u,model} = 0,87 \cdot 2 \cdot c_u \cdot t_{wc} \cdot f_u \cdot \sqrt{1 - \left(\frac{\sigma_{N,web}}{f_u}\right)^2} \tag{14}$$

where:

- $2c_u$ is the width of the buckling plate according to Ungermann (1990), eq. (2.3.27)

$$2 \cdot c_u = 2,359 \cdot \sqrt[7]{\frac{I_{fl} \cdot t_{wc}^{\ 3} \cdot E^2}{f_u^2}} \tag{15}$$

- $t_{wc}$ is the thickness of the web
- $f_u$ is the ultimate strength of the web
- $\sigma_{N,web}$ is the stress in the column web caused by an axial force N
- 0,87 is the calibration factor for HEA 240 profiles
- $I_{fl}$ is the moment of inertia

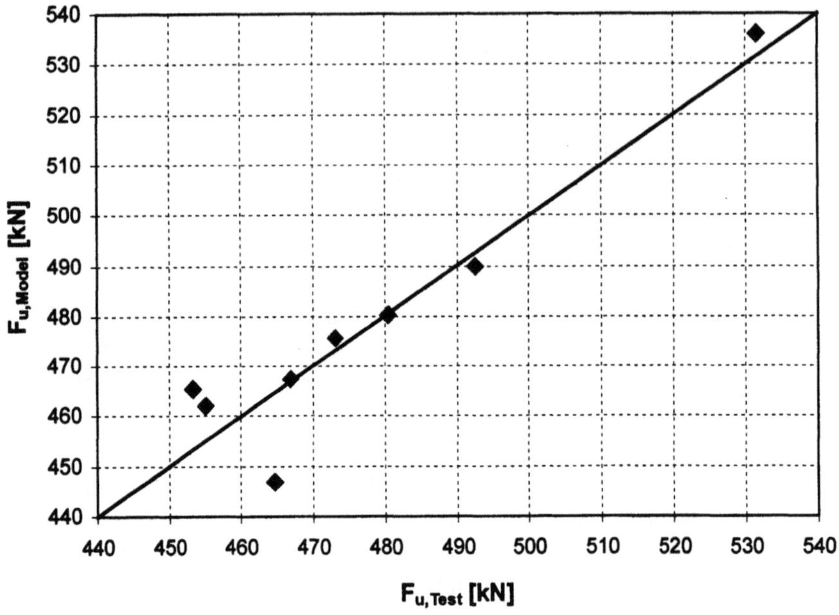


**Figure 49.** Comparison between measured ultimate loads and the ultimate loads according to the theoretical model for test specimen HEA 240

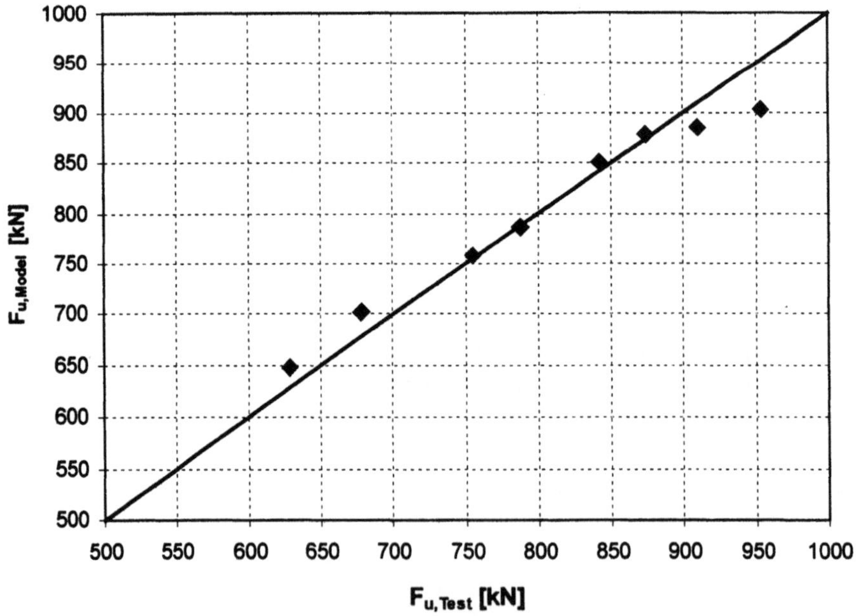


**Figure 50.** Comparison between measured ultimate loads and the ultimate loads according to the theoretical model for test specimen HEB 240

According to Eurocode 3, revised Annex J (1997) the strength of the component "column web in compression" has to be reduced for longitudinal stress values due to axial force and bending moment in the column exceeding a stress limit of $0{,}5 \cdot f_{y,wc}$. For the executed tests the interaction between the compression and the axial force is shown in Figure 51 in comparison with the graph of the Eurocode 3, revised Annex J (1997) model for interaction. The test values lie below the theoretical graph according to Eurocode 3, revised Annex J (1997). It is obvious that the theoretical limit of influence of 50% of $N_{pl}$ is not verified by this test series.

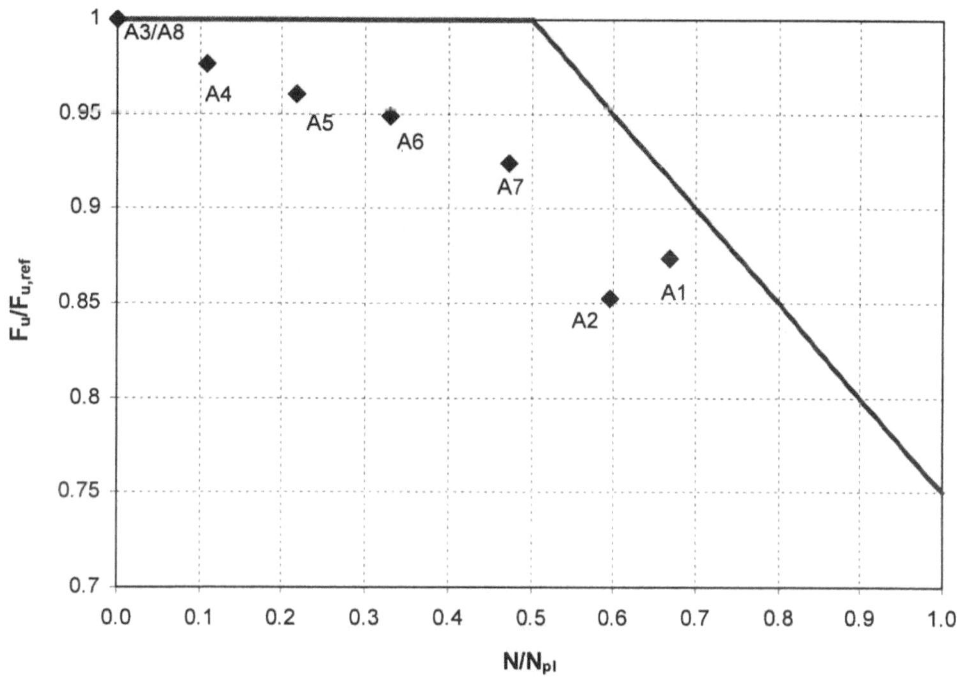


**Figure 51.** Compression force - axial force - interaction diagram

**A mechanical model to describe the decreasing branches of the load-deformation curves.** As well for the elastic and plastic part of the curves as for the ultimate load mechanical models are given in literature (see above). Since the main aim of the investigations is the determination of the deformation capacity of the component a mechanical description of the decreasing part of the curve is of great importance. In the following a model for that part of the curve on basis of a yield line mechanism, which has been developed by Kühnemund (2001) is presented.

The idea of the model is shown in Figure 52. The occurring buckle is simulated by a yield line mechanism with flat areas separated by yield lines. The shape of the model is oriented on the behaviour of the web recognised in the tests.

For the inner forces and deformations as well as for the outer ones virtual work equations were formulated. Equating of both leads to a formulation for the force F in dependence of the deformation w and the applied axial force N. A similar procedure is used for the flange behaviour. Here also a yield line model is applied. The sum of these equations - for the web and the two flanges - leads to an equation for the decreasing branch behaviour. For test specimen A8 Figure 53 shows the graph of the equation in comparison with the results. Good correspondence between test behaviour and model is reached.

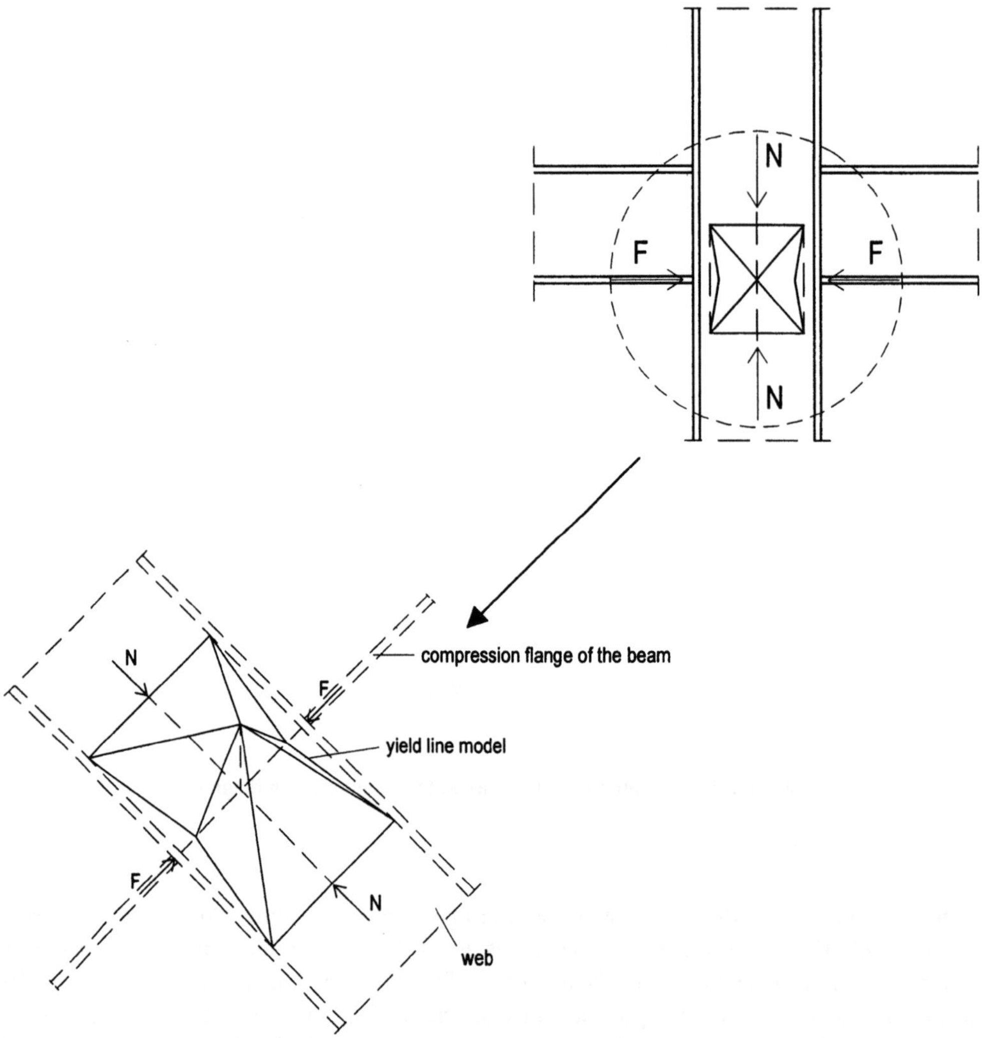


**Figure 52.** Yield line model in the column web

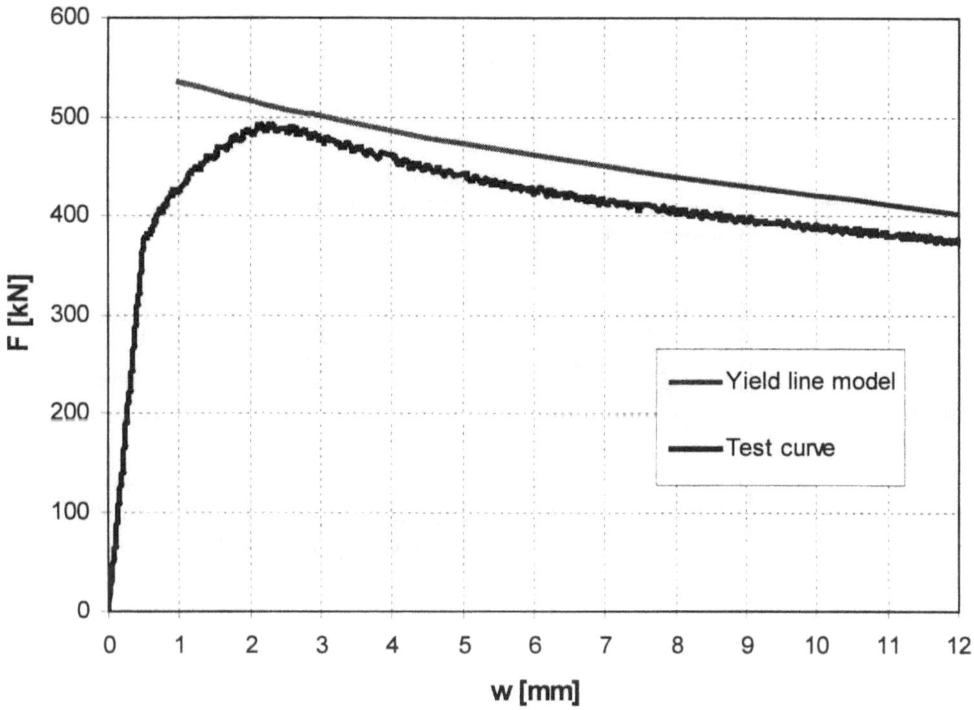


**Figure 53.** Graph of yield line model in comparison with test results A8

**Contribution of the component "column web in compression" to the rotation capacity of the whole joint.** As stated above two components especially contribute to a satisfying value of the available rotation capacity: any kinds of plate in bending which are identified as T-stub models and the component "column web in compression" in the compression area. The mentioned components are acting in series with a lever arm z (see Figure 54). To gain an estimation of the contribution of the component "column web in compression" to the rotation capacity of the joint, a welded steel joint configuration is chosen. Whereas for bolted end plate connections the point of rotation is to be found by an iterative calculation procedure (see Huber, 1999, and Tschemmernegg et al., 1997) for the model considered the point of rotation is to be assumed in the centre line of the joint for welded configurations. This means that the deformation contribution of the tension component of the welded joint is about the same as that of the component "column web in compression". Consequently the rotation capacity is found by the following formula:

$$\Phi = \frac{w_{tension} + w_{compression}}{z} = \frac{2 \cdot w_{compression}}{z} \tag{16}$$

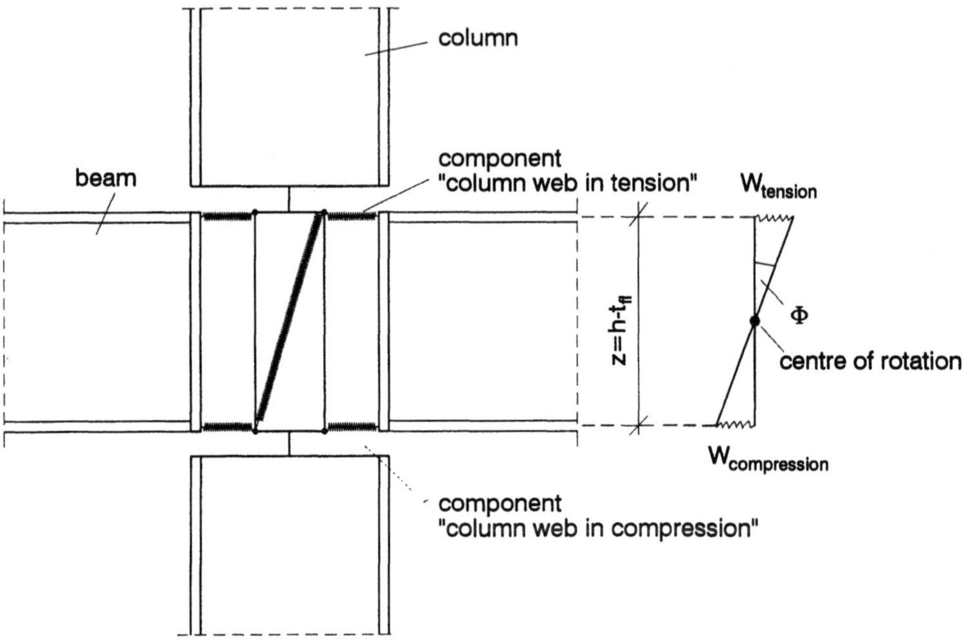


**Figure 54.** Component model for a welded steel joint

Table 9 shows the results of a study for welded joints. The column profiles are chosen corresponding to the executed tests (profiles HEA 240 and HEB 240). For the beams four different types of IPE profiles are taken into account. The lever arm z is derived from the following expression:

$$z = h_{beam} - t_{fl} \tag{17}$$

As can be seen from Table 9 all joint types without an axial force in the column show a high rotation capacity. But even with an axial force of about 30% of the plastic axial force the available rotation capacity is still of a large amount, also for the lever arm reaching values up to about 600 mm. Joints with an ratio of $N/N_{pl}$ of more than or equal to 0,6 dispose of a small rotation capacity especially for adjacent beams with large height. Because the latter configuration is of practical importance, extensive investigations for the rotation capacity of joints still have to be done.

A comparison between the two column profile types show the decisive influence of the flange and web slenderness. The compact profiles HEB 240 provide about 2.5 times larger rotation capacity values than the slender HEA 240 profiles.

**Table 9.** Estimation of rotation capacity values based on the test component results

| Column profile | ratio $N/N_{pl}$ | $w_{avail}$ [mm] | Beam profile | Lever arm z [mm] | Rotation capacity Φ [mrad] |
|---|---|---|---|---|---|
| HEA 240 | 0,00 | 10,05 | IPE 200 | 191,5 | 105 |
| | | | IPE 300 | 289,3 | 69 |
| | | | IPE 450 | 435,4 | 46 |
| | | | IPE 600 | 581,0 | 35 |
| HEA 240 | 0,33 | 5,63 | IPE 200 | 191,5 | 59 |
| | | | IPE 300 | 289,3 | 39 |
| | | | IPE 450 | 435,4 | 27 |
| | | | IPE 600 | 581,0 | 19 |
| HEA 240 | 0,60 | 2,14 | IPE 200 | 191,5 | 22 |
| | | | IPE 300 | 289,3 | 15 |
| | | | IPE 450 | 435,4 | 10 |
| | | | IPE 600 | 581,0 | 7 |
| HEB 240 | 0,00 | 26,34 | IPE 200 | 191,5 | 275 |
| | | | IPE 300 | 289,3 | 182 |
| | | | IPE 450 | 435,4 | 121 |
| | | | IPE 600 | 581,0 | 91 |
| HEB 240 | 0,36 | 12,61 | IPE 200 | 191,5 | 132 |
| | | | IPE 300 | 289,3 | 87 |
| | | | IPE 450 | 435,4 | 58 |
| | | | IPE 600 | 581,0 | 43 |
| HEB 240 | 0,65 | 3,05 | IPE 200 | 191,5 | 32 |
| | | | IPE 300 | 289,3 | 21 |
| | | | IPE 450 | 435,4 | 14 |
| | | | IPE 600 | 581,0 | 10 |

### 3.5 Conclusions

The investigations presented in this chapter aimed in the determination of load-deformation curves for the component "column web in compression". A test series of 16 tests for profiles HEA 240 and HEB 240 was executed. As the main result of the investigations was the derivation of the deformation capacity of the component one can state that the deformation capacity tends to decrease nearly linearly with an increasing axial load.

Furthermore a mechanical model was developed to describe the decreasing part of the curves which is quite important for the derivation of the deformation capacity. Together with models from literature this theoretical model leads to an analytical formulation of the load-deformation behaviour by adding component curves to the moment-rotation curve of the joint.

For the executed tests the achieved deformation capacity values are given. They underline the decisive influence of the decreasing part of the load-deformation curve . On basis of the test results rotation capacity values for welded joints were calculated. They strongly depend on the axial force in the column. The results strongly support the importance of further research work on the rotation capacity of partial strength or semi-rigid joints.

### 3.6 Acknowledgements

The authors like to thank the "Arbeitsgemeinschaft industrieller Forschungsvereinigungen "Otto von Guericke" e.V. (AiF)" for the financial support of the second tests series within the frame of a research project (see Kuhlmann et al., 2000), which the authors undertake together with Prof. Dr.-Ing. G. Sedlacek of RWTH Aachen and his team. The authors thank them and the accompanying group of practitioners for the valuable discussions.

### References

Ahmed, I.; Kirby, P. A. (1997). Maximum Connection Rotations in Non-Sway Semi-rigid Frames. In *Journal of Constructional Steel Research*, Vol. 40.

Boender, E. H. (1995). Het Randportaalprogramma. TU Delft.

Boender, E. H.; Stark, J.; Steenhuis, M.(1996). The Required Rotation Capacity of Joints in Braced Steel Frames. In *IABSE Colloquium Istanbul.*

Eurocode 3 (1992) - Design of steel structures. Part 1-1: General rules and rules for buildings. European prestandard.

Eurocode 4 (1992) - Design of composite steel and concrete structures - Part 1-1: General rules and rules for buildings. European prestandard.

Huber, G. (1999). Non-linear calculations of composite sections and semi-continuous joints. Dissertation Universität Innsbruck.

Jaspart, J. P. (1991). Etude de la semi-rigidité des noeuds poutre-colonne et son influence sur la résistance et la stabilité des ossatures en acier. Ph-D Thesis. Université Liège.

Jaspart, J. P. (1996). Plastic hinge idealisation of structural joints. EC 3 philosophy. Document COST C1/WD2/96-02, Aachen.

Jaspart, J. P. (1999). Recent advances in the field of structural steel joints and their representation in the building frame analysis and design process. COST C1 WG2 Publication, Brussels and Luxembourg.

Jaspart, J.P. (1996/1997). Contributions to recent advances in the field of steel joints. Column bases and further configurations for beam-to-column joints and beam splices. Thèse d'agrégé de l'enseignement supérieur. Université Liège.

Kattner; M. (1999). Beitrag zum Entwurf von Rahmen mit Verbundknoten im Hochbau. *Thesis No. 2055*, Ecole Polytechnique Fédérale, Lausanne.

Klein, H. (1985). Das elastisch-plastische Last-Verformungsverhalten M - υ steifenloser, geschweißter Knoten für die Berechnung von Stahlrahmen mit HEB-Stützen. Dissertation Universität Innsbruck.

Kuhlmann, U. (1986). Rotationskapazität biegebeanspruchter I-Profile unter Berücksichtigung des plastischen Beulens. Dissertation Ruhr-Universität Bochum.

Kuhlmann, U. (1989). Definition of Flange Slenderness Limits on the Basis of Rotation Capacity Values. In *Journal of Constructional Steel Research*. Vol. 14, No. 1.

Kuhlmann, U. (1997). Verification procedure for rotation capacity of joints. Document COST C1/WD2/97-24. Innsbruck.

Kuhlmann, U.; Fürch, A. (1997): Deformation capacity of the component „column web in compression". Internal test-report. University of Stuttgart.

Kuhlmann, U.; Kühnemund, F. (1999). Deformation capacity of the component „column web in compression". Internal test report. University of Stuttgart.

Kuhlmann, U.; Sedlacek, G.; Kühnemund, F.; Stangenberg, H. (2000). Verformungsverhalten der Komponenten von wirtschaftlichen steifenlosen Anschlußkonstruktionen für die Anwendung plastischer Bemessungskonzepte im Stahlbau. Research Project, Arbeitsgemeinschaft industrieller Forschungsvereinigungen "Otto von Guericke" e.V., not yet finished.

Kühnemund, F. (2001). *Zum Rotationsnachweis nachgiebiger Anschlüsse im Stahlbau*. Doctoral thesis in preparation. Universität Stuttgart.

Li, T. Q.; Choo, B. S.; Nethercot, D. A. (1995). Determination of Rotation Capacity Requirements for Steel and Composite Beams. In *Journal of Constructional Steel Research*. Vol. 32.

Nethercot, D. A.; Li, T. Q.; Choo, B. S. (1995). Required Rotations and Moment Redistribution for Composite Frames and Continuous Beams. In *Journal of Constructional Steel Research*. Vol. 35, Number 2.

Reichert, F. (1998). Investigations on the component „column web in compression" for the determination of rotation capacity of semi-rigid joints. Diploma Thesis. University of Stuttgart.

Revised Annex J of Eurocode 3 (1997). Joints in Building Frames. Edited approved draft. CEN Document CEN/TC 250/SC 3 - N 671 E.

Sibai, A. W. (1991). Semi-Rigid Joint Modelling For Nonlinear Analysis of Flexibly Connected Frames. Thesis No. 967, Ecole Polytechnique Fédérale, Lausanne.

Tschemmernegg, F (1982). Zur Entwicklung der steifenlosen Stahlauweise. In *Stahlbau 51*, Vol. 7.

Tschemmernegg, F.; Huber, K. (1987). Rahmentragwerke in Stahl unter besonderer Berücksichtigung der steifenlosen Bauweise. Österreichischer Stahlbauverband und Schweizerische Zentralstelle für Stahlbau.

Tschemmernegg, F.; Huber, G.; Huter, M.; Rubin, D. (1997). Komponentenmethode und Komponentenversuche zur Entwicklung von Baukonstruktionen in Mischbauweise. In *Stahlbau 66*, Vol. 9.

Tschemmernegg, F.; Huber, G.; Rubin, D. (1997). Classification of steel and composite joints. Document COST C1/WD2/97-25, Innsbruck.

Ungermann, D. (1990). Bemessungsverfahren für Vollwand- und Kastenträger unter besonderer Berücksichtigung des Stegverhaltens. Dissertation RWTH Aachen.

Vandegans, D. (1995). Application of the component method according to Eurocode 3 to connections with threaded studs. Document COST C1/WD2/95-21. Graz.

Vandegans, D.; Jaspart, J. P. (1996). Influence of the post-limit stiffness of joints on the frame behaviour: (Plastic mechanism collapse mode) - first draft. Document COST C1/WD2/96-03. Aachen.

Wald, F.; Villanova, F.; Mazura, V. (1997). The rotational capacity prediction, an application of the component method, Document COST C1/WD2/96-08. Helsinki.

Xiao, Y. R. (1996). Available and Required Rotation Capacities for Composite Beams and Frames. Document COST C1/WD2/96-10. Aachen.



# PART IV

# COLUMN BASE MODELLING

**F. Wald**
**Czech Technical University in Prague, Prague, Czech Republic**

**Abstract.** The present chapter aims to give an overview of the recent progress made on the investigation of column bases with end plates and embedded column bases. The column base detailing, tolerances and modelling by application of the component method to column bases are herein introduced. The particular components are also described including their influence on the connection behavior. The basic principles in background of the column base classification and the strength design are presented. The procedure is illustrated on worked examples of frame design and a column base with base plate.

## 1 Introduction

The column bases are one of the last studied structural element in European scale. Compared to beam-to-column connection, the number of available tests is limited to about 200 with different complexity of data description. In national rules the elastic prediction models were replaced by inelastic and the model of resistance of concrete foundation in crushing of concrete under the flexible plate was precised (Stockwell (1975)). The traditional elastic models of column bases gave a safe conservative solution with rather thick base plate and expensive anchoring. Some types of structures show high sensitivity of redistribution of internal forces from bending stiffness of column bases.

The structural design supported by software is nowadays less structuralized into overall structural design, structural analysis, code-check and detailing. The engineering analytical prediction models are precised to be able to describe the structural behavior in a more complex way. For connections, the component method allows for very precise description of behavior including semi-rigidity - stiffness, resistance and rotational capacity description. The support by experiments and F.E. simulation is used for special cases of design and for research verification and sensitivity studies, but is and will be limited due to the high need of inputs into the model.

The herein presented topics are mainly based on the latest European developments in column base modelling prepared under the umbrella of COST C1 project (COST C1 (1999)).

## 2 Detailing

A typical anchoring of the structure to the concrete is constructed as a column base with a base plate (see Fig.1) or as embedding of the column into a pocket in the foundation of the concrete block (see Fig.2). A combination of both details is used for fixing of heavy columns in seismic regions. Every structure has to transfer in a safe way vertical and lateral loads to the supports. In some cases, beams or other members may be supported directly, though the most common system is for columns to be supported by a concrete foundation (SCI (1995)). The column will be then connected to a base plate, which will be attached to the concrete by some form of the so-called „holding down“ assembly. Foundations themselves are supported by the substructure. The foundation may be supported directly on the existing ground, or may be supported by piles, or the foundation may be part of a slab. The influence of the support to the foundation, which may be considerable in certain ground conditions, is not covered in this document.

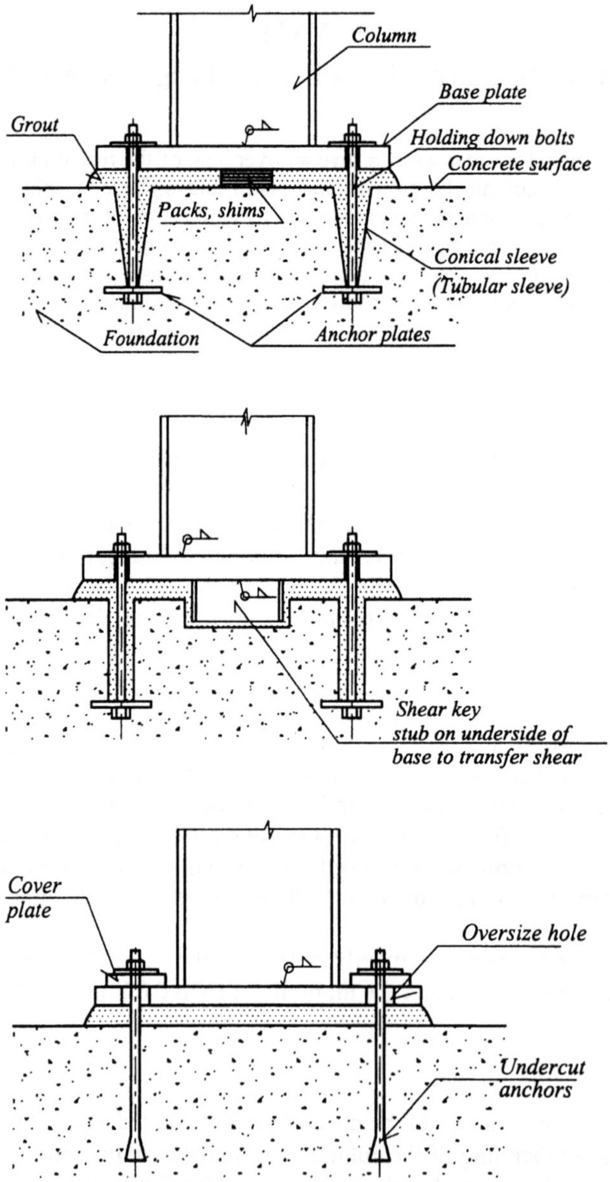


**Figure 1.** Detailing of column base with base plates (COST C1 (1999)).

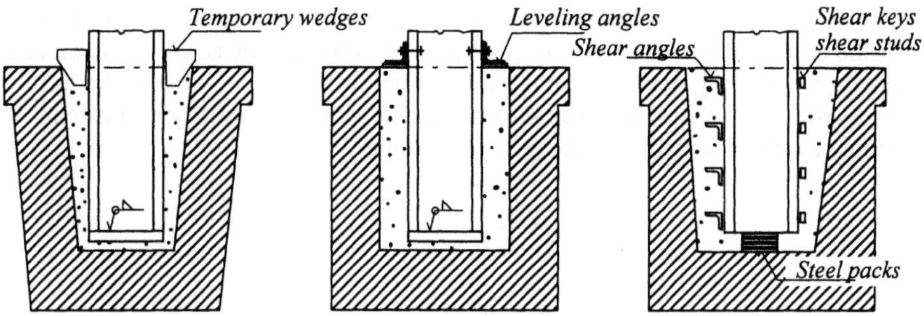


**Figure 2.** Detailing of embedded column bases.

Embedded column bases are frequently used in construction when the action of combined high bending moments and axial forces are present.

Concrete foundations are usually reinforced. The reinforcement may be nominal in the case of pinned bases, but will be significant in bases where bending moment is to be transferred. The holding down assembly comprises two, but more commonly four (or more) holding down bolts. These may be cast *in situ*, or post-fixed to the completed foundation. Cast in situ bolts usually have some form of tubular or conical sleeve, so that the top of the bolts are free to move laterally, to allow the base plate to be accurately located. Other forms of anchor are commonly used, as shown in Fig.3. Base plates for cast-in assemblies are usually provided with oversize holes and thick washer plates to permit translation of the column base. Post-fixed anchors may be used, being positioned accurately in the cured concrete. Other assemblies involve loose arrangements of bolts and anchor plates, subsequently fixed with cementicious grout or fine concrete. While loose arrangements allow for considerable translation of the base plate, the lack of initial fixity can mean that the column must be propped or guyed while the holding down arrangements are completed. Anchor plates or similar embedded arrangements are attached to the embedded end of the anchor assembly to resist pull-out. The holding down assemblies protrude from the concrete a considerable distance, to allow for the grout, the base plate, the washer, the nut and a further threaded length to allow for some vertical tolerance. The projection from the concrete is typically around 100 mm, with a considerable threaded length.

Post-fixed assemblies include expanding mechanical anchors, chemical anchors, undercut anchors and grouted anchors. Various types of anchor are illustrated in Fig.3.

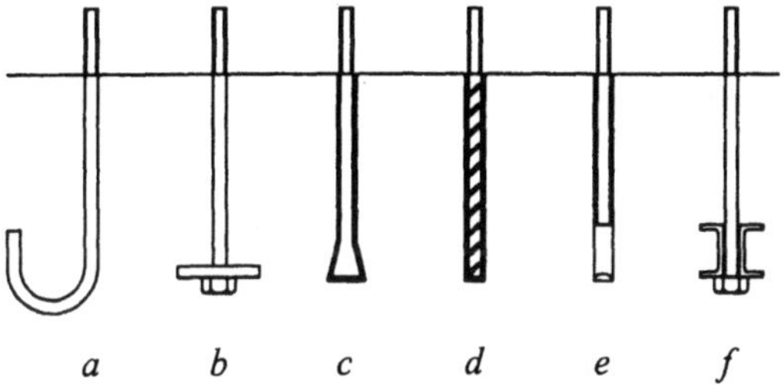


**Figure 3.** Anchor bolts, a) and b) cast in place, post fixed, c) under cat, d) post fixed chemical or cementicious grout, e) post fixed expanding anchor, f) fixed to grillage and cast in situ.

The space between the foundation and the base plate used to ensure the base plate is located at the correct absolute level. On smaller bases, this may be achieved by an additional set of nuts on the holding down assemblies, as shown in Fig.4. Commonly, the base plate is located on a series of thin steel packs as shown in Fig.5, which are usually permanent. Wedges are commonly used to assist the plumbing of the column.

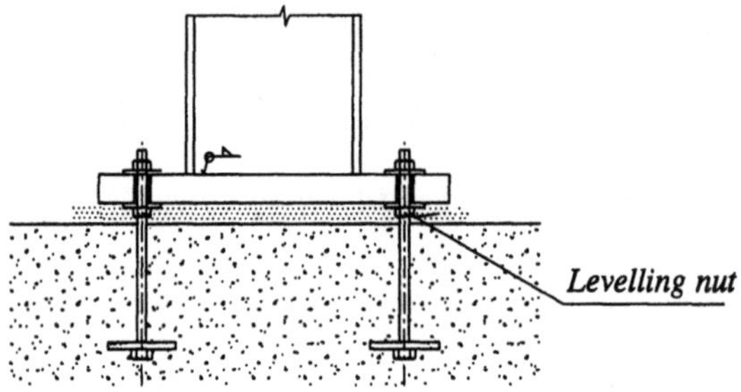


**Figure 4.** Base plate with levelling nuts (COST C1 (1999)).

The remaining void is filled with fine concrete, mortar, or more commonly, non-shrink cementicious grout, which is poured under and around the base plate. Large base plates generally have holes to allow any trapped air to escape when the base plate is grouted.

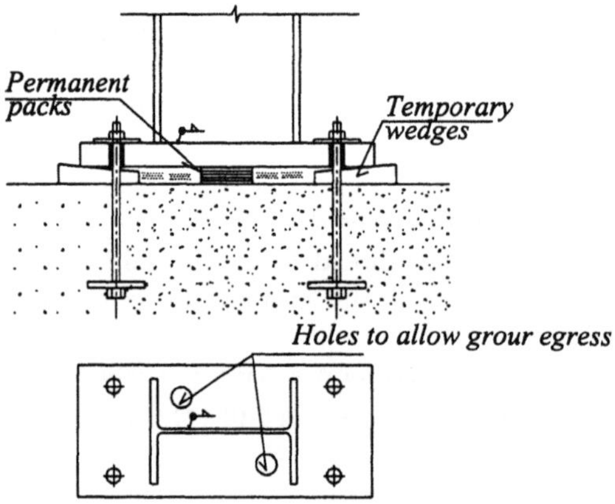


**Figure 5.** Base plate located on steel packs, see (COST C1 (1999)).

The plate attached to the column is generally rectangular. The dimensions of the plate are as required by design, though practical requirements may mean that the base is larger than necessitated by design. Steel erectors favour at least four bolts, since this is a more stable detail when the column is initially erected. Four bolts also allow the base plate to be adjusted to ensure verticality of the column. Bolts may be located within the profile of the I or the H section, or outside the profile, or both, as depicted in Fig.1. Closely grouped steel bolts with tubular or conical sleeves are to be avoided, as the remaining concrete may not be able to support the column and superstructure in the temporary condition. Bases may have stubs or other projections on the underside, which are designed to transfer horizontal loads to the foundations. However, such stubs are not appreciated by steelwork erectors and should be avoided if possible. Other solutions may involve locating the base in a shallow recess or anchoring the column directly to, for example, the floor slab of the structure.

Columns are generally connected to the base plate by welding around part or all of the section profile. Where corrosion is possible a full profile weld is recommended. Fixed (or moment-resisting) bases are assumed in analysis to be entirely rigid. Compared to pinned bases, fixed bases are likely to have a thicker base plate, and may have a larger number of higher strength holding down assemblies. Occasionally, fixed bases have stiffened base plates, as those shown in Fig. 6. The stiffeners may be fabricated from plate, or from steel members such as channels.

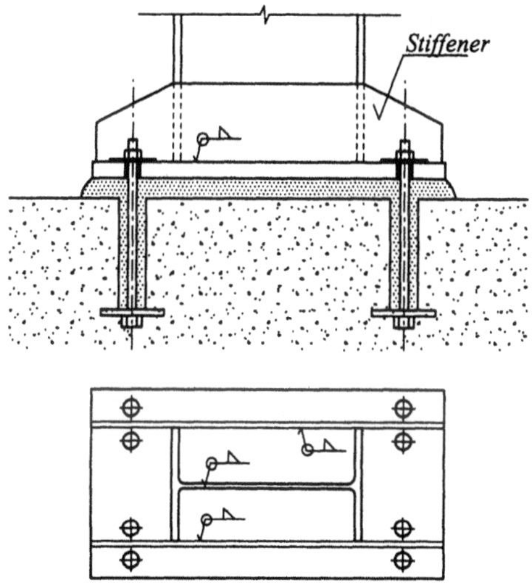


**Figure 6.** Typical stiffened column base detail, see (COST C1 (1999)).

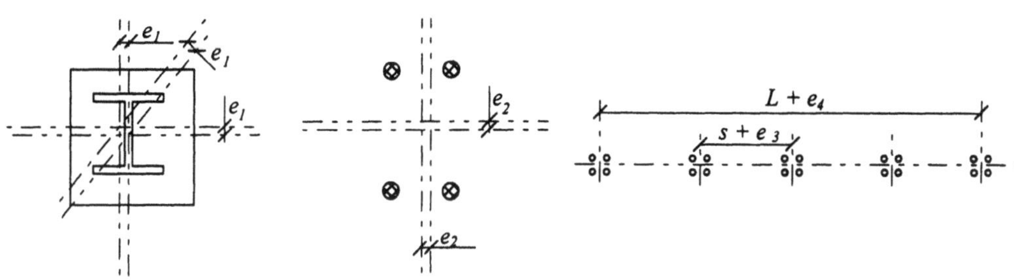


**Figure 7.** The permitted horizontal deviation for the base plates and for the anchor bolts groups.

**Tolerances.** The fabrication and erection tolerances are given in "European Prestandard Execution of Steel Structures - Part 1: General Rules and Rules for Buildings prEN 1090-1".

The eccentricity of the column at the base plate should not exceed $e_1 = 5$ mm, see Fig. 1. The position of the group of anchor bolts should not deviate by more than $e_2 = 6$ mm from the established column grid intersection, as given in Fig. 7. The deviation of the distance between the different holding down bolt groups in each row should not be more than $e_3 = 5$ mm or $e_4$, see Table 1 and Fig.7.

| $L$ [m] | $e_4$ [mm] |
|---|---|
| ≤ 30 | ± 15 |
| 30 < L < 250 | ± 0,25 ($L$ + 30) |
| ≥ 250 | ± 0,10 ($L$ + 450) |

**Table 1.** Permitted deviations on the location of anchor bolts group.

The allowed deviation from the required site level of foundation is $e_5 = +5$ mm (above) and -15 mm (bellow), see Fig.8. The tolerance from the required location level for pre-set bolts prepared for adjustment is $e_6 = +25$ mm (high) and -5 mm (low), and minimum movement in the pocket is $e_7 = \pm 10$ mm. The tolerance for pre-set bolts not prepared for adjustment is $e_8 = +45$ mm (high) and -5 mm (low), and minimum deviation in horizontal direction $e_9 = \pm 3$ mm.

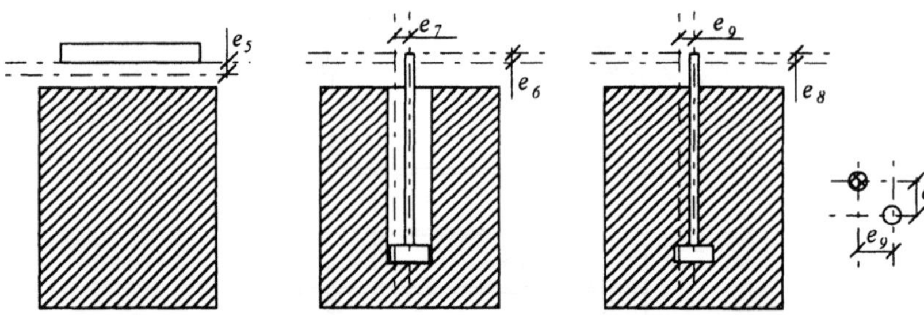


**Figure 8.** The permitted vertical deviation for the base plates and for the anchor bolts.

The squareness of the column end required for full contact between the column and base plate should be less than $e_{10} = \pm a/1000$, see Fig.9. The gap between bearing surfaces should not exceed *1 mm* anywhere, and should be less than *0,5 mm* over at least two thirds of the nominal contact area.

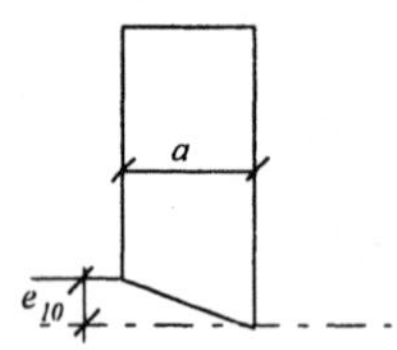


**Figure 9.** The accuracy of contact bearing surfaces.

# 3 Design of Base Plates

Eurocode 3, Section 6 and Annex L contain guidance on the strength of column bases with base plates. Section 6 contains principles, and Annex L contains detailed application rules, though these are limited examples of bases subject to axial loads only.

Traditional approaches (Dewolf and Ricker (1990), Hogan and Thomas (1990), Lescouarc` (1988), AISC (1978)) to the design of moment-resisting bases involve an elastic analysis, based on the assumption that plane sections remain plane. By solving equilibrium equations, maximum stress in the concrete (based on a triangular distribution of stress), the extent of the stress block and the tension in the holding down assemblies may be determined. While this procedure has been proved satisfactory in service over many years, the approach ignores the flexibility of the base plate in bending, the holding down assemblies and the concrete (Wald (1995), Treiberg (1987)).

Traditionally, column bases are modelled as either pinned or as fixed, while acknowledging that the reality lies somewhere within the two extremes. The opportunity to either calculate or to model the base stiffness in analysis was not available. Some national application standards recommend that the base fixity be allowed for in design. The base fixity has an important effect on the calculated frame behaviour, particularly on frame deflections.

The check of tension part is possible to provide according to Annex J of Eurocode 3 under interaction of normal force and bending. European project COST C1 (1999) was streamed to the solution of column base behavior using component method. The first step of the component method is the determination of the components, see Fig.10. The description of the component behaviour is solved in the next two lectures. The final step is the assembling of the component into the connection, see Fig.10, for resistance and stiffness calculation. The ductile behavior, the rotation capacity of connection, is ensured by adequate length of anchor bolts, resistance of anchorage into concrete and base plate ductility.

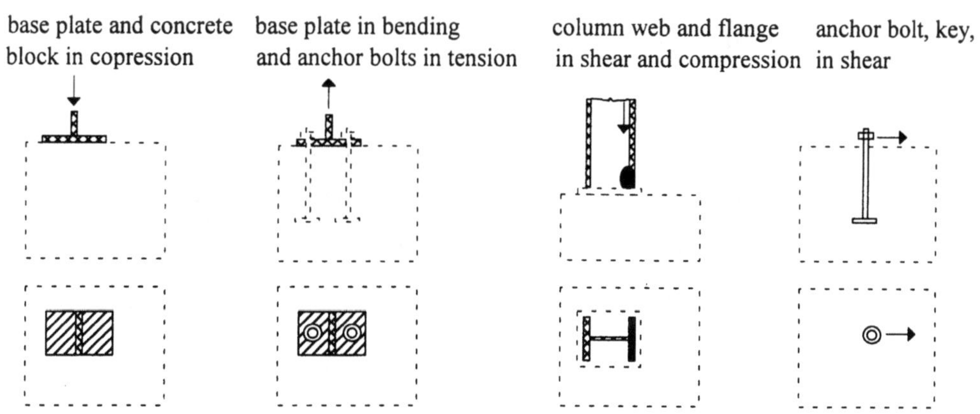


**Figure 10.** Components of column base with base plate

## 3.1 Resistance

The equilibrium of internal forces is possible to establish in elastic or in plastic stage. The effective contact area $A_{eff}$ can be evaluated based on full use of tension part $F_{t.Rd}$, see Fig.11,

$$N_{Rd} = A_{eff} f_j - \sum F_{t.Rd} , \tag{1}$$

$$M_{Rd} = \sum F_{t.Rd} r_t + A_{eff} f_j r_c . \tag{2}$$

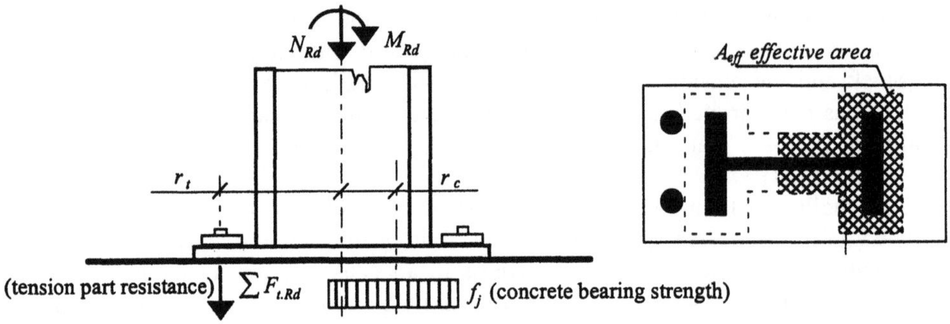


**Figure 11.** Internal forces equilibrium.

The neutral axes of the compressed part is possible to place into the axes of compressed flange for simplification, see Fig. 12. (COST C1 (1999)).
The moment resistance is in this case

$$M_{Rd} = min\left( \frac{F_{T.Rd}\ z}{1 - \frac{z_c}{M_{Sd} / N_{Sd}}} ; \frac{F_{c.Rd}\ z}{1 + \frac{z_T}{M_{Sd} / N_{Sd}}} \right) \tag{3}$$

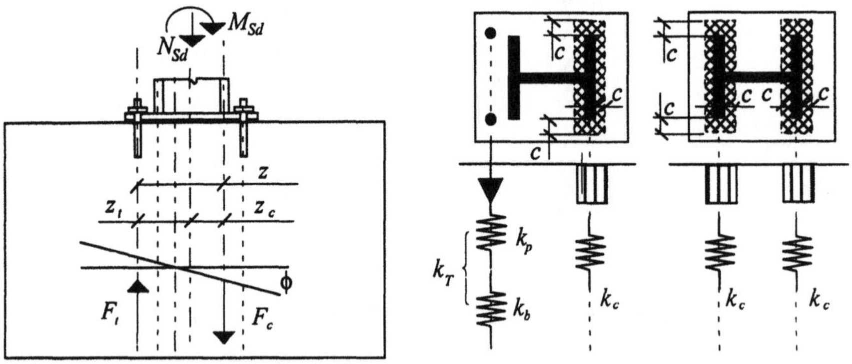


**Figure 12.** Simplified model of the compressed part.

### 3.2 Bending stiffness

From the known bilinear behaviour of components and above described simplification, Eq. (3), is expressed the bending stiffness under proportional loading, for $M_{Sd} / N_{Sd} = konst.$ In area $x_c < N_{Sd} / M_{Sd} < \alpha$ it will be

$$S_j = \frac{M_{Sd} / N_{Sd}}{M_{Sd} / N_{Sd} - \alpha} \frac{E\, z^2}{\mu \sum \frac{1}{k_i}} , \tag{4}$$

for the spring stiffness as

$$\alpha = \frac{z_c\, k_c - z_T\, k_t}{k_c + k_t} . \tag{5}$$

The curve shape factor is similar for all end plate connection

$$\mu = (1{,}5\gamma)^{2,7} \tag{6}$$

and the bending moment ration can be introduces as

$$\gamma = \frac{1 + \dfrac{z/2}{M_{Sd} / N_{Sd}}}{M_{Rd} / N_{Sd} + \dfrac{z/2}{M_{Sd} / N_{Sd}}} . \tag{7}$$

The above described model was compared to set of experiments. The results of analytical simulation is shown for one example including working diagrams of experiments, see Fig.14. The sensitivity study of base plate thickness shows the influence of tension part resistance and stiffness.

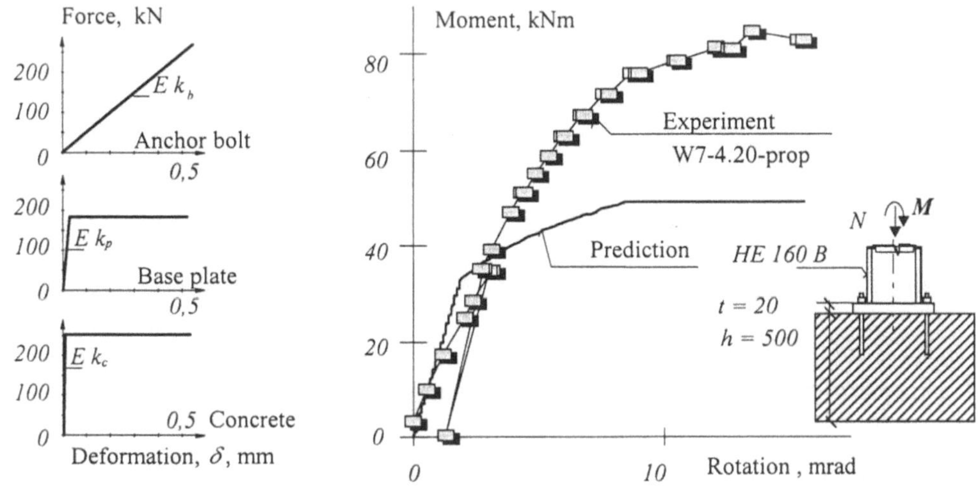


**Figure 13.** Comparison of prediction model to W7-4.20-prop (Sokol and Wald (1998)), component modelling, final moment rotational curve.

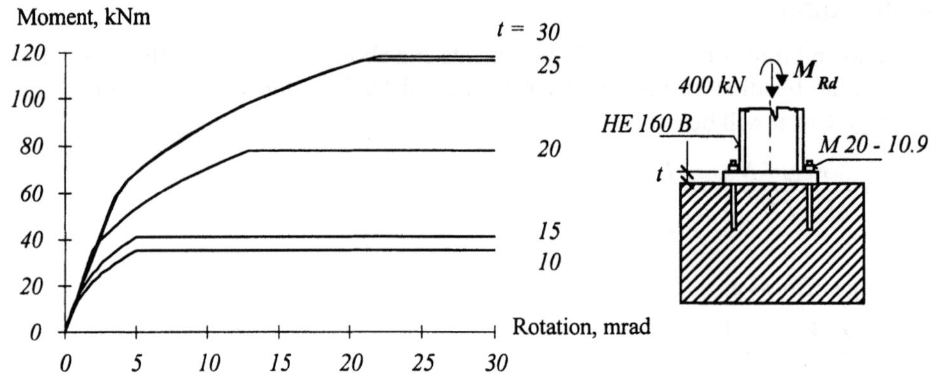


**Figure 14.** The moment-rotational diagram of the column base plate with different base plate thickness.

### 3.3 Pre-design of column base stiffness

For pre-design of bending stiffness of beam-to-column connections a simplified method based on summary of experiments and prediction models has been developed (Steenhuis, (1994)). The method was extended also for the column bases with base plate (Wald at al. (1997)). The estimation of stiffness can be written in the following form

$$S_{j.ini.app} = \frac{E\ z^2 t}{20}, \tag{8}$$

where $E$ is modulus of elasticity of steel and $t$ is the base plate thickness. The lever arm of internal forces $z$ is taken according to Fig.15, as a distance between anchor bolts and centre of compressed flange.

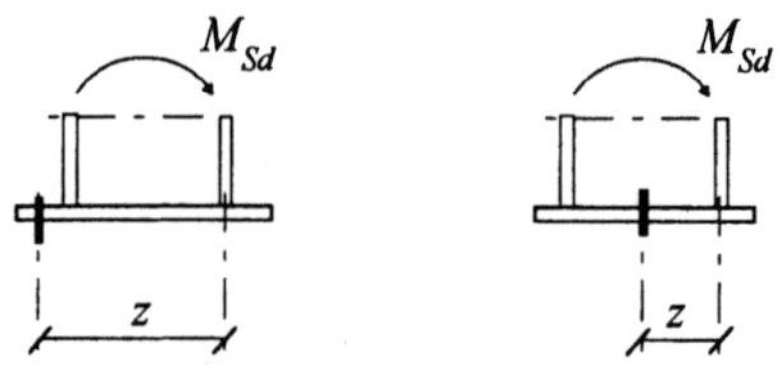


**Figure 15.** Assumption of the lever arm of internal forces z for pre-design of column base stiffness.

# 4 Concrete in Compression and Base Plate in Bending

### 4.1 Introduction

The components concrete in compression and base plate in bending including the grout represent the behavior of the compressed part of a column base with a base plate. The strength of these components depends primarily on the bearing resistance of the concrete block. The grout is influencing the column base bearing resistance by improving the joint resistance due to application of high quality grout or by decreasing the resistance due to poor quality of the grout material or due to poor detailing.

The deformation of this component is relatively small. The description of the behavior of this component is required for the prediction of column bases stiffness loaded by normal force primarily.

The technical literature concerned with the bearing strength of the concrete block loaded through a plate may be treated in two broad categories. Firstly, investigations focused on the bearing stress of rigid plates, most were concerned the prestressed tendons. Secondly, studies were concentrated on flexible plates loaded by the column cross section due to an only portion of the plate.

The experimental and analytical models for the components concrete in compression and plate in bending included the ratio of concrete strength to plate area, relative concrete depth, the location of the plate on the concrete foundation and the effects of reinforcement. The result of these studies on foundations with punch loading and fully loaded plates offers qualitative information on the behavior of base plate foundations where the plate is only partially loaded by the column. Failure occurs when an inverted pyramid forms under the plate. The application of limit state analysis on concrete can include the three-dimensional behavior of materials, plastification and cracking. Experimental studies (Shelson (1957), Hawkins (1968), DeWolf (1978)) led to the development of an appropriate model for column base bearing stress estimation that was adopted into the current codes.

The separate check of the concrete block itself is necessary to provide to check the shear resistance of the concrete block as well as the bending or punching shear resistance according to the concrete block geometry detailing.

The influence of a flexible plate was solved by replacing the equivalent rigid plate (Stockwell (1975)). This reasoning is based on recognition that uniform bearing pressure is unrealistic and that maximum pressure would logically follow the profile shape. This simple practical method was modified and checked against the experimental results (Bijlaard (1982), Murray (1983)). Eurocode 3 (Annex L (1990)) adopted this method in conservative form suitable for standardisation using an estimate including the dimensions of the concrete block cross-section and its height. It was also found (DeWolf and Sarisley (1980), and Wald (1993)) that the bearing stress increases with larger eccentricity of normal force. In this case the base plate is in larger contact with the concrete block due to its bending. In case, when the distance between the plate edge and the block edge is fixed and the eccentricity is increased, the contact area is reduced and the value of bearing stress increases. In case of the crushing of the concrete surface under the rigid edge is necessary to apply the theory of damage. These cases are unacceptable from a design point of view and are determining the boundaries of above described analysis.

### 4.2 Component Resistance

The proposed design model resistance of the components concrete in compression and base plate in bending is given in Eurocode 3 Annex L (1990). The resistance of these components is determined with help of an effective rigid plate concept.

The concrete block size has an effect on the bearing resistance of the concrete under the plate. This effect can be conservatively introduced for the strength design by the concentration factor

$$k_j = \sqrt{\frac{a_1\ b_1}{a\ b}} \tag{9}$$

where the geometry conditions, see Fig.16, are introduced by

$$a_1 = min\begin{Bmatrix} a+2a_r \\ 5a \\ a+h \\ 5b_1 \end{Bmatrix},\ a_1 \geq a\ , \tag{10}$$

$$b_1 = min\begin{Bmatrix} b+2b_r \\ 5b \\ b+h \\ 5a_1 \end{Bmatrix},\ b_1 \geq b\ , \tag{11}$$

This concentration factor is used for evaluation of the design value of the bearing strength as follows

$$f_j = \frac{\beta_j\, k_j\ f_{ck}}{\gamma_c}\quad , \tag{12}$$

where joint coefficient is taken under typical conditions with grout as $\beta_j = 2/3$. This factor $\beta_j$ represents the fact that the resistance under the plate might be lower due to the quality of the grout layer after filling.

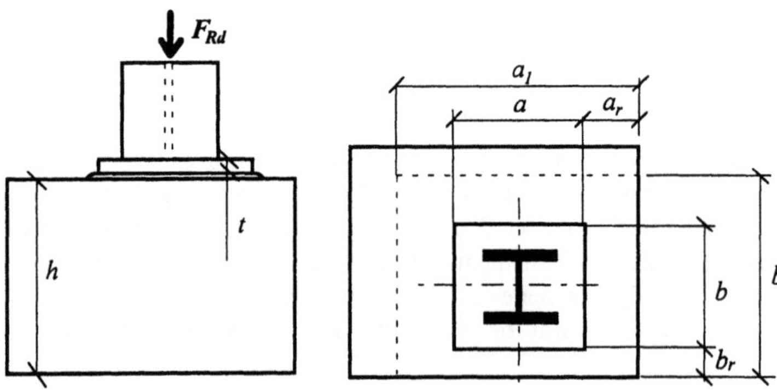


**Figure 16.** Evaluation of the concrete block bearing resistance.

The flexible base plate, of the area $A_p$, can be replaced by an equivalent rigid plate with area $A_{eq}$, see Fig.17. The formula for calculation of the effective bearing area under the flexible base plate around the column cross section can be based on estimation of the effective width $c$. The prediction of this width $c$ can be based on the T-stub model. The calculation secures that the yield strength of base plate is not exceeded. Elastic bending moment resistance of the base plate per unit length should be taken as

$$M' = \frac{1}{6} t^2 f_y \tag{13a}$$

and the bending moment per unit length on the base plate acting as a cantilever of span $c$ is, see Fig.18,

$$M' = \frac{1}{2} f_j c^2 \tag{13b}$$

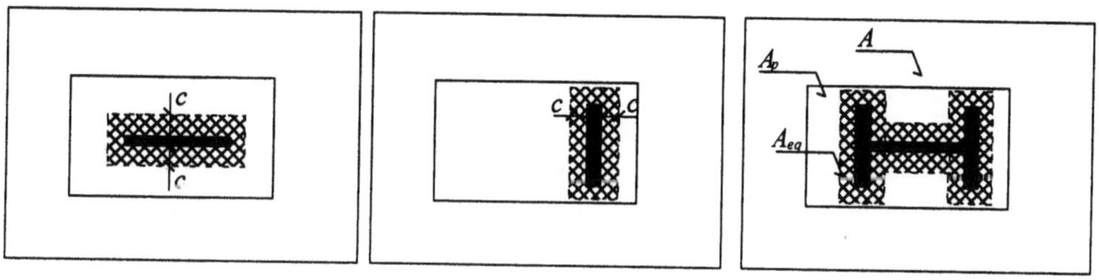


**Figure 17.** Flexible base plate modelled as a rigid plate of effective area with effective width c.

In the case that these moments are equal, the bending moment resistance is reached and the formula evaluating $c$ can be obtained

$$\frac{1}{2} f_j c^2 = \frac{1}{6} t^2 f_y \tag{13c}$$

as

$$c = t \sqrt{\frac{f_y}{3\, f_j \gamma_{M0}}}\,. \tag{13d}$$

The component is loaded by normal force $F_{Sd}$. The strength, expecting the constant distribution of the bearing stresses under the effective area, see Fig.18, is possible to evaluate for a component by

$$F_{sd} \le F_{Rd} = A_{eq} f_j = (2c + t_w) L f_j \tag{14}$$

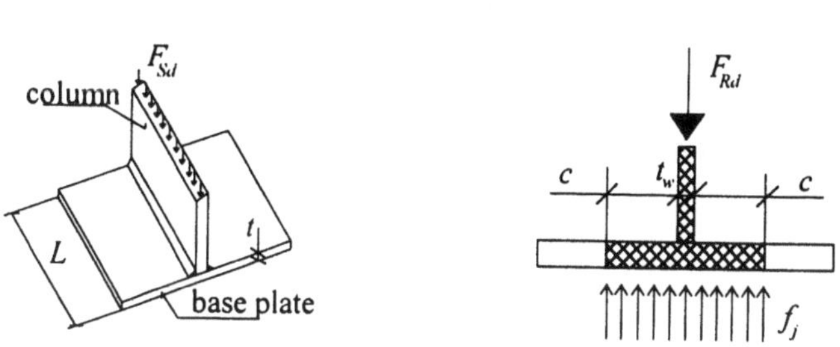


**Figure 18.** T-stub in compression, the effective width calculation.

The improvement of effective area due to the plate behavior for plates fixed on three or four edges can be based on elastic resistance of plates (Wald (1995)) or more conservatively can be limited by the deformations of plate as is reached for cantilever prediction. This improvement is not significant for open cross sections, till about 3%. For tubular columns the plate behaviour increase the strength up to 10% according to the geometry.

The practical conservative estimation of the concentration factor, see Eq. (9), can be precised by introduction of the effective area into the calculation; into the procedure Eqs (9) - (11). This leads however to an iterative procedure and is not recommended for practical purposes.

The grout quality and thickness is introduced by the joint coefficient $\beta_j$, see SBR (1973). For $\beta_j = 2/3$, it is expected the grout characteristic strength not less than $0,2$ times the characteristic strength of the concrete foundation $f_{c.g} \geq 0,2\, f_c$ and that the thickness of the grout is not greater than $0,2$ times the smaller dimension of the base plate $t_g \leq 0,2 \min (a\,;\, b)$. In cases of different quality or high thickness of the grout $t_g \geq 0,2 \min (a\,;\, b)$, it is necessary to check the grout separately. The bearing distribution under 45° can be expected in these cases, see Fig.19, (Bijlaard (1982)).

The influence of packing under the steel plate can be neglected for the design (Wald at al. (1993)). The influence of the washer under plate used for erection can be also neglected for design in case of good grout quality $f_{c.g} \geq 0,2\, f_c$. In case of poor grout quality $f_{c.g} \leq 0,2\, f_c$ it is necessary to take into account the anchor bolts and base plate resistance in compression separately.

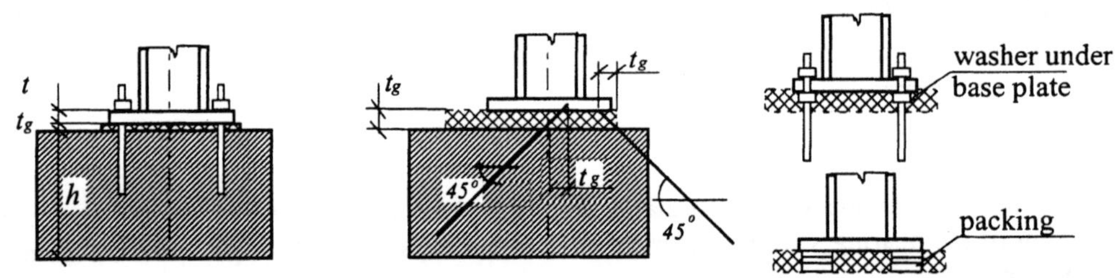


**Figure 19.** The stress distribution in the grout.

### 4.3 Component Stiffness

The elastic stiffness behavior of the T-stub components concrete in compression and plate in bending exhibit the interaction between the concrete and the base plate as demonstrated for the strength behavior. The initial stiffness can be calculated from the vertical elastic deformation of the component. The complex problem of deformation is influenced by the flexibility of the base plate and by the concrete block quality and size.

The simplified prediction of deformations of a rigid plate supported by an elastic half space is considered first including the shape of the rectangular plate. In a second step, an indication is given how to replace a flexible plate by an equivalent rigid plate. In the last step, assumptions are made about the effect of the size of the block to the deformations under the plate for practical base plates.

Different authors solved the deformation of a rectangular rigid plate in equivalent half space. Lambe and Whitman (1967) gave it in simplified form as

$$\delta_r = \frac{F\,\alpha a_r}{E_c A_r}, \tag{15}$$

where

$\delta_r$ is the deformation under a rigid plate,
$F$ the applied compressed force,
$a_r$ the width of the rigid plate,
$E_c$ the Young's modulus of concrete,
$A_r$ the area of the plate, $A_r = a_r\,L$,
$L$ the length of the plate,
$\alpha$ a factor dependent on ratio between $L$ and $a_r$.

The value of factor $\alpha$ depends on the Poison's ratio of the compressed material, see in Table 2, for concrete ($\nu \approx 0,15$). The approximation of these values as $\alpha \approx 0,85\sqrt{L\ /a_r}$ can be read from the following Table 2.

| $L / a_r$ | $\alpha$ **according to Lambe and Whitman (1967)** | **Approximation as** $\alpha \approx 0,85\sqrt{L\ /a_r}$ |
|---|---|---|
| 1 | 0,90 | 0,85 |
| 1,5 | 1,10 | 1,04 |
| 2 | 1,25 | 1,20 |
| 3 | 1,47 | 1,47 |
| 5 | 1,76 | 1,90 |
| 10 | 2,17 | 2,69 |

**Table 2.** Factor α and its approximation.

With the approximation for $\alpha$, the formula for the displacement under the plate can be rewritten

$$\delta_r = \frac{0,85F}{E_c\sqrt{L\ a_r}} \tag{16}$$

A flexible plate can be expressed in terms of equivalent rigid plate based on the same deformations. For this purpose, half of a T-stub flange in compression is modelled as shown in Fig.20.

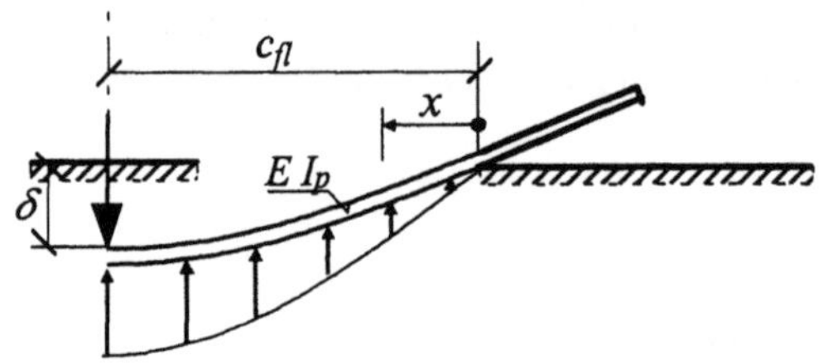


**Figure 20.** A flange of a flexible T-stub.

The flange of a unit width is elastically supported by independent springs. The deformation of the plate is a sine function, which can be expressed as

$$\delta(x) = \delta \sin (\tfrac{1}{2}\, \pi x / c_{fl}) \tag{17}$$

The uniform stress on the plate can then be replaced by the fourth differentiate of the deformation multiplied by $E\,I_p$, where $E$ is the Young's modulus of steel and $I_p$ is the moment of inertia per unit length of the steel plate with thickness $t$ ($I_p = t^3 / 12$)

$$\sigma_{(x)} = E_s\, I_p\, (\tfrac{1}{2}\, \pi / c_{fl})^4\, \delta \sin (\tfrac{1}{2}\, \pi x / c_{fl}) = E_s\, \frac{t^3}{12}\, (\tfrac{1}{2}\, \pi / c_{fl})^4\, \delta \sin (\tfrac{1}{2}\, \pi x / c_{fl}) \tag{18}$$

The concrete part should be compatible with this stress

$$\delta(x) = \sigma_{(x)}\, h_{ef} / E_c \tag{19}$$

where $h_{ef}$ is the equivalent concrete height of the portion under the steel plate. Assume that $h_{ef} = \xi\, c_{fl}$ hence

$$\delta(x) = \sigma_{(x)}\, \xi\, c_{fl} / E_c \tag{20}$$

Substitution gives

$$\delta \sin (\tfrac{1}{2}\, \pi x / c_{fl}) = E\, t^3 / 12\, (\tfrac{1}{2}\, \pi / c_{fl})^4\, \delta \sin (\tfrac{1}{2}\, \pi x / c_{fl})\, \xi\, c_{fl} / E_c \tag{21}$$

This may be expressed as

$$c_{fl} = t \sqrt[3]{\frac{(\pi/2)^4}{12}\, \xi\, \frac{E}{E_c}} \tag{22}$$

The flexible length $c_{fl}$ may be replaced by an equivalent rigid length $c_r$ such that uniform deformations under an equivalent rigid plate give the same force as the non-uniform deformation under the flexible plate

$$c_r = c_{fl}\, 2 / \pi \tag{23}$$

The factor $\xi$ represents the ratio between $h_{ef}$ and $c_{fl}$. The value of $h_{ef}$ can be expressed as $a_r$. From Tab. 1 can be read that factor $\alpha$ for practical T-stubs is about equal to $1,4$. The width $a_r$ is equal to $t_w + 2\, c_r$, where $t_w$ is equal to the web thickness of the T-stub. As a practical assumption it is now assumed that $t_w$ equals to $0,5\, c_r$ which leads to

$$h_{ef} = 1,4 \cdot (0,5 + 2)\, c_r = 1,4 \cdot 2,5\, c_{fl}\, 2 / \pi = 2,2\, c_{fl} \tag{24}$$

hence $\xi = 2,2$.

For practical joints can be estimated by $E_c \cong 30\,000\ N / mm^2$ and $E \cong 210\,000\ N / mm^2$, which leads to

$$c_{fl} = t\sqrt[3]{\frac{(\pi/2)^4}{12}\xi\frac{E}{E_c}} \approx t\sqrt[3]{\frac{(\pi/2)^4}{12}2{,}2\frac{210000}{30000}} \cong 1{,}98 \tag{25}$$

or $c_r = 1{,}26\,t \approx 1{,}25\,t$ , (26)

which gives for the effective width calculated based on elastic deformation

$$a_{eq.el} = t_w + 2{,}5\,t\ . \tag{27}$$

The influence of the finite block size compared to the infinite half space can be neglected in practical cases. For example, the equivalent width $a_r$ of the equivalent rigid plate is about $t_w + 2\,c_r$. In case $t_w$ is $0{,}5\,c_r$ and $c_r = 1{,}25\,t$ the width is $a_r = 3{,}1\,t$. That means, that peak stresses are present even in the elastic stage spread over a very small area.

In general, a concrete block has dimensions at least equal to the column with and column depth. Furthermore, it is not unusual that the block height is at least half of the column depth. This means that stresses under the flange of a T-stub, which represents for instance a plate under a column flange, are spread over a relative big area compared to $a_r = 3{,}1\,t$. If stresses are spread, the strains will be low where stresses are low and therefore, these strains will not contribute significantly to the deformations of the concrete just under the plate. Therefore, for simplicity it is proposed to make no compensation for the fact that the concrete block is not infinite.

From the strength procedure the effective width of a T-stub is calculated as

$$a_{eq.str} = t_w + 2\,c = t_w + 2\,t\sqrt{\frac{f_y}{3 f_j \gamma_{M0}}} \tag{28}$$

Based on test, see Figs22-23 and FE simulation, see Fig.24, it may occur that the value of $a_{eq.str}$ is also a sufficient good approximation for the width of the equivalent rigid plate as the expression based on elastic deformation only. If this is the case, it has a practical advantage for the application by designers. However, in the model $a_{eq.str}$ will become dependent on strength properties of steel in concrete, which is not the case in the elastic stage. In Fig.21 the influence of the base plate steel quality is shown for particular example on the concrete quality - deformation diagram for flexible plate $t = t_w = 20\ mm$, $L_{eff} = 300\ mm$, $F = 1000\ kN$. From the diagram it can be seen that the difference between $a_{eq.el} = t_w + 2{,}5\,t$ and $a_{eq.str} = t_w + 2\,c$ is limited.

The concrete surface quality is affecting the stiffness of this component (see the tests by Alma and Bijlaard (1980), and Sokol and Wald (1997)). So, the reduction of the modulus of elasticity of the upper layer of concrete of thickness of *30 mm* was proposed (Sokol and Wald, (1997)) according to the observation of experiments with concrete surface only, with poor grout quality and with high grout quality. For analytical prediction the reduction factor of the surface quality was observed from *1,0* till *1,55*. For the proposed model the value *1,50* had been proposed, see Eq. (29).

The simplified procedure to calculate the stiffness of the component concrete in compression and base plate in bending can be summarised in Eurocode 3, Annex J form as

$$k_c = \frac{F}{\delta E} = \frac{E_c\sqrt{a_{eq.el}\,L}}{1{,}5 * 0{,}85 E} = \frac{E_c\sqrt{a_{eq.el}\,L}}{1{,}275\ E}, \tag{29}$$

where

$a_{eq.el}$ the equivalent width of the T-stub, $a_{e.el} = t_w + 2{,}5\,t$,

| | |
|---|---|
| $L$ | the length of the T-stub, |
| $t$ | the flange thickness of the T-stub, the base plate thickness, |
| $t_w$ | the web thickness of the T-stub, the column web or flange thickness. |

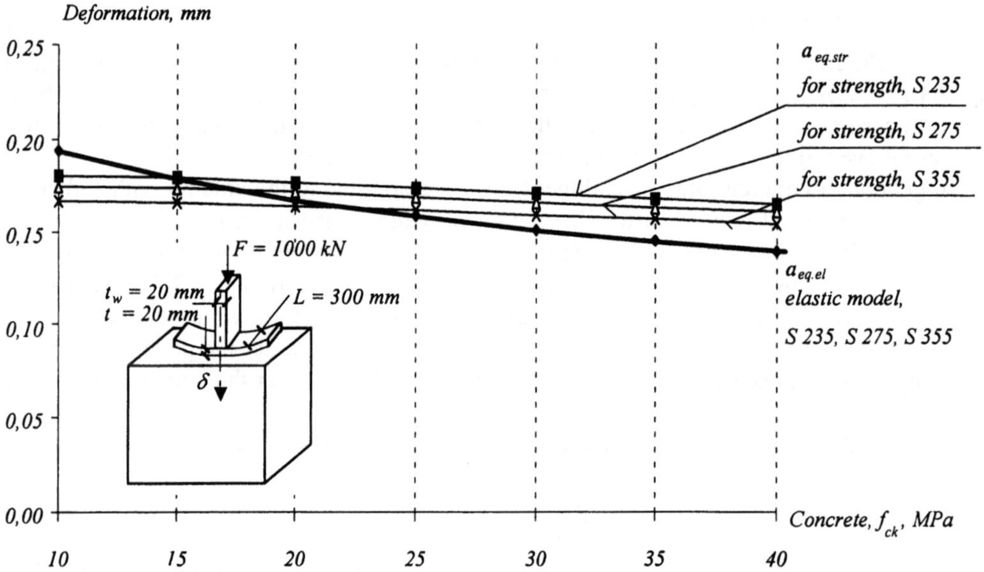


**Figure 21.** Comparison of the prediction of the effective width on concrete - deformation diagram for particular example for unlimited concrete block $k_j = 5$, base plate and web thickness 20 mm, L = 300 mm, force F = 1000 kN.

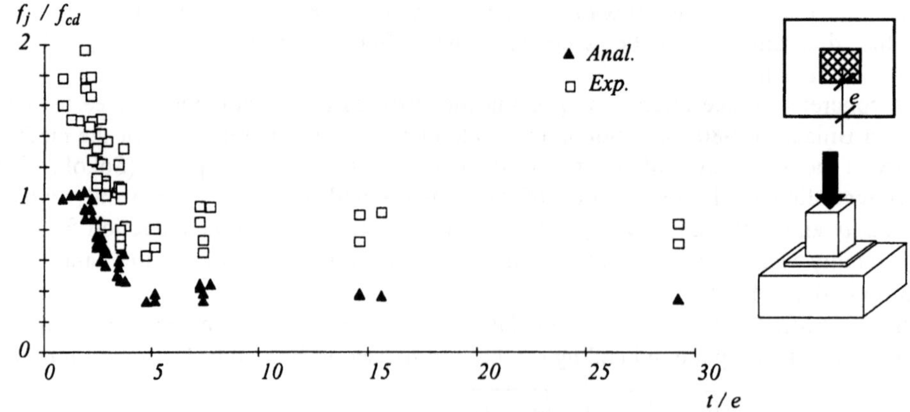


**Figure 22.** Relative bearing resistance-base plate slenderness relationship (experiments DeWolf (1978), and Hawkins (1968)).

### 4.4 Validation

The proposed model is validated against the tests for strength and for stiffness separately. 50 tests in total were examined in this part of study to check the concrete bearing resistance (DeWolf (1978), Hawkins (1968)). The test specimens consist of a concrete cube of size from 150 to 330 *mm* with centric load acting through a steel plate. The size of the concrete block, the size and thickness of the steel plate and the concrete strength are the main variables.

Fig.22 shows the relationship between the slenderness of the base plate, expressed as a ratio of the base plate thickness to the edge distance and the relative bearing resistance. The design approach given in Eurocode 3 is in agreement with the test results, but conservative. The bearing capacity of test specimens at concrete failure is in the range from 1,4 to 2,5 times the capacity calculated according to Eurocode 3 with an average value of 1,75.

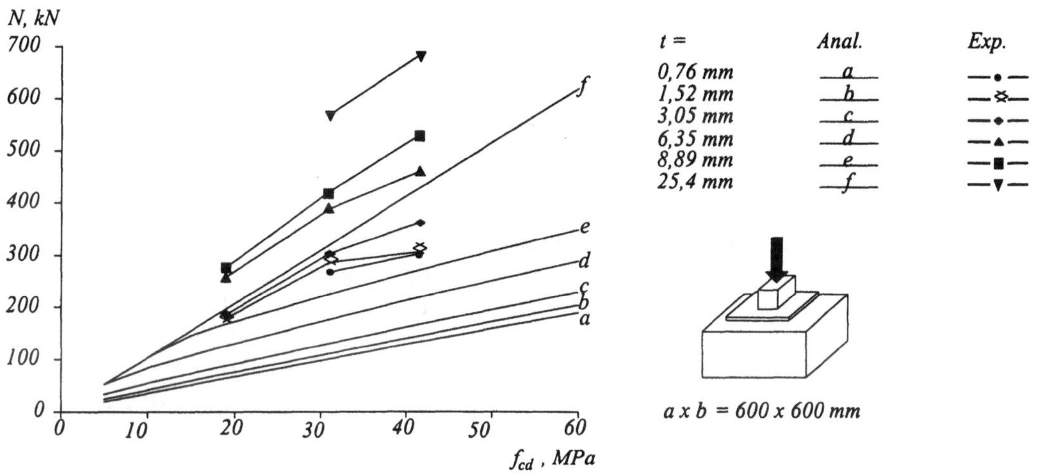


**Figure 23.** Concrete strength - ultimate load capacity relationship (Hawkins (1968)).

The influence of the concrete strength is shown in Fig.23, where the validation of the proposal based on proposal $t_w + 2\ c$ is shown. A set of 16 tests with similar geometry and material properties was used in this diagram from the set of tests (Hawkins (1968)). It worths to notice here that the only variable was the concrete strength of *19, 31* and *42 MPa*.

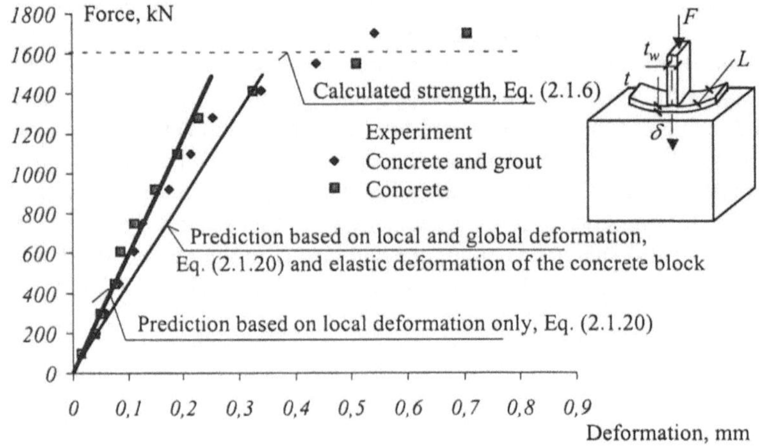


**Figure 24.** Comparison of the stiffness prediction to Test 1, (Alma and Bijlaard (1980)), concrete block 800x400x320 mm, plate thickness t = 32,2 mm, T stub length L = 300 mm.

The stiffness prediction is compared to tests Alma and Bijlaard (1980) in Fig.24. The tests of flexible plates on concrete foundation are very sensitive to boundary conditions (rigid tests frame) and measurements accuracy (very high forces and very small deformations). The predicted value based on eq. (16) is the local deformation only. The elastic global and local deformation of the whole concrete block is shown separately. Considering the spread in test results and the accuracy achievable in practice, the comparison shows a sufficiently good accuracy of prediction.

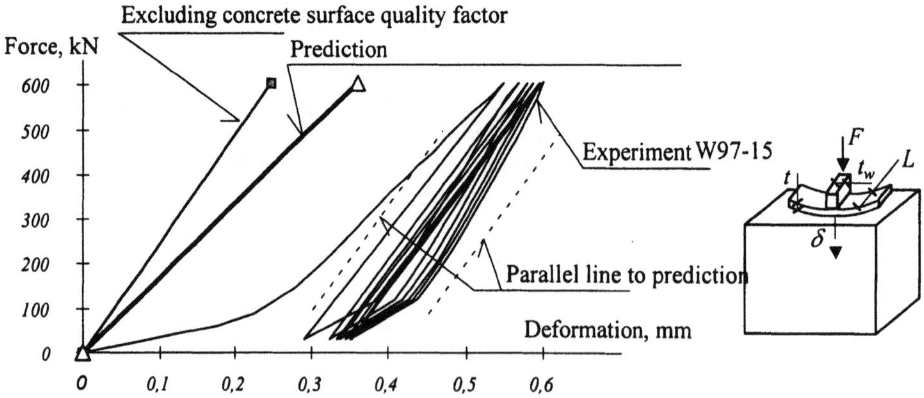


**Figure 25.** Comparison of the stiffness prediction to Test W97-15, repeated loading, cleaned concrete surface without grout only (Sokol and Wald (1997)), concrete block 550 x 550 x 500 mm, plate thickness t = 12 mm, T stub length L = 335 mm.

The comparison of local and global deformations can be shown on Finite Element (F.E.) simulation. In Fig.26 the prediction of elastic deformation of rigid plate 100 x 100 *mm* on concrete block 500 x 500 x 500 *mm* is compared to calculation using the F.E. model.

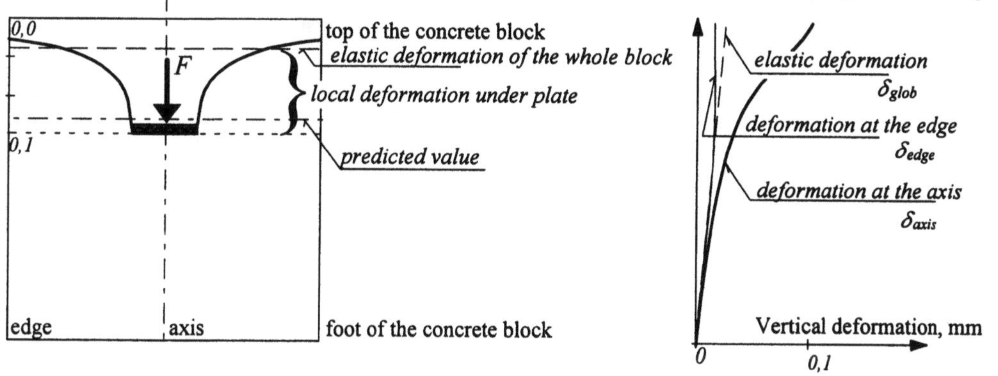


**Figure 26.** Calculated vertical deformations of a concrete block 0,5 x 0,5 x 0,5 m loaded to a deflection of 0,01 mm under a rigid plate 0,1 x 0,1 m; in the figure on the right, the deformations along the vertical axis of symmetry $\delta_{axis}$ are given and the calculated deformations at the edge $\delta_{edge}$, included are the global elastic deformations according to $\delta_{glob} = F\,h\,/(E_c\,A_c)$, where $A_c$ is full the area of the concrete block.

Based on these comparisons, the recommendation is given that for practical design, besides the local effect of deformation under a flexible plate, the global deformation of the supporting concrete structure must be taken into consideration.

# 5 Base Plate in Bending and Anchor Bolt in Tension

## 5.1 Introduction

When the anchor bolts are activated in tension, the base plate is subjected to tensile forces and deforms in bending while the anchor bolts elongate. The failure of the tensile zone may result from the yielding of the plate, from the failure of the anchor bolts, or from a combination of both phenomena.

Compared to the end plate beam-to-column connections the base plates have different configuration. The bolts are longer due to washers, thick base plate, grout and embedded length compared to beam-to-column end plate connections. The base plates are designed thick to resist compression contact forces and are stiffened more by the column itself. This difference is necessary to introduce into the strength, stiffness and rotational capacity calculation of the base plate in tension. For the component method (Wald at al. (1992)) the resistance and stiffness calculation is proposed. The quality of behavior is guided by contact between the base plate and concrete surface.

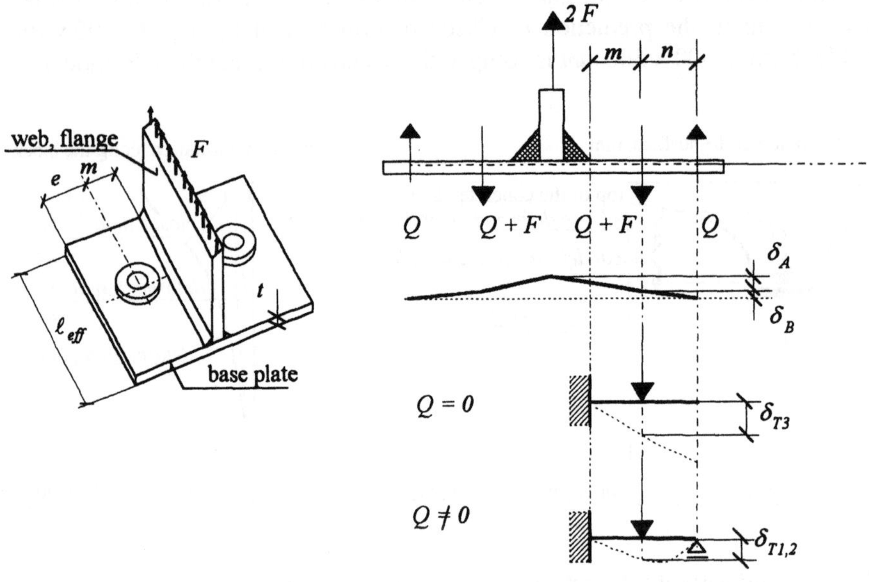


**Figure 27.** The T-stub deformation.

### 5.2 Component Stiffness

To calculate the deformation curve of T-stub by beam theory, the following parameters corresponding to the geometry are introduced

$$\beta = t / m \,, \tag{30}$$

$$\lambda = n / m \,. \tag{31}$$

The bolt stiffness can be expressed as

$$k_b = 2{,}0 \frac{A}{L_b} \tag{32}$$

and the plate stiffness as

$$k_p = \ell_{eff}\, \beta^3 \tag{33}$$

The deformation of the plate and of the bolt in case of full separation can be written as

$$\delta_p = \frac{1}{2} * \frac{F_T\, m^3}{3EI} \,, \tag{34}$$

plus

$$\delta_b = \frac{F_T\, L_b}{A_b\, E_b} \,, \tag{35}$$

which can be rewritten in the Annex J component stiffness form

$$k_{T3} = \frac{F_T}{E(\delta_p + \delta_b)} . \tag{36}$$

The stiffness of both component in the case of no contact is

$$k_{T3} = \frac{1}{2}\left(\frac{2k_b k_p}{2k_b + k_p}\right) . \tag{37}$$

If the contact is present, the direct integration gives the deformation

$$\delta_{T1,2} = \frac{F_t}{E}\frac{2}{1}\left(\frac{2k_p(\lambda+1)^3 + k_b\lambda^2(4\lambda+3)}{k_p\left(k_p + 2k_b\lambda^2(\lambda+3)\right)}\right) = \frac{F_t}{Ek_{T.1,2}} , \tag{38}$$

which is in the Annex J component stiffness form

$$k_{T1,2} = \frac{1}{2}\left(\frac{k_p\left(k_p + 4k_b\lambda^2(\lambda+3)\right)}{2k_p(\lambda+1)^3 + k_b\lambda^2(4\lambda+3)}\right) \tag{39}$$

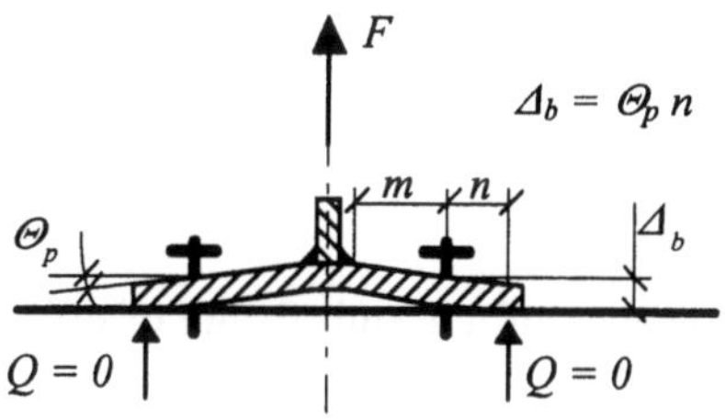


**Figure 28.** Boundary for prying action.

The boundary, where the contact occurs (Wald (1992)), see Fig.28, are based on contact limit

$$\lambda_{lim} = \frac{k_p}{3k_b} \tag{40}$$

or

$$L_{b.boundary} = \frac{7m^2 n\, A_s}{\ell_{eff}\, t^3} . \tag{41}$$

If, as a further assumption, $n$ is defined as equal to $1,25\ m$, see Eurocode 3, Annex J, then

$$L_{b.boundary} = \frac{8,82\ m^3\ A_s}{\ell_{eff}\, t^3} . \tag{42}$$

The simplified solution, based on Annex J structure of the independent components is proposed in form if $$\frac{A_s}{L_b} \geq \frac{\ell_{eff}\, t^3}{8{,}82\, m^3} \tag{43}$$

$$k_p = \frac{\ell_{eff}\, t^3}{m^3} = \frac{0{,}85\, L_{eff}\, t^3}{m^3} \tag{44}$$

$$k_b = 1{,}6\,\frac{A_s}{L_b} \tag{45}$$

and if $$\frac{A_s}{L_b} \leq \frac{\ell_{eff}\, t^3}{8{,}82\, m^3} \tag{46}$$

$$k_p = \frac{\ell_{eff}\, t^3}{2m^3} = \frac{0{,}425\, L_{eff}\, t^3}{m^3} \tag{47}$$

$$k_b = 2\,\frac{A_s}{L_b} \tag{48}$$

### 5.3 Component Resistance

In Eurocode 3, the design resistance of a T-stub flange of effective length $\ell_{eff}$ is derived as follows for each failure mode

Mode 3: bolt fracture, Fig.29a,

$$F_{Rd,3} = \Sigma B_{t.Rd} \tag{49}$$

Mode 1: plastic mechanism, Fig.29b,

$$F_{Rd,1} = \frac{4\ell_{eff}\, m_{pl,Rd}}{m} \tag{50}$$

Mode 2: mixed failure, Fig.29c,

$$F_{Rd,2} = \frac{2\ell_{eff}\, m_{pl,Rd} + \Sigma B_{t.Rd}\, n}{m+n} \tag{51}$$

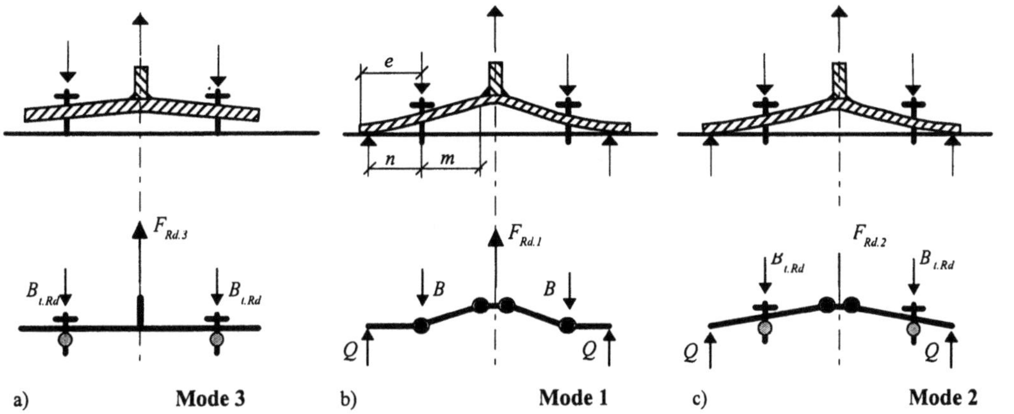


**Figure 29.** Failure modes in a T-stub.

In these expressions $m_{pl,Rd}$ is the plastic moment of the T-stub flange per unit length ($\frac{1}{4}t^2 f_y / \gamma_{M0}$) with $t$ flange thickness, $f_y$ yield stress of the flange, $\gamma_{M0}$ partial safety factor, $m$ and $e$ are geometrical characteristics defined in Fig.29, $\Sigma\, B_{t.Rd}$ is the sum of the design resistances $B_{t.Rd}$ of the bolts connecting the T-stub to the foundation, $B_{t.Rd} = 0{,}9\, A_s f_{ub} / \gamma_{Mb}$, where $A_s$ is the tensile stress area of the bolts, $f_{ub}$ the ultimate stress of the bolts and $\gamma_{Mb}$ a partial safety factor, $n$ designates the place where the prying force $Q$ is assumed to be applied, as shown in ($n = e$, but its value is limited to 1,25 $m$). $\ell_{eff}$ is derived at the smallest value of the effective lengths corresponding to all the possible yield lines mechanisms in the specific T-stub flange being considered.

The design strength $F_{Rd}$ of the T-stub is derived as the smallest value got from expressions (50) to (51) :

$$F_{Rd} = min(F_{Rd,1}, F_{Rd,2}, F_{Rd,3}) \qquad (52)$$

The influence on Mode 1 failure of backing plates aimed at strengthening the column flanges in beam-to-column bolted joints is also considered. A similar influence may result in column bases from the use of washer plates. The effect of the latter on the base plate resistance will be taken into consideration in a similar way than it is done in Annex J for backing plates.

In the particular case of base plates, it may happen that the elongation of the anchor bolts in tension is such, in comparison to the flexural deformability of the base plate that no prying forces develop at the extremities of the T-stub flange. In this case, the failure results either from that of the anchor bolts in tension (Mode 3) or from the yielding of the plate in bending, see Fig.30, where a "two hinges" mechanism develops in the T-stub flange. This failure is not likely to appear in beam-to-column joints and splices because of the limited elongation of the bolts in tension. This particular failure mode is named "Mode 1*", see Fig.31.

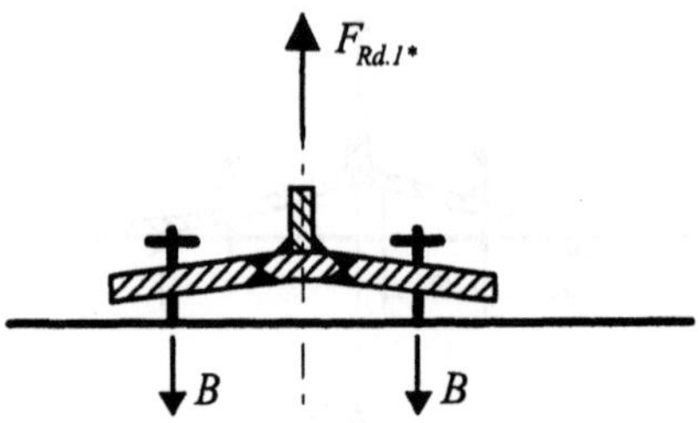


**Figure 30**. Mode 1* failure.

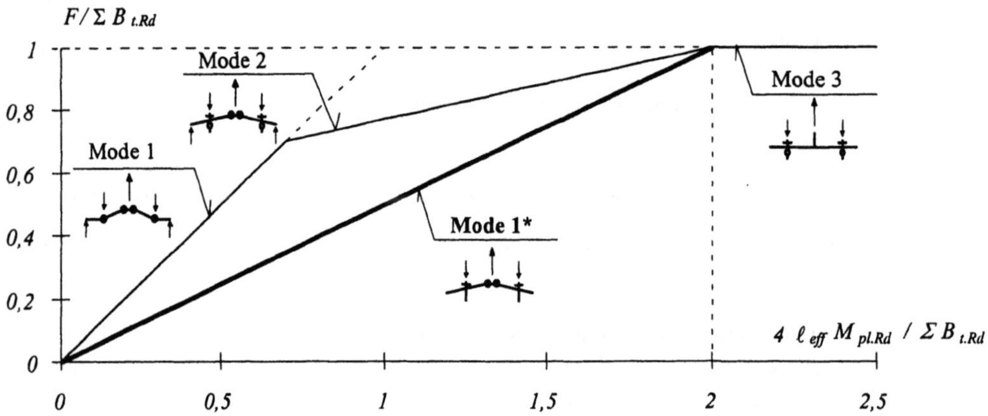


**Figure 31.** T-stub design resistance according to Annex J.

The corresponding resistance writes

$$F_{Rd.1*} = \frac{2\,\ell_{eff}\, m_{pl.Rd}}{m} \tag{53}$$

As soon as the Mode 1* mechanism forms, large base plate deformations develop; they may result in contacts between the concrete block and the extremities of the T-stub flange, i.e. in prying forces. Further loads may therefore, be applied to the T-stub until failure is obtained through Mode 1 or Mode 2. But to reach this level of resistance, large deformations of the T-stub are necessary, which is not acceptable in design conditions. The extra strength which separates Mode 1* from Mode 1 or Mode 2 in this case is therefore, disregarded and formula (53) is applied despite the discrepancy, which could result from comparisons with some experimental tests.

As a result, in cases where no prying forces develop, the design resistance of the T-stub is taken as equal to

$$F_{Rd} = min\left(F_{Rd,1*}, F_{Rd,3}\right), \tag{54}$$

when $F_{Rd,3}$ is given by formula (50).

Eurocode 3 distinguishes now between the so-called circular and non-circular yield lines mechanisms in T-stub flanges (see Figure **31.**). The latter differ by their shape and lead to specific values of T-stub effective lengths noted respectively $\ell_{eff,cp}$ and $\ell_{eff,np}$. But the main difference between circular and non-circular patterns is linked to the development or not of prying forces between the T-stub flange and the rigid foundation: circular patterns form without any development of prying forces $Q$, and the reverse happens for non-circular ones.

The direct impact on the different possible failure modes is as follows:

Mode 1 :

the presence or not of prying forces do not alter the failure mode which is linked in both cases to the development of a complete yield mechanism in the plate. Formula (50) applies therefore, to circular and non-circular yield patterns.

Mode 2 :

the bolt fracture clearly results here from the over-loading of the bolts in tension because of prying effects; therefore, Mode 2 only occurs in the case of non-circular yield lines patterns.

Mode 3 :

this mode does not involve any yielding in the flange and applies therefore, to any T-stub.

As a conclusion, the calculation procedure differs according to the yield line mechanisms developing in the T-stub flange (Figure **31.** )

$$F_{Rd} = min(F_{Rd,1}; F_{Rd,3}) \quad \text{for circular patterns} \tag{55}$$

$$F_{Rd} = min(F_{Rd,1}; F_{Rd,2}; F_{Rd,3}) \quad \text{for non-circular patterns} \tag{56}$$

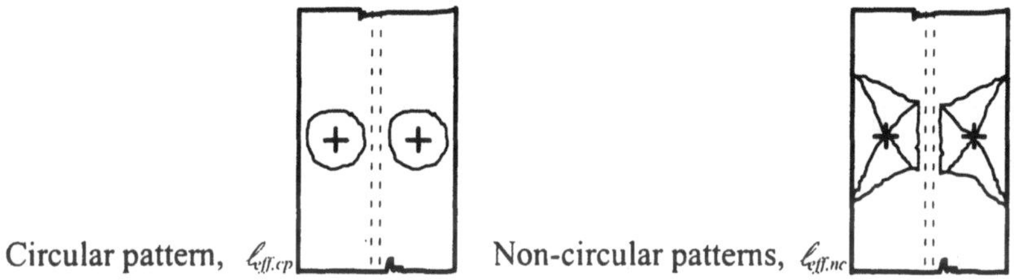


**Figure 32.** Patterns of failure.

All the possible yield line patterns are considered through recommended values of effective lengths grouped into two categories : circular and non-circular ones. The minimum values of the effective lengths - respectively termed $\ell_{eff,cp}$ and $\ell_{eff,np}$ - are therefore, selected for category. The failure load is then derived, by means of formula (52), by considering successively all the three possible failure modes, but with specific values of the effective length

$$\text{Mode 1 :} \quad \ell_{eff,1} = min(\ell_{eff,cp}; \ell_{eff,nc}) \tag{57}$$

$$\text{Mode 2 :} \quad \ell_{eff,2} = \ell_{eff,nc} \tag{58}$$

$$\text{Mode 3 :} \quad \ell_{eff,3} = \ell_{eff,1} = min(\ell_{eff,cp}; \ell_{eff,nc}) \tag{59}$$

The non-circular patterns referred to in revised Annex J of Eurocode 3 cover cases where prying forces develop at the extremities of the plated component. The circular patterns develop without any prying. Concerning Mode 1* failure, the circular and non-circular patterns have to be taken into consideration. Concerning Mode 1* failure only circular patterns are taken into account and the non-circular patterns (Eurocode 3) have to be disregarded. Mode 1* identifies exactly Mode 1 and, in order to ensure that design resistances (eqs. (50)-(51))are equal, the effective lengths for circular patterns defined in Annex J have to be multiplied by a factor 2 before being implemented in Formula (53).

Besides that, non-circular patterns not involving prying forces in the bolts may occur. These ones may be considered through Formula (56), but by introducing appropriate effective length characteristics. The lowest of the effective lengths between those derived for circular and non-circular patterns respectively is this one that will determine the design resistance of the T-stub.

Table 3 indicates how to select the values of $\ell_{eff}$ for two classical base plate configurations, in cases where prying forces develop and do not develop.

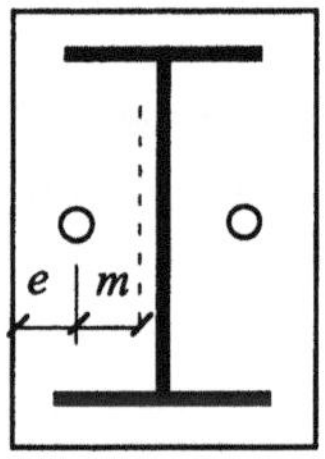


**Figure 33.** Effective length of a T stub $\ell_{eff}$ for base plate with bolts inside flanges.

| Prying | No prying |
|---|---|
| $\ell_1 = 2\,\alpha\,m - (4\,m + 1{,}25\,e)$ | $\ell_1 = 2\,\alpha\,m - (4\,m + 1{,}25\,e)$ |
| $\ell_2 = 2\,\pi\,m$ | $\ell_2 = 4\,\pi\,m$ |
| $\ell_{eff,1} = min(\ell_1;\ell_2)$ | $\ell_{eff,1} = min(\ell_1;\ell_2)$ |
| $\ell_{eff,2} = \ell_1$ | $\ell_{eff,2} = \ell_1$ |

**Table 3.** Effective length of a T stub $\ell_{eff}$ for base plate with bolts inside flanges.

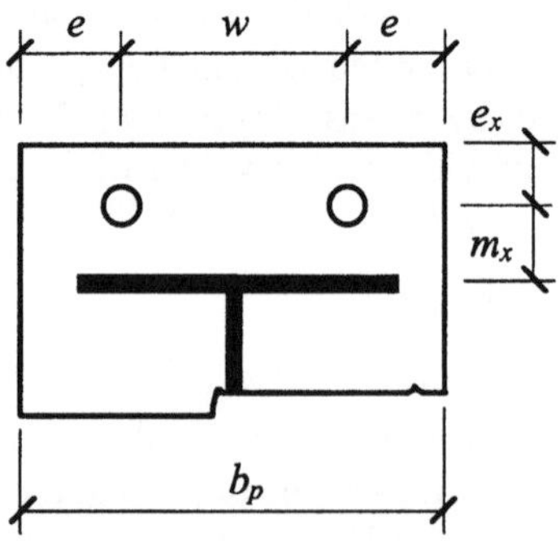


**Figure 34.** Effective length of a T-stub $\ell_{eff}$ for base plate with bolts outside flanges.

| Prying | No prying |
|---|---|
| $\ell_1 = 4.m_x + 1{,}25\ e_x$ | $\ell_1 = 4.m_x + 1{,}25\ e_x$ |
| $\ell_2 = 2\ \pi\ m_x$ | $\ell_2 = 4 \pi m_x$ |
| $\ell_3 = 0{,}5\ b_p$ | $\ell_3 = 0{,}5\ b_p$ |
| $\ell_4 = 0{,}5\ w + 2\ m_x + 0{,}625\ e_x$ | $\ell_4 = 0{,}5\ w + 2\ m_x + 0{,}625\ e_x$ |
| $\ell_5 = e + 2\ m_x + 0{,}625\ e_x$ | $\ell_5 = e + 2\ m_x + 0{,}625\ e_x$ |
| $\ell_6 = \pi\ m_x + 2\ e$ | $\ell_6 = 2\ \pi\ m_x + 4e$ |
| $\ell_{eff,1} = min(\ell_1;\ell_2;\ell_3;\ell_4;\ell_5;\ell_6)$ | $\ell_{eff,1} = min(\ell_1;\ell_2;\ell_3;\ell_4;\ell_5;\ell_6)$ |
| $\ell_{eff,2} - min(\ell_1;\ell_3;\ell_4;\ell_5)$ | $\ell_{eff,2} = min(\ell_1;\ell_3;\ell_4;\ell_5)$ |

**Table 4.** Effective length of a T stub $\ell_{eff}$ for base plate with for bolts flanges.

The anchor bolt effective free length $L_b$ consists of physical free length $L_{bf}$ and embedded free length in case of embedded anchor bolts $L_b = L_{bf} + L_{be}$, see Fig.35. The embedded free length can be for typical structural bolts predicted as $L_{be} \cong 8\ d$, see (Wald (1995)). The anchor bolt area can be taken as net area $A_s$ for simplicity.

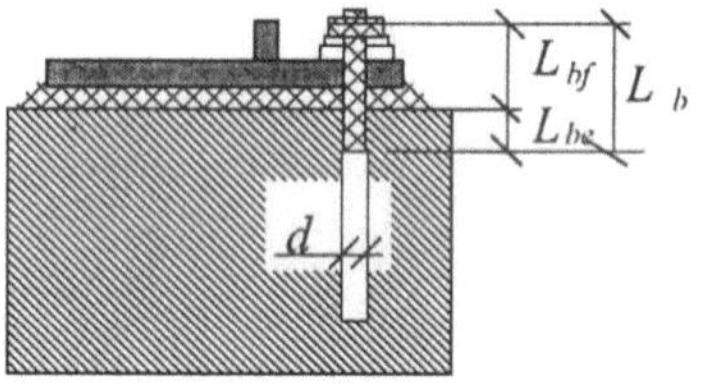


**Figure 35.** The anchor bolt effective free length.

## 5.4 Anchorage of the bolts in the concrete block

Different types of anchor bolts are used as shown in Fig.36.

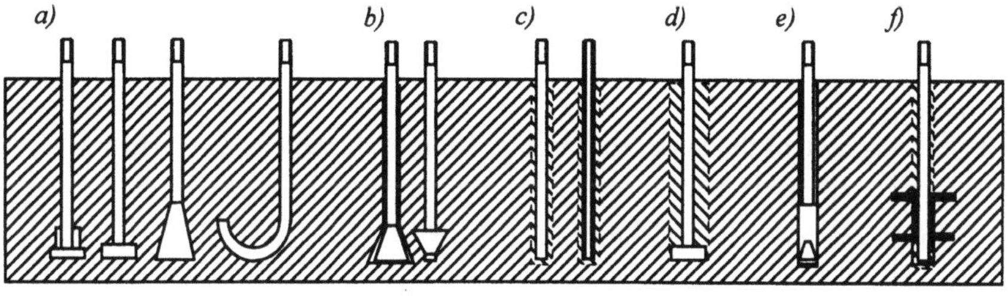


**Figure 36.** Basic types of anchoring; cast-in-place (a), undercut (b), adhesive (c), grouted (d), expansion (e), anchoring to grillage beams (f).

The most economical solutions for anchoring are, for instance, hooked bars for light anchoring, cast-in-place headed anchors and bounded anchors to drilled holes. The more expensive anchoring systems such as "grillage beams embedded in concrete" are designed for large frames. Models for the anchoring design resistance compatible with Eurocodes have been published in CEB Guide (1991).

The anchoring resistance is provided by CEB rules based on the ultimate limit state concept. As already said, this resistance has to be such that the anchor bolts fail in tension before the anchorage (pull-out of the anchor, failure of the concrete, ...) reaches its own resistance.

For a single anchor, the following failure modes have to be considered:
Pull-out failure $N_{Rd.p}$,
Concrete cone failure $N_{Rd.c}$,
Splitting failure of the concrete $N_{Rd.sp}$,
Similar verifications are required for anchor groups.

The calculation of the anchoring design resistance of cast-in-situ headed anchor bolts loaded in tension is presented here below.

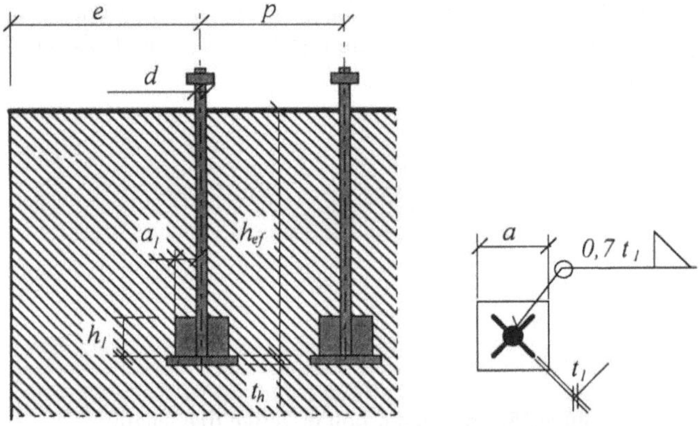


**Figure 37.** The geometry of the *in-situ*-cast headed anchor bolts.

The pull-out failure design resistance may be obtained as

$$N_{Rd.p} = p_k A_h / \gamma_{Mp} , \tag{60}$$

where $p_k$ is taken in non-cracked concrete as

$$p_k = 11,0 f_{ck} \tag{61}$$

and $A_h$ is the bearing area of the head; for circular head of diameter $d_h$, it writes

$$A_h = \pi (d_h^2 - d^2) / 4 \tag{62}$$

The concrete cone failure design resistance is given as

$$N_{Rd.c} = N^0_{Rd.c} \frac{A_{c.N}}{A^0_{c.N}} \Psi_{s.N} \Psi_{ec.N} \Psi_{re.N} \Psi_{ucr.N} \tag{63}$$

where

$$N^{0}_{Rk.c} = k_1 f_{ck}^{0,5} h_{ef}^{1,5} / \gamma_{Mc} \tag{64}$$

is the characteristic resistance of a single fastener. The coefficient $k_1$ could be taken for non cracked concrete as

$$k_1 = 11,0 \; [\sqrt{N/mm}] \tag{65}$$

The geometric effect of spacing $p$ and edge distance $e$ is included in the calculation of the area of the cone, see Fig.38, as

$$A^{0}_{c.N} = p_{cr.N}^{2} \tag{66}$$

$$A_{c.N} = \left(p_{cr.N} + p_1\right)\left(p_{cr.N} + p_2\right) \tag{67}$$

for examples in Figure 10a and 10b or as

$$A_{c.N} = \left(e + 0,5 \; p_{cr.N}\right) p_{cr.N} \tag{68}$$

for Fig.10c. It is possible to approximately consider

$$p_{cr.N} \cong 2,0 \; e_{cr.N} \cong 3,0 \; h_{ef} \tag{69}$$

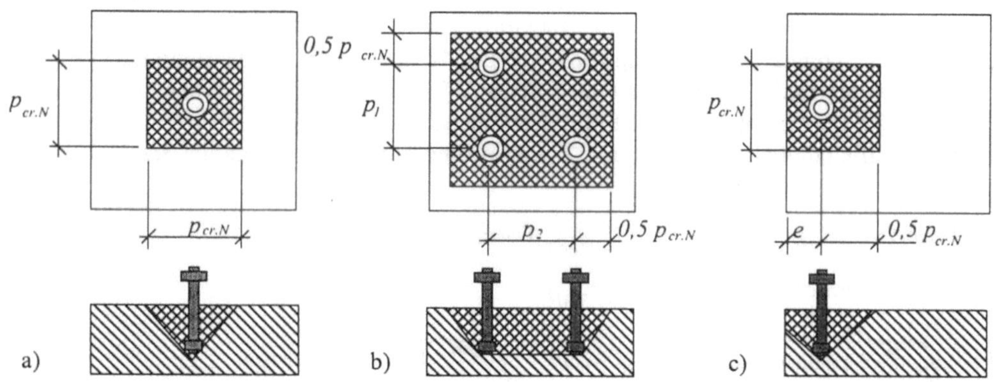


**Figure 38.** An idealised concrete cone, individual anchor (a), anchor group (b), single anchor at edge (c).

The disturbance of the stress distribution in the concrete may be introduced through the following parameter

$$\Psi_{s.N} = 0,7 + 0,3 \frac{e}{e_{cr.N}} \leq 1 \tag{70}$$

The parameter $\Psi_{ec.N}$ takes into account the group effect. Parameter $\Psi_{re.N}$ is used for small embedded depths ($h_{ef} \leq 100 \; mm$). The resistance is increased in non-cracked concrete by parameter $\Psi_{ure.N} = 1,4$.

The splitting failure for the in-situ-cast anchors is prevented if the concrete is reinforced or by limiting

- the spacing

$$p_{min} = 5\ d_h \geq 50\ mm \qquad (71)$$

- the edge distances

$$e_{min} = 3\ d_h \geq 50\ mm \qquad (72)$$

- and the height of the concrete block

$$h_{min} = h_{ef} + t_h + c_{\varnothing} \qquad (73)$$

where $t_h$ is thickness of the anchor bolt head and $c_{\varnothing}$ the required concrete cover for reinforcement.

For fastenings with an edge distance $e > 0{,}5\ h_{ef}$ in all directions, a check of the characteristic pull-out resistance may be omitted.

The detailed complex description of the evaluation formulae for the design resistance of different types of fastenings in tension is included in the CEB Guide (1991). When calculating the anchoring resistance, the tolerances for the position of the bolts should be taken into account according to Eurocode 3, clause 7.7.5 (ENV 1992-1-1, Part 1.1).

### 5.5 Validation

The proposed analytical formulas for complex component calculation, base plate in bending and anchor bolt in tension, were checked against test (Sokol and Wald (1997)) with good agreement. The comparisons of simplified, see Eq. 42, and complex calculation, see Eq. 41, give very similar results except for experiments W97-01 and W97-02 where the different boundary for prying / no prying leads to different resistance for both models. According to the curve of simplified model, there is prying, but according to the complex model, there is not.

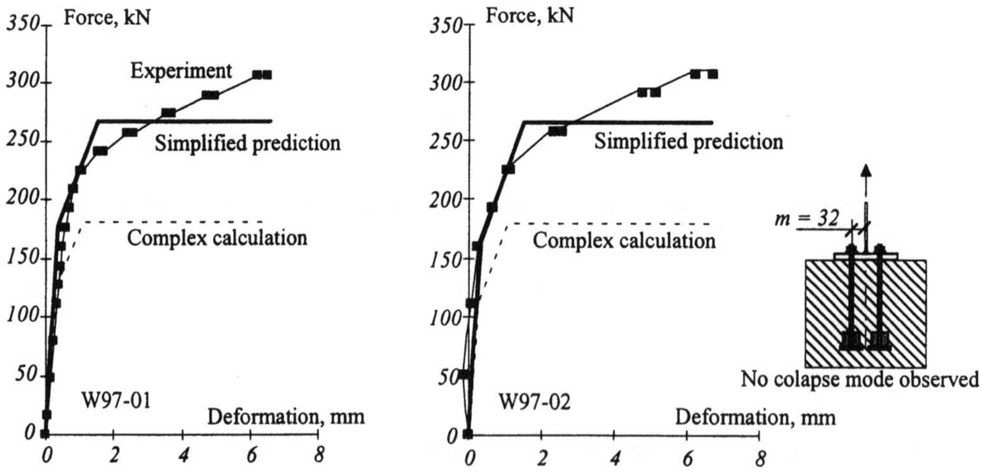


**Figure 39.** Load deflection diagram for experiment W 97 - 01 and W 97 – 02 (Sokol and Wald (1997)) bolts M24, plate 20 mm, m = 32 mm, n = 40 mm, concrete block 550 x 550 x 500 *mm*, no collapse was reached due to loading cell limitation.

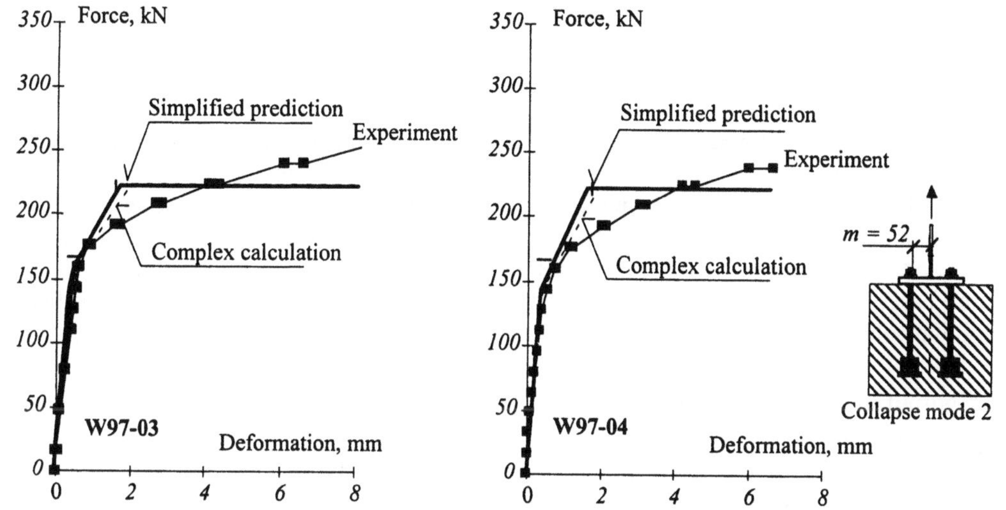


**Figure 40.** Load deflection diagram for experiment W 97 - 03 and W 97 – 04 (Sokol and Wald (1997)) bolt M24, plate 20 mm, m = 52 mm, n = 40 mm, concrete block 550 x 550 x 500 mm, the collapse mode 2, collapse breaking of bolts and plate mechanisms.

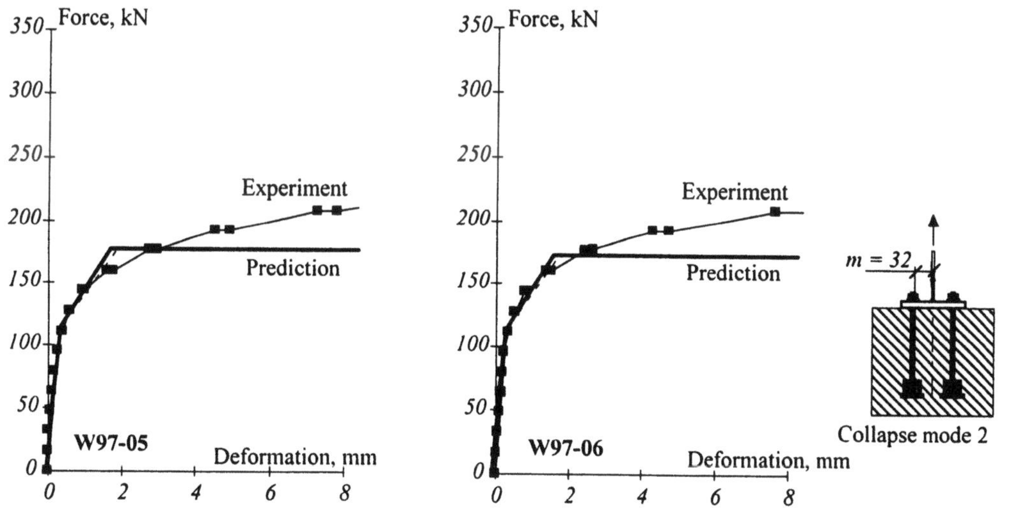


**Figure 41.** Load deflection diagram for experiment W 97 - 05 and W 97 – 06 (Sokol and Wald (1997)) bolts M24, plate 12 mm, m = 32 mm, n = 40 mm, concrete block 550 x 550 x 500 mm, the collapse mode 2 was observed, collapse breaking of bolts and plate mechanisms.

# 6 Classification

## 6.1 Introduction

The classification of a connection may by based on its resistance, its stiffness or its ductility. The column bases are designed for column resistance for the acting internal forces. The stiffness of column bases affects the internal force distribution. The boundaries of separation of rigid column bases from semi-rigid ones can be evaluated based on engineering skills or more exactly, on design calculation accuracy for resistance 5% and for serviceability 10%. For column bases two cases of sway and non-sway frames can be taken into account as for beam-to-column connections.

## 6.2 Column bases in non-sway frames

A modification of the actual moment-rotation characteristic of column bases is likely to affect the whole response of non-sway frames and in particular, the lateral displacements of the beams and the buckling resistance of the column. This second aspect - the buckling resistance of the columns - is the one for which the influence is rather important, as seen in Fig.42. (see Wald and Seifert (1991)). It shows how the buckling length coefficient of a column pinned at the upper extremity is affected by the variation of the column base rotational stiffness. The buckling length coefficient $K$ is reported on the vertical axis and is expressed as the ratio between the elastic critical load ($F_{cr,pin}$) of the column pinned at both extremities and that of the same column but restrained by the column base at the lower extremity ($F_{cr,res}$); it is shown to vary from *1,0* (pinned - pinned support conditions) to *0,7* (pinned - fixed support conditions).

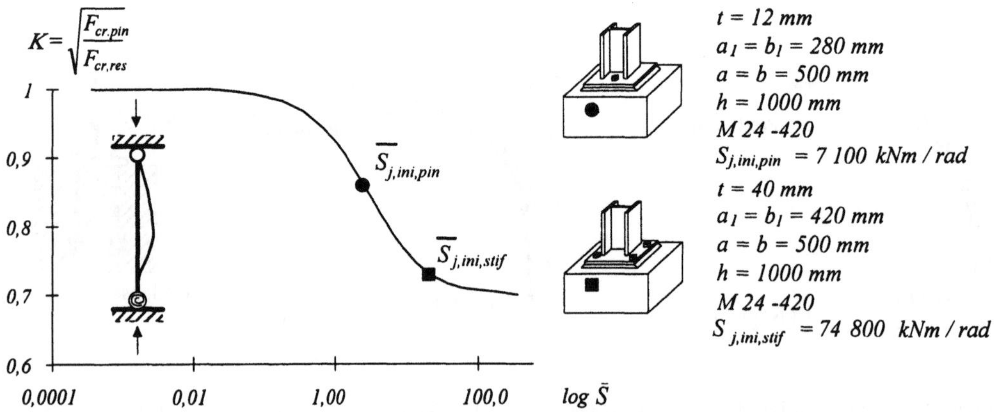


**Figure 42.** Elastic critical buckling load versus column base initial stiffness Wald and Seifert (1991).

$$K = \sqrt{\frac{F_{cr,pin}}{F_{cr,res}}} \tag{74}$$

$$F_{cr,pin} = \frac{\pi^2 EI_c}{L_c^2} \tag{75a}$$

$$F_{cr,res} = \frac{\pi^2 EI_c}{(KL_c)^2} \tag{75b}$$

where $E$ is the modulus of elasticity of steel; $L_c$ and $I_c$ are respectively the system length and the moment of inertia of the column. In Fig.42, the non-dimensional stiffness $\overline{S}$ of the column base is reported in a logarithmic scale on the horizontal axis.

$$\overline{S} = \frac{S_{j.ini}\, L_c}{EI_c} \tag{76}$$

$S_{j,ini}$ is the initial elastic stiffness in rotation of the column base. The numerical values indicated in Fig.42 have been obtained by considering the particular case of a *4 m* length column with a *HE 200 B* cross-section.
The actual initial stiffness of two typical column bases is reported in Fig.42:

- column base with a base plate and two anchor bolts inside the *H* cross-section; this configuration is traditionally considered as pinned, but possesses an initial stiffness $S_{j,ini,pin}$ equal to *7 100 kNm / rad*;
- column base with a base plate and four bolts outside the column cross-section; such a column base is usually considered as rigid, even if its initial stiffness $S_{j,ini,stif}$ = *74 800 kNm / rad* is not infinite.

As for beam-to-column joints, it may be concluded that stiff column bases always deforms slightly in rotation while presumably pinned ones exhibit a non-zero rotational stiffness. Some column bases are however, so flexible or so rigid that the structural frame response obtained by considering the actual column base characteristics in rotation is not significantly different from that obtained by modelling respectively the column bases as perfectly pinned or rigid. For beam-to-column joints, this has led to the concept of stiffness classification into pinned, semi-rigid and rigid joints (see Eurocode 3 (1992)).

The stiffness classification in Eurocode 3, Annex J is achieved by comparing the initial stiffness of the beam-to-column joints to boundary values. For instance, rigid joints are characterised by a stiffness higher than $8\, E\, I_b / L_b$ where $I_b$ and $L_b$ are respectively the moment of inertia and the length of the beam. This rigidity check is based on the so-called "5% criterion". It says that a joint may be considered as rigid if the ultimate resistance of the frame in which it is incorporated is not affected by more than 5% in comparison with the situation where fully rigid joints are considered (see Wald and Jaspart (1998)).

Adopting the same basic principle may derive a rigid classification boundary for column bases may be derived. To achieve it, the single storey - single bay non-sway frame shown in Fig.43 is considered. The study of the sensitivity of the frame to a variation of the column base stiffness properties is influenced by the beam and column characteristics in bending, $E\, I_b / L_b$ and $E\, I_c / L_c$ respectively. Two limit cases are however, obtained in the case that

- the beam is rather stiff as shown in Fig.43b ($E\, I_b / L_b = \infty$);

– the beam is rather flexible (or when a pinned joint connect the beam to the column); this situation is illustrated in Fig.43c ($E\,I_b / L_b = 0$).

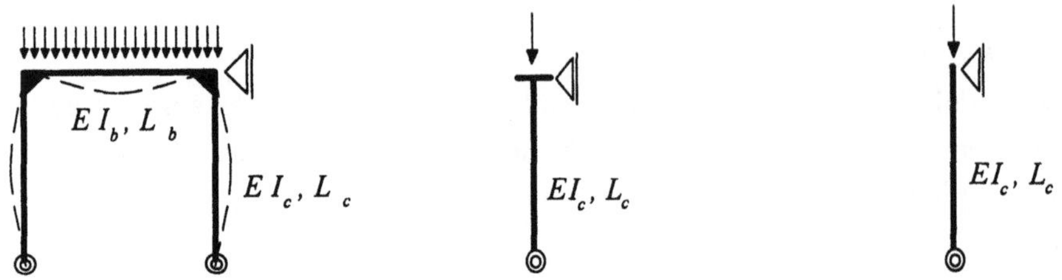


a) Portal frame (non-sway) b) Isolated column (top-fixed) c) Isolated column (top-pinned)

**Figure 43.** Portal frame and isolated columns for classification study.

The application of the "5% criterion" to the first limit case (column fixed at top extremity) writes as follows

$$\frac{\dfrac{\pi^2 EI_c}{(KL_c)^2}}{\dfrac{\pi^2 EI_c}{(0{,}5L_c)^2}} \geq 0{,}95 \; . \tag{77}$$

For sake of simplicity, it is applied to the critical elastic loads and not to the ultimate ones (integrating the effects of plasticity, imperfections, etc.); it may be demonstrated that by doing so a safe value of the rigid stiffness boundary is obtained.

From Equation (77), the minimum value of the buckling length coefficient may be derived

$$K \leq 0{,}513 \; . \tag{78}$$

According to Eurocode 3, Annex E on the «effective buckling length of members in compression », the K coefficient is expressed as a function of restraint coefficients ($k_l$, $k_u$) at both column ends and writes in this specific case

$$K = \frac{1 + 0{,}145(k_l + k_u) - 0{,}265 k_l k_u}{2 - 0{,}364(k_l + k_u) - 0{,}247 k_l k_u} \tag{79}$$

with

$$k_l = \frac{EI_c / L_c}{EI_c / L_c + S_{j,ini}} \quad \text{at lower extremity} \tag{80a}$$

$$k_u = 0 \quad \text{at upper extremity} \tag{80b}$$

The application of the "5% criterion" to the first limit case (column fixed at top extremity) writes as

$$\frac{1}{N_u} = \frac{1}{N_p} + \frac{1}{N_{cr}} \tag{81}$$

where $N_p$ and $N_{cr}$ designate respectively the squash load and the critical elastic buckling load of the column.

As $N_{cr} = N_p / \bar{\lambda}^2$, $\bar{\lambda}$ being the reduced slenderness of the column, Equation (81) writes also

$$N_u = N_p \frac{1}{1+\bar{\lambda}^2} \ . \tag{82}$$

With reference to Fig.43b, the reduced slenderness of the column with a fully stiff column base at lower end equal

$$\bar{\lambda}_{rigid\ column\ base} = 0{,}5\ \bar{\lambda}_o \ , \tag{83a}$$

while that of the column with a semi-rigid column base writes

$$\bar{\lambda}_{semi\text{-}rigid\ column\ base} = K\ \bar{\lambda}_o \ , \tag{83b}$$

where $\bar{\lambda}_o$ is the reduced slenderness of the column assumed as pinned at both extremities (K = 1).

The application of the "5 % criterion" to the ultimate column resistance therefore, gives $N_p$ being considered as constant

$$\frac{1+(0{,}5)^2\bar{\lambda}_o^2}{1+K^2\bar{\lambda}_o^2} \geq 0{,}95 \ , \tag{84a}$$

or $$K \leq 0{,}513\sqrt{1+\frac{1}{5\bar{\lambda}_o^2}} \ . \tag{84b}$$

Expression (84b) may be compared to the expression (78). For high $\bar{\lambda}_o$ values, both expressions are similar. In such cases, the ultimate resistance $N_u$ equals $N_{cr}$ and a high boundary value of $S_{j,ini}$ (see Formula (81)) is required. For low values of $\bar{\lambda}_o$, the condition (84b) relaxes and, as a consequence, less severe boundary values of $S_{j,ini}$ are required, the influence of the cross-section yielding becoming then more predominant than the instability. For $\bar{\lambda}_o$ = 0,48, Equation (84b) writes K ≤ 0,7, which means that any column base, even a perfectly pinned one, will be considered as rigid. By integrating Formula (84b) in Formulae (79) to (80b), the following stiffness boundaries are obtained

if $\bar{\lambda}_o \leq 0{,}48$ then $$S_{j,ini} \geq 0 \ , \tag{85a}$$

if $\bar{\lambda}_o > 0{,}48$ then $$S_{j,ini} \geq \frac{1{,}145-0{,}838\mu}{1{,}026\mu-1}\frac{4EI_c}{L_c} \ , \tag{85b}$$

with

$$\mu = \sqrt{1+\frac{1}{5\bar{\lambda}_o^2}} \tag{86}$$

For practical applications, simpler expressions are proposed, which fit rather well, as seen in Fig.44, with the exact ones (Formulae 88) in the usual range of application ($\bar{\lambda}_o \leq 2$ to 3). These are

if $\bar{\lambda}_o \leq 0,5$ then $S_{j,ini} \geq 0$ , (87a)

if $0,5 < \bar{\lambda}_o < 3,93$ then $S_{j,ini} \geq 7\,(2\,\bar{\lambda}_o - 1) E\, I_c / L_c$ , (87b)

if $\bar{\lambda}_o \geq 3,93$ then $S_{j,ini} \geq 48\, E\, I_c / L_c$ (87c)

as a safe approximation, Formula (87c) may obviously be applied for any column slenderness.

### 6.3 Column bases in sway frames

The sway frames are more sensitive than non-sway ones to the variation of the rotational properties of column bases, mainly because of their high sensitivity to lateral deflections as well as to changes of the overall stability conditions when the lateral flexibility increases.

To illustrate this statement, a single-bay single-storey sway frame is considered in Fig.44. The diagram indicates the evolution with increasing values of $\bar{S}$ of the ratio $\beta_s = y_S / y_P$ between the lateral deflection $y_S$ of the frame with actual column base stiffness and the deflection $y_P$ of the frame with assumed ideally pinned column bases. The non-dimensional stiffness $\bar{S}$ defined by Eq. (76) is again reported in a logarithmic scale. First order elastic theory is used to compute the $y$ values.

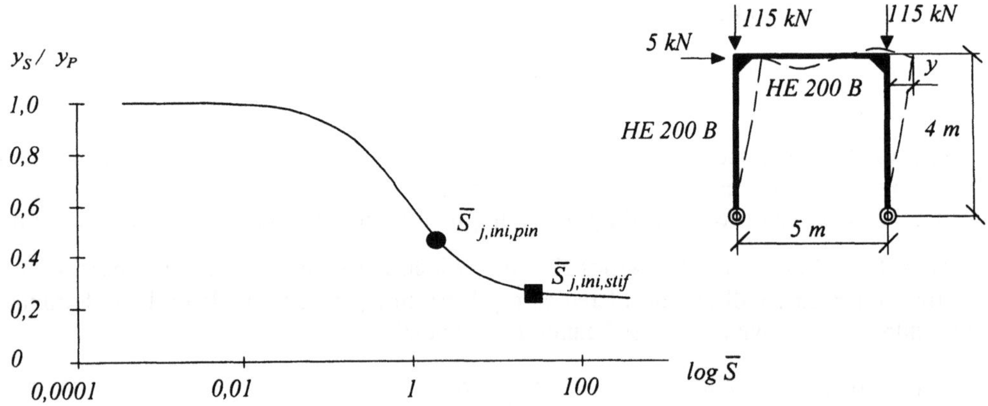


**Figure 44.** Sensitivity of the sway deflection to a variation of the column base stiffness in a portal frame.

A stiffness classification boundary similar to that one expressed in the case of non-sway frames may again be derived here on the basis of a "5% resistance criterion". For sway frames it may also be demonstrated that the more restrictive situation corresponds to the limit case where the beam flexural stiffness is rather high in comparison with that of the columns. Referring to the isolated column represented in Fig.45b, it therefore, carries out the derivation of the classification boundary.

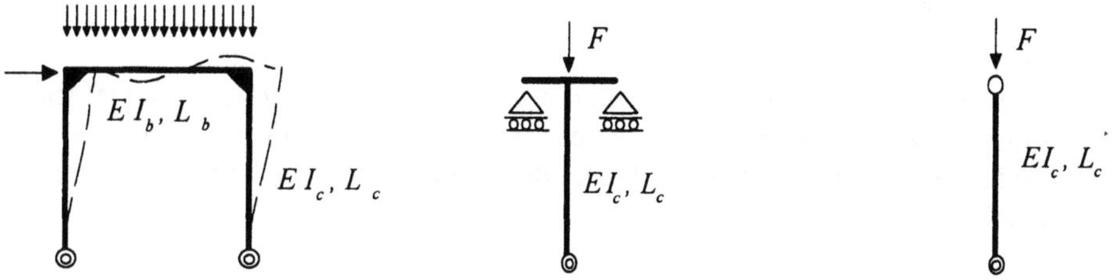


a) Portal frame (non-sway) b) Isolated column (top-fixed) c) Isolated column (top-pinned)

**Figure 45.** Sway portal frame and isolated columns for classification study.

The application of the "5% resistance criterion" writes in this case

$$\frac{\dfrac{\pi^2 EI_c}{(KL_c)^2}}{\dfrac{\pi^2 EI_c}{(L_c)^2}} \geq 0{,}95 \tag{88}$$

As a result

$$K \leq 1{,}026 \ . \tag{89}$$

For sway frames, the $K$ - $k$ relationship given by Eq. (79) has to be replaced by the following

$$K = \sqrt{\frac{1 - 0{,}2(k_l + k_u) - 0{,}24 k_l k_u}{1 - 0{,}8(k_l + k_u) + 0{,}6 k_l k_u}} \ , \tag{90}$$

while Eqs (80a) and (80b) remain unchanged. The combination of these equations leads to the following expression of the stiffness classification boundary

$$S_{j,ini} \geq 11 EI_c / L_c \tag{91}$$

However the 5% resistance criterion fully disregards the aspects of lateral frame deflections which have been pointed out as important. It may be shown (Wald and Sokol (1997)) that the lateral deflection $y_s$ of the portal frame, illustrated in Fig.45a, may by written as

$$y_S = \frac{F L_c^{\ 3}}{2EI_c} \frac{1}{12} \frac{4(3+\overline{S}) + 6(4+\overline{S})\varsigma}{\overline{S} + 6(1+\overline{S})\varsigma} \ , \tag{92}$$

where

$$\overline{S} = \frac{S_{j,ini}}{EI_c / L_c} \tag{93}$$

$$\varsigma = \frac{EI_b / L_b}{EI_c / L_c} \tag{94}$$

For $\overline{S} \Rightarrow \infty$ , the deflection for the frame with rigid column bases may be derived from (92)

$$y_R = \frac{FL_c^{\,3}}{2EI_c} \frac{1}{12} \frac{4+6\varsigma}{1+6\varsigma} \tag{95}$$

In comparison to the case where rigid column bases are used, Eq. (91), the actual frame - where the column bases possesses some degree of flexibility - will experience a larger deflection, formula (88); this increase of the lateral displacement may be expressed in terms of percentage $w$ as follows

$$\frac{y_S}{y_R} = 1 + \omega \ . \tag{96}$$

As far as classification is concerned, an $\omega$ % resistance criterion may be suggested with the objective to limit the increase of the lateral displacement of the actual frame to $\omega$% of the deflection evaluated in the case of rigid column bases. Combining the expressions (92), (91) and (96), the value of the minimum rotation that the column bases should exhibit to be considered as rigid from a displacement point of view is derived

$$\overline{S} \geq \frac{12 + 24\varsigma - 6\varsigma(1+\omega)\dfrac{4+6\varsigma}{1+6\varsigma}}{(4+6\varsigma)\omega} \ . \tag{97}$$

This condition is illustrated in Figure 46, (see Wald and Sokol (1997)). The required stiffness is seen to be rather insensitive to the values of $\varsigma$ for significant values of $w$. Conservatively the values obtained for $\varsigma = 0,1$ may be selected, i.e.:

- for $\omega = 20\%$, the following stiffness boundary is obtained $\overline{S} \geq 15$,
- for $\omega = 10\%$ $\overline{S} \geq 30$,
- for $\omega = 5\%$ $\overline{S} \geq 60$.

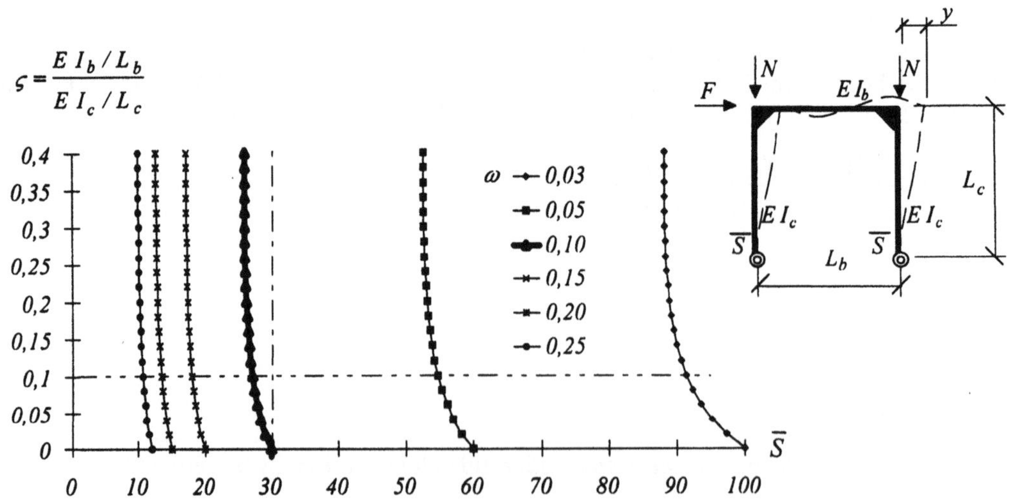


**Figure 46.** Displacement classification criteria for column bases.

As a consequence, the displacement classification criterion is seen to be much more restrictive than the resistance one given by Equation (17). The selection of the value for the boundary is obviously strongly related to the level of accuracy, which is thought to be necessary for the evaluation of the lateral frame deflection. A value of 10% appears to be quite realistic and the following stiffness classification boundary for presumably rigid column bases may be therefore, proposed

- column bases

$$S_{j,ini} \geq 30\ EI_c\ /\ L_c \tag{98a}$$

- semirigid column bases

$$S_{j,ini} < 30\ EI_c\ /\ L_c \tag{98b}$$

For similar reasons than those given for non-sway frames no classification boundary for presumably pinned column bases is suggested.

### 6.4 Summary

For particular structure is possible to evaluate its boundaries of stiffness classification. The save conservative boundaries proposed for structural frames are shown in Figure 47. Two cases of sway and non-sway frames can be evaluated based on calculation accuracy for resistance 5% and for serviceability 10% (Wald and Jaspart (1998)). The above values are based on Eurocode assumptions and can be changed according to different standard rules, as well as importance of structure.

For typical non sway frames based on ultimate limit state and limits of relative slenderness as $\bar{\lambda}_o = 1{,}36$ is possible to take $12\ E\ I_c\ /\ L_c$, (COST C1 (1999)).

The sway frames are more sensitive to column base stiffness. Based on sway limits under serviceability limit state the boundaries are $S_{j,ini} \geq 30\ EI_c\ /\ L_c$.

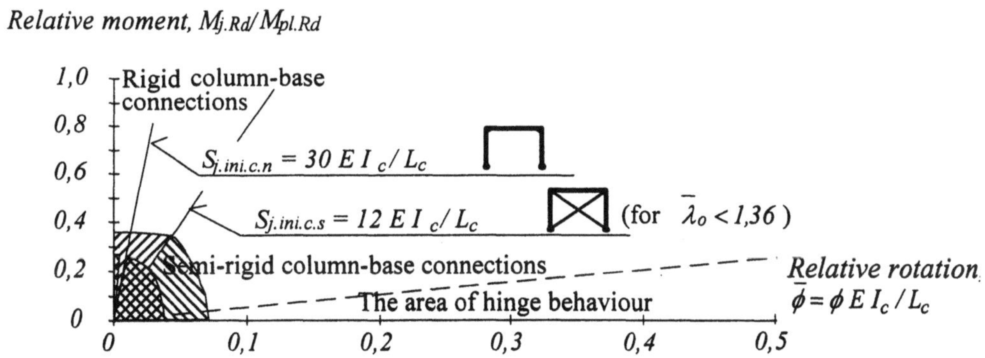


**Figure 47.** Proposed classification system according to the initial stiffness.

# 7 Worked Examples

## 7.1 Frame Design

Design a frame of Fig.48. The column is supported against the lateral torsional buckling along its full lengths by corrugated sheeting. The beam is supported in connections by purlins. Steel S 235.

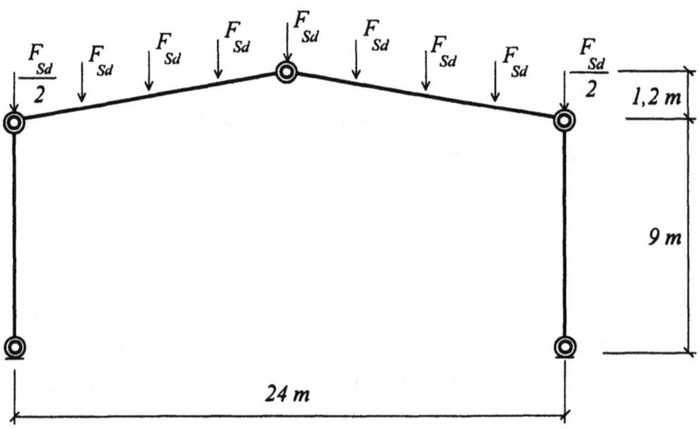


**Figure 48**. Portal frame of worked example.

1. Two load combination are guiding the frame design. The first one (dead load, snow and parallel to facade wind load) is major for bending moments in beam to column connections and horizontal drift. The second one (dead load and snow load) gives major moments in beam, column normal force and beam deflections. Both load combinations are shown on Fig.49.

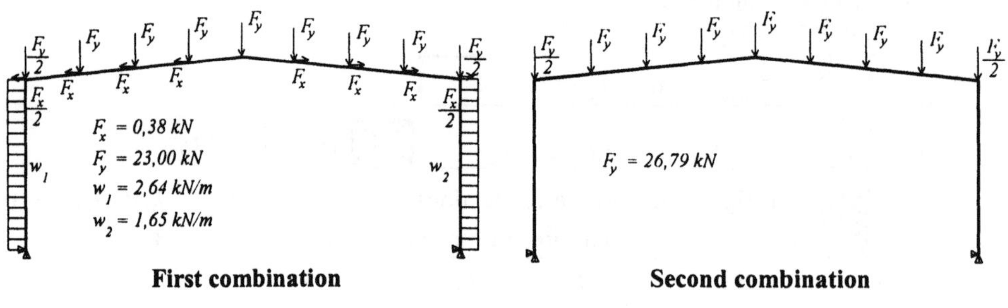


**Figure 49.** Major load combinations.

2. Frame imperfections are included by initial sway of columns $\phi$, which is for portal frames $\phi = \frac{1}{200}$. The effect is possible to be replaced by a pair of horizontal forces.

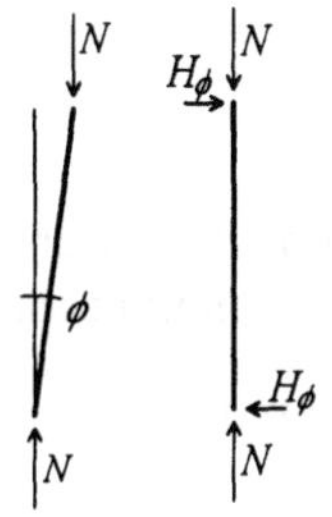


**Figure 50.** Frame imperfections as equivalent forces.

The column normal force by load combination is

$$N^{(1)} = 4 * F_y = 4 * 23{,}0 = 92{,}0\ kN\ ,$$

$$N^{(2)} = 4 * F_{Sd} = 4 * 26{,}79 = 107{,}2\ kN\ .$$

Equivalent forces are

$$H_\phi^{(1)} = \phi \sum N^{(1)} = \frac{1}{200} * 2 * 92{,}0 = 0{,}92\ kN\ ,$$

$$H_\phi^{(2)} = \phi \sum N^{(2)} = \frac{1}{200} * 2 * 107{,}2 = 1{,}07\ kN\ .$$

3. For global analysis of statically indetermine structure the preliminary judge of cross-sections is necessary. The columns HE 340 B and beam IPE 550 is chosen. The prediction of bending stiffness of connections is possible to be preliminary predicted by the simple formula

$$S_{j.ini} = \frac{E\, z^2 t}{k_f}\ .$$

For beam-to-column connection with end plate of thickness *20 mm*, with stiffeners and haunch on beam is the bending stiffness can be preliminary predicted

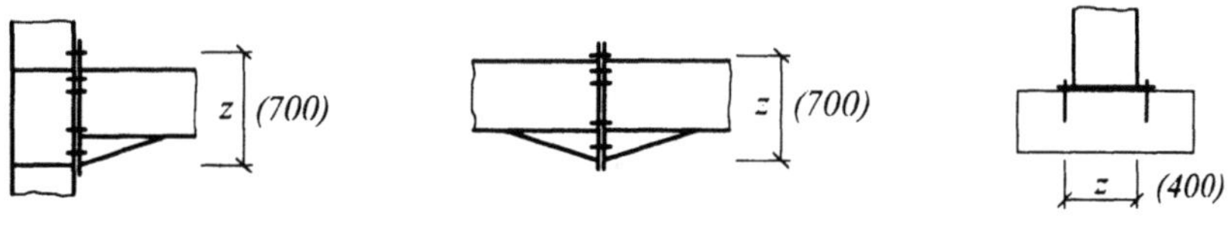


**Figure 51.** Connections lever arm.

$$S_{j.ini.b-c} = \frac{E\, z^2 t_{fl}}{k_f} = \frac{210\,000 * 700^2 * 21{,}5}{8{,}5} = 260\,300\ kNm\,/\,rad$$

The beam-to-beam connection with the similar end plate and haunch

$$S_{j.ini.b-b} = \frac{E\, z^2 t_p}{k_f} = \frac{210\,000 * 700^2 * 20}{6} = 343\,000\ kNm\,/\,rad$$

The bending stiffness of column base with four bolts and base plate of *30 mm* will be

$$S_{j.ini.cb} = \frac{E\,z^2 t}{k_f} = \frac{210\,000 * 400^2 * 30}{20} = 50\,400 \ kNm\,/\,rad$$

The value of second stiffness is possible to predicteconservatively using coefficient $\kappa$. The coefficient $\kappa$ is *2* for beam-to-column connections and *3* for another connections:

$$S_{j.b-c} = \frac{260\,300}{2} = 130\,150 \ kNm\,/\,rad\ ,$$

$$S_{j.b-b} = \frac{343\,000}{3} = 114\,330 \ kNm\,/\,rad\ ,$$

$$S_{j.cb} = \frac{50\,400}{3} = 16\,800 \ kNm\,/\,rad\ .$$

4. The internal forces are calculated by any standard Finite Element code for both load combinations. The semi-rigid connections ware introduced by linear elements with one degree of freedom - rotation.

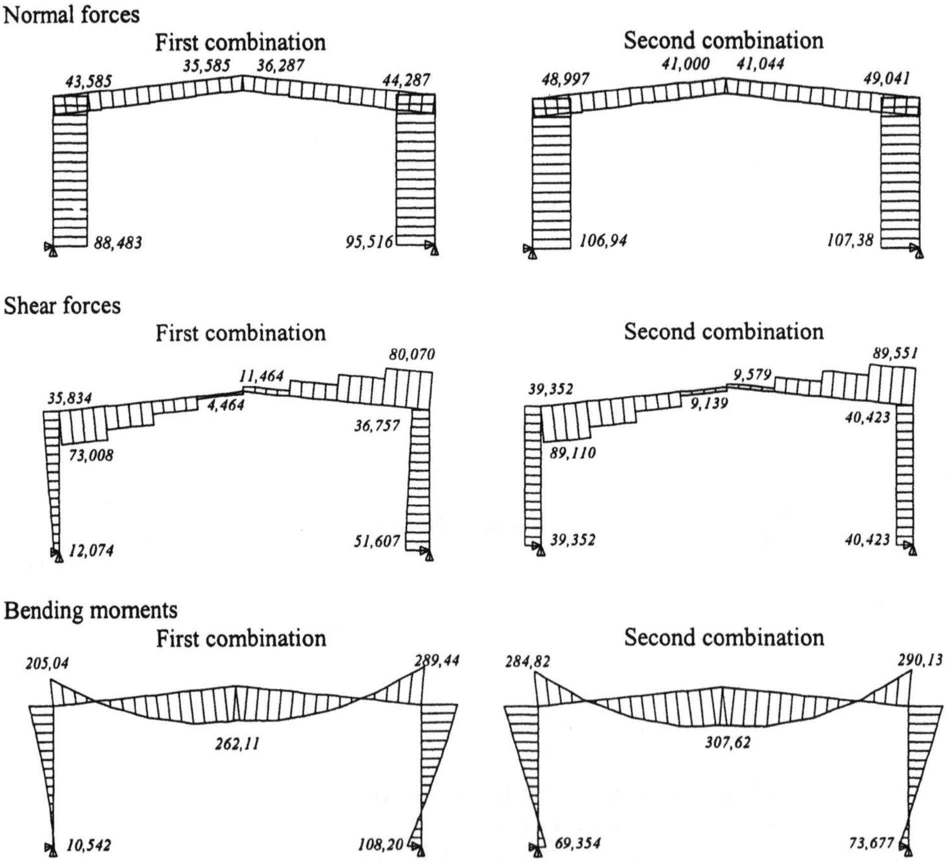


**Figure 52**. Internal forces and moments for global analysis with semi-rigid connections and column bases.

5. Preliminary designed elements are checked for calculated internal forces. The re-calculation of global analysis is not necessary in case of small changes of cross sections under design process because the portal frame has a small level of fixed freedom of structure.
6. For approved cross sections of elements, the connections are designed and its resistance is checked.
7. The global analysis for serviceability limit states is calculated for initial stiffness of connections with the values:

   $S_{j.ini.b-c} = 260\,300\ kNm / rad$,

   $S_{j.ini.b-b} = 343\,000\ kNm / rad$,

   $S_{j.ini.cb} = 50\,400\ kNm / rad$.

   The characteristic load is used. The maximal horizontal drift of frame is due to the first combination in the left corner

   $$\delta_x = 27{,}4\ mm < 60\ mm = \frac{h}{150}.$$

   The maximal deflection is due to the second load combination

   $$\delta_y = 109{,}8\ mm < 120\ mm = \frac{l}{200}.$$

**Note**

The results of global analysis of three examples of frames shows that the influence of column base connections stiffness is shown on the previous example in the table below (with rigid connections and pinned column bases a) with above designed semi-rigid connections and column bases b) as well as with rigid connections and column bases c).

| | Maximal moment in column base, *kNm* | Maximal moment in corner, *kNm* | Maximal moment in beam, *kNm* | Maximal beam deflection, *mm* | Maximal horizontal drift of left corner, *mm* |
|---|---|---|---|---|---|
| a)  | *0* | *337,85* | *318,10* | *113,68* | *73,70* |
| b) 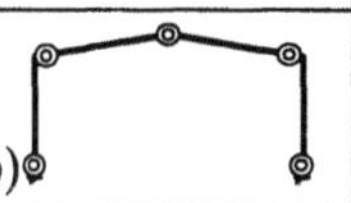 | ***108,20*** | ***290,13*** | ***307,62*** | ***109,80*** | ***27,43*** |
| c)  | *214,09* | *305,90* | *274,73* | *95,54* | *19,42* |

**Table 5.** Review of global analysis for different connection representation.

From the Tables 5 and 6 is visible that the influence of connection stiffness is not so important for bending moments (excluding the moments a column base is hinged or fixed), but the influence of connection stiffness on frame deformation is very high. The assumption of column base hinge behaviour goes to an uneconomical element design. The fully rigid connection with base plate with stiffener gives an uneconomical design of column bases. A realistic approach is nominally pinned column bases for resistance (ULS) and semi-rigid for serviceability (SLS).

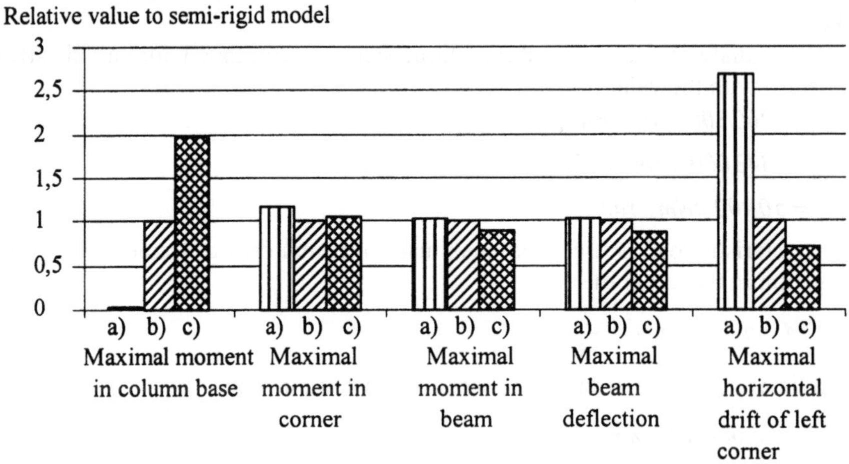


**Table 6**. Relative bending moments and horizontal deformation.

### 7.2 Column Base with Base Plate

Calculate moment resistance and bending stiffness of column base on Fig.53. Column *HE 200 B* is loaded by normal force $F_{Sd} = 500\ kN$. Concrete block *C16 / 20* of size *1 600 * 1 600 * 1 000 mm* is designed for particular soil conditions. Base plate of thickness *30 mm*, steel *S235*, $\gamma_c = 1{,}5$, $\gamma_{M0} = 1{,}15$, $\gamma_{Mb} = 1{,}25$.

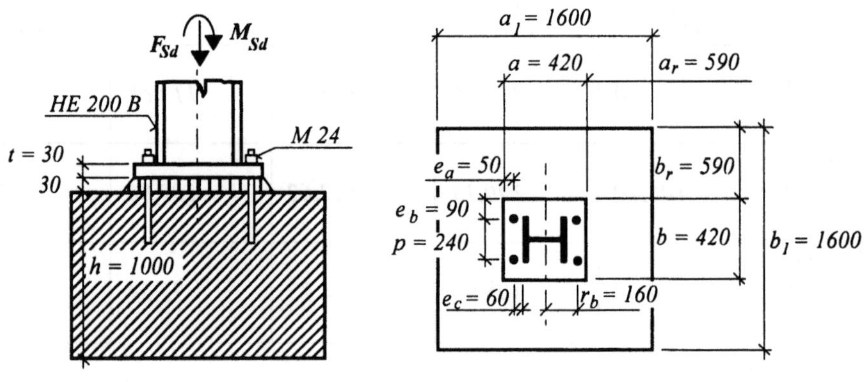


**Figure 53.** Column base of worked example.

1. Resistance of component base plate in bending and anchor bolts in tension. Anchor bolt lever arm is for fillet weld $a_{wf} = 6\ mm$.

$$m = 60 - 0{,}8 * a_w * \sqrt{2} = 60 - 0{,}8 * 6 * \sqrt{2} = 53{,}2\ mm\ .$$

The T - stub length for no prying mode is:

$$L_{eff.1} = min\begin{Bmatrix} 4m + 1{,}25e_a = 4 * 53{,}2 + 1{,}25 * 50 = 275{,}3 \\ 4\pi m = 4\pi * 53{,}2 = 668{,}6 \\ 0{,}5b = 0{,}5 * 420 = \underline{210} \\ 2m + 0{,}625e_b + 0{,}5p = 2 * 53{,}2 + 0{,}625 * 90 + 0{,}5 * 240 = 282{,}7 \\ 2m + 0{,}625e_b + e_a = 2 * 53{,}2 + 0{,}625 * 90 + 50 = 212{,}7 \\ 2\pi m + 4e_b = 2\pi * 53{,}2 + 4 * 90 = 694{,}2 \\ 2\pi m + 2p = 2\pi * 53{,}2 + 2 * 240 = 814{,}2 \end{Bmatrix} = 210\ mm$$

The effective length of anchor bolt $L_{bef}$ can be taken as

$$L_{bef} = 8 * d + t_g + t + t_n/2 = 8 * 24 + 30 + 30 + 19/2 = 261{,}5\ mm\ .$$

The check of prying of anchor bolts. Limit thickness of base plate $t_p$, with no prying is

$$t_p = 2{,}07m\ \sqrt[3]{\frac{A_s}{L_{bef}\ L_{eff.1}}} = 2{,}07 * 53{,}2\ \sqrt[3]{\frac{353}{261{,}5 * 210}} = 20{,}5\ mm\ .$$

The no-prying occurs, because $t > t_p$. The resistance of T - stub with two anchor bolts is

$$F_{Rd.1*} = \frac{2\ L_{eff.1}\ t^2 f_y}{4\ m\ \gamma_{M0}} = \frac{2 * 210 * 30^2 * 235}{4 * 60 * 1{,}10} = 336{,}5 * 10^3\ N\ .$$

The resistance is limited by tension resistance of two anchor bolts M 24 in tension (the area of threaded part of bolt $A_s = 353\ mm$ )

$$F_{Rd.3} = 2B_{t.Rd} = 2\,\frac{0{,}9 f_{ub} A_s}{\gamma_{Mb}} = 2\,\frac{0{,}9 * 360 * 353}{1{,}25} = 183{,}0 * 10^3\ N\ .$$

2. To evaluate the compressed part, resistance is calculated by the connection concentration factor as

$$a_l = b_l = min\begin{Bmatrix} a + 2a_r = 420 + 2 * 590 = 1600 \\ 5a = 5 * 420 = 2100 \\ a + h = 420 + 1000 = \underline{1420} \\ 5b_l = 5 * 1420 = 7100 \end{Bmatrix} = 1420\ mm\ ,$$

and

$$a_l = b_l = 1420 > a = b = 420\ mm\ .$$

The above condition is fulfilled and

$$k_j = \sqrt{\frac{a_1\, b_1}{a\, b}} = \sqrt{\frac{1420 * 1420}{420 * 420}} = 3{,}38\,.$$

The grout is not affecting the concrete bearing resistance because

$$0{,}2 \min(a;b) = 0{,}2 * \min(420;420) = 84 \ mm > 30 \ mm = t\,.$$

The concrete bearing resistance can be calculated as

$$f_j = \frac{2}{3}\frac{k_j\, f_{ck}}{\gamma_C} = \frac{2}{3}\frac{3{,}38 * 16}{1{,}50} = 24{,}0 \ MPa\,.$$

From the force equilibrium in the vertical direction

$$F_{Sd} = A_{eff}\, f_j - F_{t.Rd}\,,$$

the area of concrete in compression $A_{eff}$ under the full resistance of tension part can be calculated

$$A_{eff} = \frac{F_{Sd} + F_{t.Rd}}{f_j} = \frac{500 * 10^3 + 183{,}0 * 10^3}{24{,}0} = 28\,458 \ mm^2\,.$$

3. The flexible base plate is transferred into a rigid plate of equivalent area. The with of the strip $c$ around the column cross section, see Fig.54, is calculated from

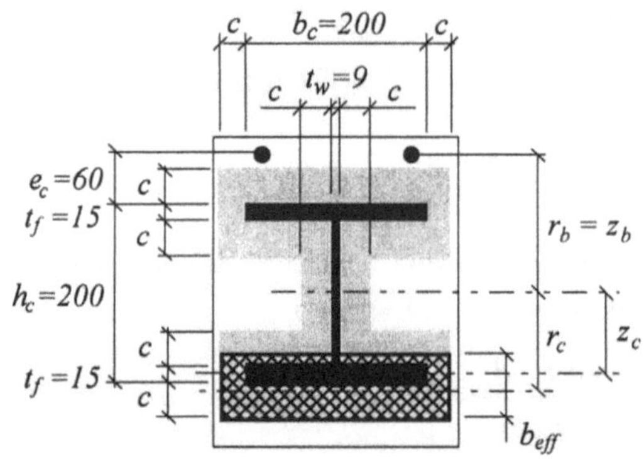


**Figure 54.** The effective equivalent area under base plate.

$$c = t\sqrt{\frac{f_y}{3\, f_j \gamma_{M0}}} = = 30 * \sqrt{\frac{235}{3 * 24{,}0 * 1{,}10}} = = 51{,}7 \ mm$$

4. The active effective with can be calculated from a known area in compression of width

$$b_{eff} = \frac{A_{eff}}{b_c + 2c} = \frac{28\,458}{200 + 2 * 51{,}7} = 93{,}8 \ mm < t_f + 2\,c = 15 + 2 * 51{,}7 = 118{,}4 \ mm\,.$$

The lever arm of the concrete to the column axes of symmetry can be calculated as

$$r_c = \frac{h_c}{2} + c - \frac{b_{eff}}{2} = \frac{200}{2} + 51,7 - \frac{93,8}{2} = 104,8\ mm$$ and lever arm of bolts as

$$r_b = \frac{h_c}{2} + e_c = \frac{200}{2} + 60 = 160\ mm\,.$$

Moment resistance of column base is

$$M_{Rd} = F_{t.Rd}\, r_b + A_{eff}\, f_j\, r_c$$
$$= 183,0 * 10^3 * 160 + 28\,458 * 24,0 * 104,8 = 100,9 * 10^6\ Nmm = 100,9\ kNm$$

Under acting normal force $N_{Sd} = 500kN$ the moment resistance in bending is $M_{Rd} = 100,9\ kNm$.

5. The end of column needs to be checked. The design resistance in poor compression is

$$N_{pl.Rd} = \frac{A f_y}{\gamma_{M0}} = \frac{7808 * 235}{1,10} = 1\,668 * 10^3\ N$$

and column bending resistance is

$$M_{pl.Rd} = W_{pl}\, f_y / \gamma_{M0} = 642,5 * 10^3 * 235 / 1,10 = 137,3 * 10^6\ Nmm\,.$$

The interaction of normal force reduces the moment

$$M_{Ny.Rd} = M_{pl.Rd} \frac{1 - \dfrac{N_{Sd}}{N_{pl.Rd}}}{1 - 0,5\dfrac{A - 2\, b\, t_f}{A}} = 137,3 \frac{1 - \dfrac{500}{1\,668}}{1 - 0,5\dfrac{7808 - 2 * 200 * 15}{7808}} = 108,7\ kNm\,.$$

The column base is designed on acting force only (not for column resistance).

6. To evaluate the bending stiffness, the particular component stiffness is calculated

$$k_b = 2,0 \frac{A_s}{L_{bef}} = 2,0 \frac{353}{261,5} = 2,700\ mm\,,$$

$$k_p = \frac{0,425\, L_{eff}\, t^3}{m^3} = \frac{0,425 * 210 * 30^3}{53,2^3} = 16,004\ mm\,.$$

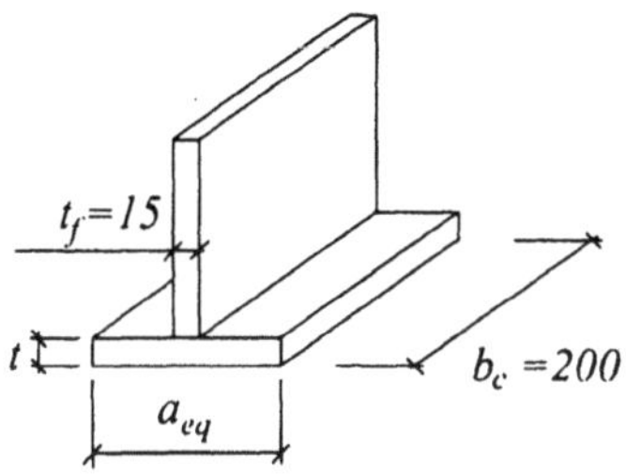


**Figure 55.** The T-stub in compression for stiffness calculation.

The concrete block stiffness can be evaluated based on a T-stub in compression, see Fig.55

$$a_{eq} = t_f + 2{,}5\,t = 15 + 2{,}5 * 30 = 90\ mm\,,$$

$$k_c = \frac{E_c}{1{,}275\ E_s}\sqrt{a_{eq}\ b_c} = \frac{27\,500}{1{,}275 * 210\,000}\sqrt{90 * 200} = 13{,}780\ mm\,.$$

7. The lever arm of component in tension $z_t$ and in compression $z_c$ to the column base neutral axes is

$$z_t = r_c = \frac{h_c}{2} + e_c = \frac{200}{2} + 60 = 160\ mm\,,$$

$$z_c = \frac{h_c}{2} - \frac{t_f}{2} = \frac{200}{2} - \frac{15}{2} = 92{,}5\ mm\,.$$

The stiffness of tension part, bolts and T-stub, can be calculated as

$$k_t = \frac{1}{\dfrac{1}{k_{bolt}} + \dfrac{1}{k_{T-stub}}} = \frac{1}{\dfrac{1}{2{,}700} + \dfrac{1}{16{,}004}} = 2{,}310\ mm\,.$$

8. For the calculation of the initial stiffness of the column base the lever arm is evaluated

$$z = z_t + z_c = 160 + 92{,}5 = 252{,}5\ mm$$

and

$$a = \frac{k_c\ z_c - k_t\ z_t}{k_c + k_t} = \frac{13{,}780 * 92{,}5 - 2{,}310 * 160}{13{,}780 + 2{,}310} = 56{,}2\ mm\,.$$

The bending stiffness is calculated for particular constant eccentricity

$$e = \frac{M_{Rd}}{F_{Sd}} = \frac{100{,}9 * 10^6}{500 * 10^3} = 201{,}8\ mm \text{ as}$$

$$S_{j.ini} = \frac{e}{e+a}\,\frac{E_s\,z^2}{\mu \sum_i \dfrac{1}{k_i}} = \frac{201{,}8}{201{,}8 + 56{,}2} * \frac{210\,000 * 252{,}5^2}{1 * \left(\dfrac{1}{2{,}310} + \dfrac{1}{13{,}780}\right)} =$$

$$= 20{,}718 * 10^9\ Nmm/rad = 20\,718\ kNm/rad\,.$$

**Notes:**

- ✓ The classification of the column base according to its bending stiffness is evaluated in comparison to column bending stiffness. For column length $L_c = 4{,}0\ m$ and its cross-section *HE 200 B* relative bending stiffness is

$$\overline{S}_{j.ini} = S_{j.ini}\,\frac{L_c}{E_s\,I_c} = 2{,}0470 * 10^{10}\,\frac{4\,000}{210\,000 * 56{,}96 * 10^6} = 6{,}85$$

- ✓ The designed column base is sway for braced, as well as non-sway frames because

$$\overline{S}_{j.ini} = 6{,}93 < 12 = \overline{S}_{j.ini.EC3.n} \, ,$$

$$\overline{S}_{j.ini} = 6{,}93 < 30 = \overline{S}_{j.ini.EC3.s} \, .$$

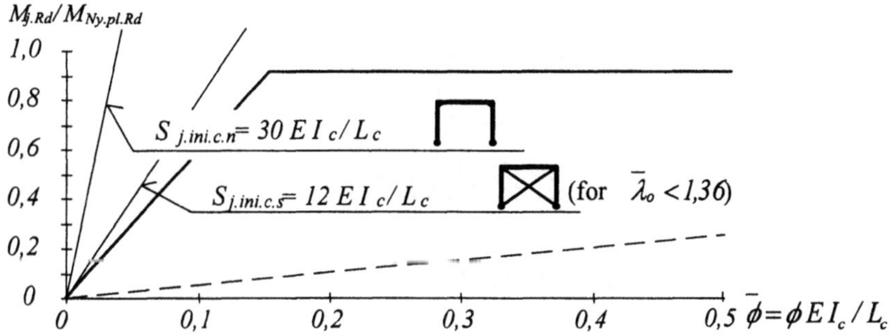


**Figure 56.** The column base classification.

- ✓ The influence of tolerances, see Eurocode 3, Annex L, and size of welds is not covered in the above calculation.
- ✓ A conservative simplification can be applied by placing the concrete reaction on the axes of compressed flange for resistance calculation.

# 8 Embedded Column Base

## 8.1 Introduction

Embedded column bases are frequently used in construction when the action of combined high bending moments and axial forces are present. They are often adopted in the steel frames structure design. Another example is to support windshield walls. However, existing design models for embedded column bases are quite different in practices from different countries. Previous research work has been mainly conducted in Japan (Akyiama (1984), Nakashima, (1994)), Germany (Koch et al. (1993), and Schleich and Witny (1999)) and France (Lescoure' (1988)). Recent experiments have provided extensive test data to derive a rational design approach. This includes tests conducted on vertical resistance of the column base and on the bending resistance. The proposed design model following the existing procedures is based on these test results and the numerical modelling validation. It gives a method to calculate the required length of column embedment in order to transfer bending moment, shear and vertical forces as shown in Fig.57. The model provides the calculation methods for shear resistance, axial capacity and bending moment for column bases. After extensive validation of experimental and numerical results it has been proved that this model is safe and relatively simple to use.

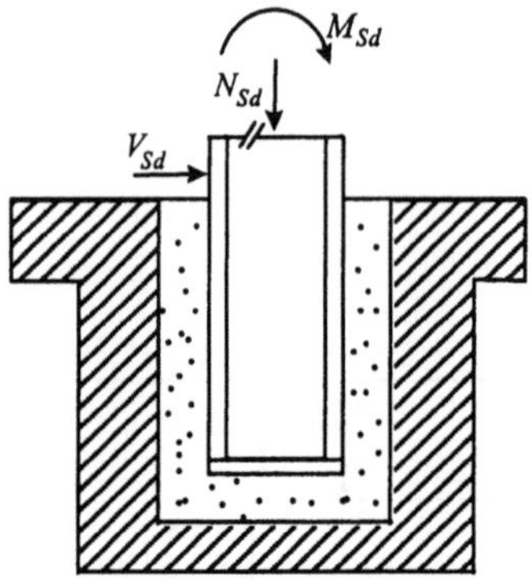


**Figure 57.** Embedded column base.

### 8.2 Moment Capacity

In the embedded column base design, the moment capacity calculation is very important. From the tests and numerical analyses the stress distribution pattern has been identified between steel column and concrete when subject to the combined shear force and bending moment. Simplification can be applied by using the rigid-plastic stress theory. Plastic stress distribution is assumed in the concrete along both flanges of steel column with certain criteria introduced. Involvement of both flanges in the column for stress transfer is measured by the effective column width $b_{eff}$. The concrete stress in horizontal direction in the analysis is set to 0,67 $f_{cd}$ according to numerical results (Pertold (1999)). This has taken into account the orientation of the principal stresses not being horizontal. The length of embedded steel column is limited in the range from $b_{eff}$ to 2 $b_{eff}$. This is the common practice in design.

The calculation of the effective width of the column flange $b_{eff}$ is described in the following formula. Involvement of the exterior and interior part of flanges has been considered in the stress transfer from the steel column to the concrete base. The effective width can be presented as

$$b_{eff} = min \begin{Bmatrix} b_c + 0{,}5\, h_c \\ 2b_c - t_{wc} \end{Bmatrix} . \tag{99}$$

The slenderness ratio of the steel column will be limited to

$$\frac{b_c}{t_{fc}} \leq 20 . \tag{100}$$

As discussed in the previous section, the length of embedment must fall in the range of two boundaries as shown here

$$b_{eff} \leq h_{emb} \leq 2b_{eff} , \tag{101}$$

where

$t_{fc}$ is the thickness of the column flange,
$t_{wc}$ is the thickness of the column web,
$h_c$ is the depth of the column section,

$b_c$ is the width of the column section,
$b_{eff}$ is the effective width of the column flange,
$h_{emb}$ is the embedment depth.

Following these assumptions the stress distribution in the concrete in contact with the column flanges can be simplified as shown in Fig.58. This has been validated by test results and numerical simulation.

Horizontal forces in the top position $F_t$ and in the bottom position $F_b$ transmitted from the column flanges to the concrete can be calculated (see Fig.58) as

$$F_t = 0{,}8 * 0{,}67\ f_{cd}\ b_{eff} \tag{102}$$

$$F_b = 0{,}8\left(h_{emb} - x\right)0{,}67\ f_{cd}\ b_{eff} \tag{103}$$

Establishing the horizontal equilibrium the position of the neutral axis can be derived as

$$x = \frac{0{,}93\ V_{Sd} + 0{,}5\ b_{eff}\ f_{cd}\ h_{emb}}{b_{eff}\ f_{cd}} \tag{104}$$

Taking the moments about the lower face of the column end, the following equation can be obtained

$$M_{Sd} + V_{Sd}\ h_{emb} + F_b\ 0{,}4\left(h_{emb} - x\right) - F_t\left(h_{emb} - 0{,}4\ x\right) = 0 \tag{105}$$

Combining equations (102), (103), (104) and (105), the embedded length $h_{emb}$ of the steel column can be calculated

$$h_{emb} = \frac{1{,}56\ V_{Sd} + \sqrt{4{,}74\ V_{Sd}^{\ 2} + 6{,}22\ M\ f_{cd}\ b_{eff}}}{b_{eff}\ f_{cd}} \tag{106}$$

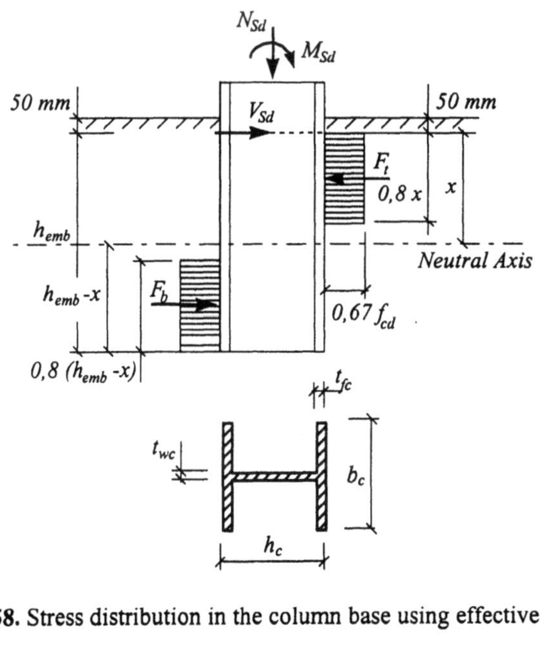


**Figure 58.** Stress distribution in the column base using effective width $b_{eff}$.

In the case that the actual depth of embedment in design is smaller than the calculated one, the maximum horizontal stress in the concrete will be smaller than $0,67 f_{cd}$. In this case the concrete will not govern the failure of the embedded column.

It must be noticed that it is necessary to check the moment resistance of the steel column at the same time. When a slender column is used, e.g. aspect ratio of $b_c/t_{fc}$ greater than $20$, the stress in the concrete in contact with the area of the column flanges outside the effective width will become nearly zero due to high flexibility of the column flange. This model assumes that stress is constantly distributed along the effective width. This has already been proved in numerical modelling (Pertold (1999)).

The concrete will be insignificantly stressed when the length of embedded column exceeds $2\ b_{eff}$. This is due to the flexibility of the whole column. When the length of embedment is smaller than $b_{eff}$, the steel column is going to be pulled out of the concrete base. This has been demonstrated in previous experimental work (Nakashima (1994)). Both flanges of the steel column will be involved in stress transfer from the steel column to the concrete base, as shown in the numerical simulation (Pertold (1999)). The ratio of the stress transferred by the two flanges is dependent on the strength of the concrete filling within the column void area. This condition can be quantified by the effective column width $b_{eff}$. This in turn, will be determined by the column section size.

### 8.3 Shear Resistance

If the shear collapse of column bases is going to be prevented, the shear resistance of the steel column and column base must be higher than the acting shear force. The shear force acting in the concrete base is illustrated in Fig.59. This shear force scheme has been evolved from the horizontal stress distribution in Fig.58. The column base is subjected to the combination of bending moment and horizontal force. Calculation of the shear capacity can be divided into two parts. First, the steel column part above the surface of the concrete base is taken into consideration. The design of this part will depend on the steel section alone. Secondly, the column part inside the concrete base is taken into consideration. Taking into account the effect of concrete filling will enhance steel column shear resistance.

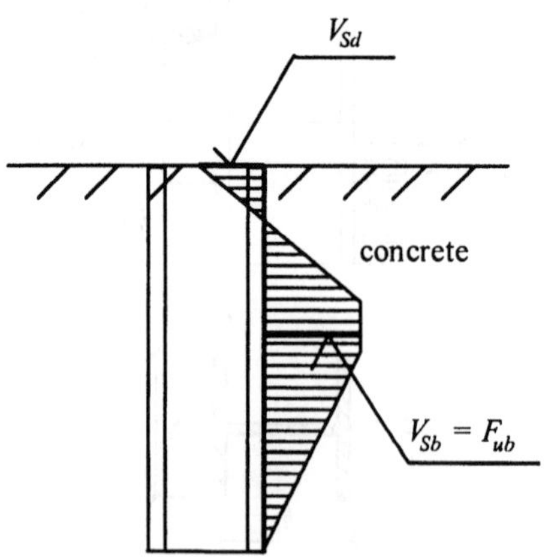


**Figure 59.** The shear force distribution in the concrete base.

The calculation of the shear resistance can be carried out in the following steps:

Steel column must transfer the maximum shear force $V_{Sd}$ acting above the surface of the concrete base

$$V_{Sd} \leq V_s \,, \tag{107}$$

where

$V_s$ is the plastic shear resistance of steel column,

$V_{Sd}$ is the maximum shear force acting above the surface of the concrete base.

Steel column with concrete filling has to transfer the maximum shear force $V_{Sb}$ acting inside the concrete base.

$$V_{Sb} \leq V_{sc} \,, \tag{108}$$

$$V_{sc} = V_s + V_c \,, \tag{109}$$

where

$V_c$ is the shear resistance of the concrete filling between the column flanges,

$V_{sc}$ is the shear resistance of steel column with concrete filling,

$V_{Sb}$ is the maximum shear force acting inside the concrete base.

The design shear capacity of the concrete filling $V_c$ will be given in the later formulation.

The concrete filling between column flanges must transfer the shear force $V_{Sbc}$ acting inside the concrete filling between column flanges

$$V_{Sbc} < V_c \,, \tag{110}$$

where

$V_{Sbc}$ is the maximum shear force acting inside the concrete filling between column flanges. Calculation of the maximum shear force inside the concrete filling $V_{Sbc}$ and the maximum shear force $V_{Sb}$ inside the concrete base is derived from stress distribution pattern in Figure 60. For the moment resistance calculation, the effective column width $b_{eff}$ has been used to describe the involvement of both flanges in stress transfer. For the calculation of maximum shear force $V_{Sbc}$ in the concrete filling, the stress distribution on both flanges is used.

It can be seen from Fig.59 that the maximum shear force $V_{Sb}$ acting in the column base equals to the value of $F_b$ (111). The maximum shear force $V_{Sbc}$ acting in the concrete filling equals to the value of $F_{b2}$

$$F_b = F_{b1} + F_{b2} \,. \tag{111}$$

$F_{b2}$ can be derived from the definition of the effective width $b_{eff}$ and from the Fig.60

$$F_{b2} = F_b \, \frac{b_{eff} - b_c}{b_{eff}} \,. \tag{112}$$

$F_{b2}$ can be written further (by using equations (103) and (112)) as

$$F_{b2} = 0{,}54 \left(h_{emb} - x\right) f_{cd} \left(b_{eff} - b_c\right) . \tag{113}$$

The shear capacity of the concrete filling between column flanges $V_c$ is derived from a truss model. This model is illustrated in Fig.60. Design parameters were adopted for shear capacity calculation of an embedded column base.

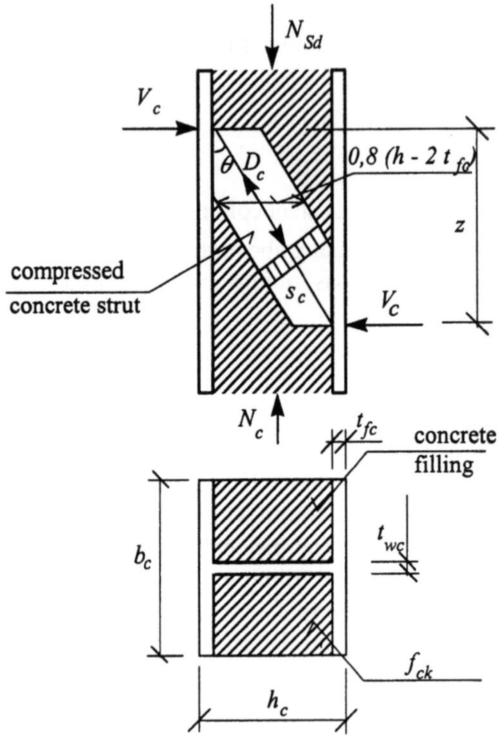


**Figure 60.** Truss model of concrete for shear resistance calculation.

The horizontal shear resistance $V_c$ can be represented by the axial capacity $D_c$ of the concrete strut in the truss model

$$V_c = D_c \sin\theta \ , \tag{114}$$

$$D_c = 0{,}468 \frac{f_{ck}}{\gamma_c} A_c \ , \tag{115}$$

where

$D_c$ is the force in the concrete strut,
$A_c$ is the area of the concrete strut,
$s_c$ is the stress in the concrete strut,
$\theta$ is the angle of the concrete strut,
z is the level arm of shear forces,
$\gamma_c$ is the partial safety factor for concrete,
$f_{ck}$ is the characteristic compressive cylinder strength of concrete at 28 days.

The effective concrete section area $A_c$ can be obtained as

$$A_c = \lfloor 0{,}8 \left(h_c - 2\, t_{fc}\right) \cos\theta \rfloor \left(b_c - t_{wc}\right) \ . \tag{116}$$

The angle of inclination of the concrete strut component is

$$\theta = arctan \frac{h_c - 2\, t_{fc}}{z} \quad . \tag{117}$$

The lever arm of the two shear forces is

$$z = 0{,}6\; h_{emb} \quad . \tag{118}$$

### 8.4 Axial Load Resistance

The experimental results (Pertold (1999)) have proved that significant axial load capacity can be produced by bond strength between a steel column and concrete base. It would be an ideal situation if dynamic and cyclic loading cases could be considered in the bond resistance calculation. However, much more experimental work is needed before bond resistance can be considered for cyclic and dynamic loading. Only static behaviour of the axial load resistance can be calculated for steel column by checking the collapse on the column section perimeter along the length $h_{emb}$ of embedded column, see Fig.58, or a combination of bond failure along the column flanges and shear failure of the concrete filling.

**Punching and crushing of steel column without base plate.** By using Eurocodes the punching and crushing strength of the concrete can be calculated. The behaviour of a concrete base in punching is similar to that of a concrete slab described in details in Eurocode 2 (1992). The contribution of the punching capacity of the concrete to the axial load resistance of the column base can be calculated as

$$V_{1.Rd} = \frac{u}{\beta}\left(v_{Rd1}, v_{Rd3}\right), \tag{119}$$

where

$\beta = 1$ is the load eccentricity coefficient,

$u$ is the perimeter of the critical section,

$v_{Rd1}$ is the design shear resistance per unit length of the critical perimeter, for a slab without shear reinforcement,

$v_{Rd3}$ is the design shear resistance per unit length of the critical perimeter, for a slab withshear reinforcement

To calculate $u$, $v_{Rd1}$ and $v_{Rd3}$, Eurocode 2 (1992), should be used. For the value of $d$ (average effective depth of the slab) the thickness of the concrete below the column end can be taken. However, this thicknes must not exceed half of the size of the concrete base.

Contribution of the crushing capacity to the axial load resistance of the column base can be calculated as

$$F_{ber1.Rd} = f_j\; A_c\,, \tag{120}$$

where

$f_j$ is the bearing strength of the concrete, Eurocode 2 (1992),

$A_c$ is the section area of the column.

**Punching and crushing of a steel column with base plate.** Contribution of the punching capacity to the axial load resistance of the column base can be calculated as

$$V_{2.Rd} = \frac{u_{bp}}{\gamma}\left(v_{Rd1}, v_{Rd3}\right) , \tag{121}$$

where

$\gamma = 1$ is the coefficient of load eccentricity,

$u_{bp}$ is the perimeter of the critical section for the column with base plate.

From the results of numerical simulations the bearing width of embedded base plates can be safely calculated based on bearing width for endplates which are connected to the base surface, see Chapter 3.

### 8.5 Design Sugesstions

From the proposed design model, it can be seen that design of embedded column base requires several steps. This involves assessing the strength and the geometry of the steel column and concrete base. Two suggestions are made to quickly select the material strength, the size of the column section and the length of embedment of the column.

- If concrete strength is higher than C 30, the embedded column base with length of embedment 2 $b_{eff}$ will transmit the full plastic moment capacity of the steel column with a grade of S 275.
- The embedded column base will not fail in shear if $h_{emb} = 2\ b_{eff}$ . The value of shear strength must be smaller than the plastic bending moment of the steel column with a grade of S 275 divided by 1m.

### 8.6 Validation of Design Model With Test Data

In this section the proposed design model is compared with experimental and numerical results. Reported experiments (Koch at al. (1993)) have been used to analyse the embedded column base collapse under the action of shear force and moment. Experiments from Part I of this paper have been used to analyse the collapse in bond, crushing and punching under the action of vertical load.

**Length of embedment governed by the shear force and bending moment.** The main design parameter for the column base is the embedment length. In design the bending moment and shear force must be considered together. Comparison of the actual length with the calculated one indicates a good agreement for different column sizes, as shown in Fig.61. The embedment length is calculated using equation (106). For comparison purposes, only tests that were due to the failure of the concrete were used. If in any test, failure of the steel column occurs, there is no need to check the embedment length. Instead designers will only need to check the steel section strength. This will not be discussed here.

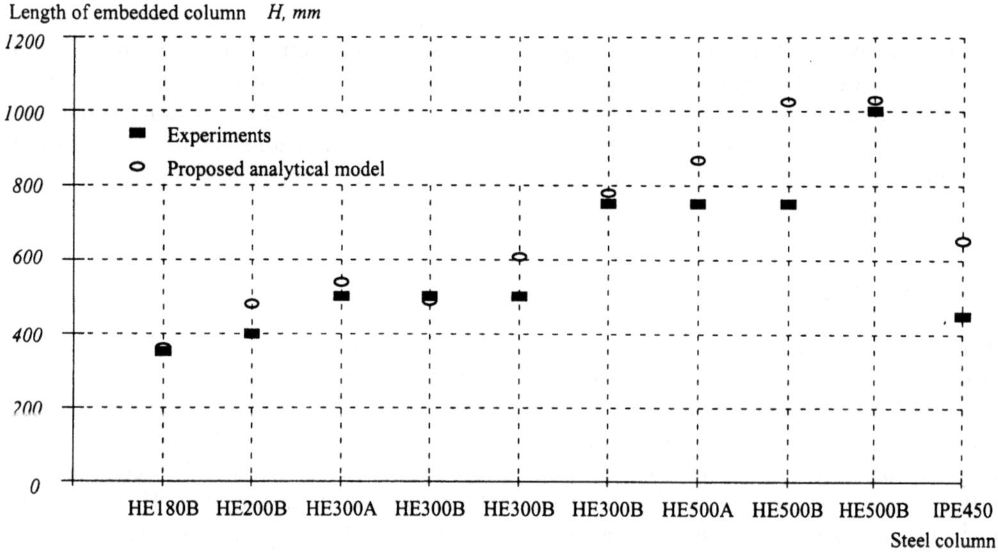


**Figure 61.** Actual and calculated length of column embedment.

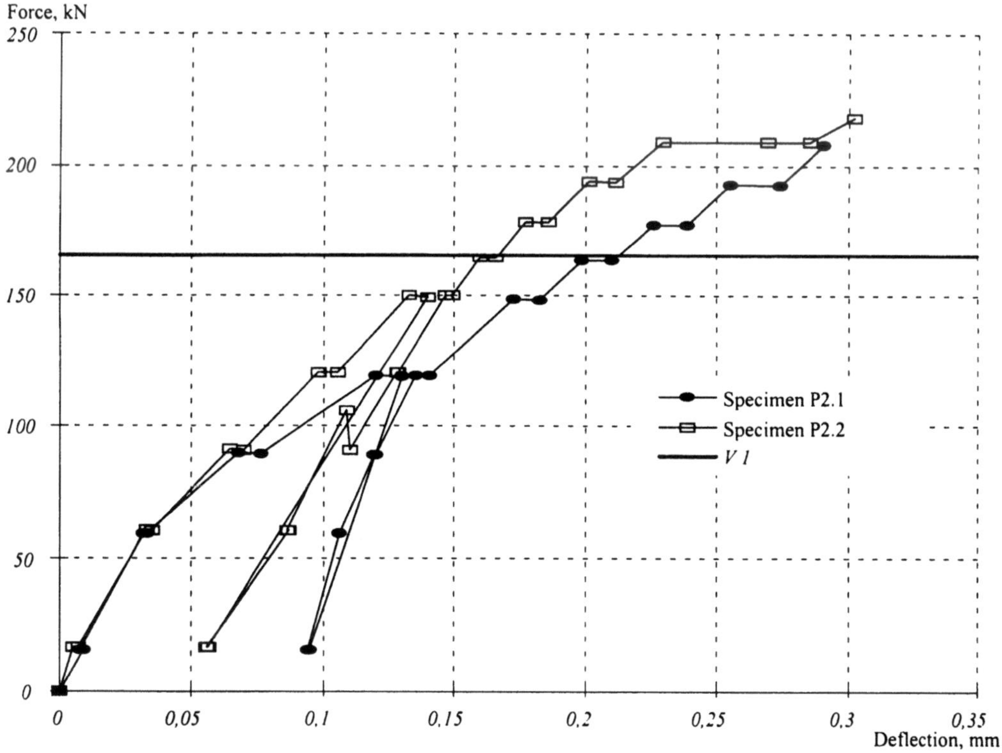


**Figure 62.** Comparison of calculated crushing and punching resistance with test results (Pertold (1998)).

**Comparison of crushing and punching with tests.** Comparison of the measured (Pertold (1999() and calculated resistance in crushing and punching is shown in Fig.62. The average calculated value of maximum load capacity in punching equals 166 kN. The maximum load capacity for crushing failure mode resulted as 734 kN. In fact failure occurred in punching. Crushing under the embedded column had not started at the moment of base failure.

## Acknowledgement

Within the framework of the activities of the COST C1 European Project and the Technical Committee 10 of ECCS, an ad-hoc working group was established to prepare a background document for European standardisation, and a European Manual for Column Bases with base plates. Members are: D. Brown, SCI London; A. M. Gresnigt, TU Delft; J. P. Jaspart, University of Liège; Z. Sokol, CTU in Prague, J. W. B. Stark, TU Delft; M. Steenhuis, TNO Delft; F. Wald, CTU in Prague (the convenor of the group), K. Weynand, RTWH Aachen. The author would like to express his thanks to all colleagues for help with preparation of all materials and for the support of Czech Republic grants COST C10 and J04/98:210000003.

## References


AISC (1978). *Specification for the Design, Fabrication and Erection of Structural Steel for Buildings*. New York: AISC.

Akiyama, H., Kurosawa, M., Wakuni, N., and Nishimura, I. (1984). Strength and deformation of column bases embedded in base concrete. *AIJ*, 335: 45-53.

Alma, J. G. J. and Bijlaard, F. S. K. (1980). *Berekening van kolomvoetplaten*. Rapport No. BI-80-46/63.4.3410. Delft: IBBC-TNO.

Bijlaard, F. S. K. (1982). *Rekenregels voor het ontwerpen van kolomvoetplaten en experimentele verificatie*. Rapport No. BI-81-51/63.4.3410. Delft: IBBC-TNO.

CEB - *Guide on the Design of Fastenings in Concrete* (1994) draft. Part 1 - Provisions. Part 2 - Characteristic Resistance of Fastening with Post - Installed Expansion- and Undercut Anchors. Part 3 - Characteristic Resistance of Fastening with Cast - in Place Headed Studs. August.

COST C1 (1999). *Column Bases in Steel Building Frames* (ed. Weynand K.), Brussels: E.U.

DeWolf, J. T. (1978). Axially loaded column base plates. *Journal of the Structural Division ASCE* 104: 781-794.

DeWolf, J. T. and Sarisley, E. F. (1980). Column base plates with axial loads and moments. *Journal of the Structural Division ASCE* 106: 2167-2185.

DeWolf, J.T. and Ricker, D.T. (1990). *Column Base Plates. Steel Design Guide - Series 1*. Chicago: AISC.

ENV 1090 (1996). *Execution of steel structures - Part 1, General rules and rules for buildings. European Prenorm*. Brussels: C.E.N.

Eurocode 2 (1992). ENV 1992-1-1, *Part 1.1, Design of Concrete Structures. European Prestandard*. European Committee for Standardization. Brussels: C.E.N.

Eurocode 3 (1992). ENV 1993-1-1, *Part 1.1, Design of Steel Structures. European Prestandard*. European Committee for Standardization. Brussels: C.E.N.

Eurocode 4 (1992). ENV 1994-1-1, *Part 1.1, Design of Composite Steel and Concrete Structures. European Prestandard*. European Committee for Standardization. Brussels:C.E.N.

Hawkins, N. M. (1968). The bearing strength of concrete loaded through rigid plates. *Magazine of Concrete Research*, 20: 31-40.

Hawkins, N. M. (1968). The bearing strength of concrete loaded through flexible plates. *Magazine of Concrete Research*, 20: 95-102.

Hogan, T. J. and Thomas, I. R. (1990). *Design of Structural Connections*. A.C.N. 000 973 839. Sydney: Australian Institute of Steel Construction.

Jaspart, J.P.(1991). *Etude de la Semi-rigidité des Noeuds Poutre-colonne et son Influence sur la Résistance et la Stabilité des Ossatures en Acier*. Ph.D. Thesis. Department MSM. Liège: University of Liège.

*Joints in Steel Construction, Moment Connections* (1995). SCI, BCSA, No.207/95: 233, ISBN 1 85942 018 4.

Koch, E., Mang, F., Schleich, J.B., Seiler, J. and Stiglat, K. (1993). Steel columns embedded in concrete foundation. *Fourth International Conference on Structural Failure, Durability and Retrofitting*. Singapore, 14-15 July 1993.

Lambe, T. W. and Whitman, R.V. (1969). *Soil Mechanics*. MIT. New York.: John Wiley & Sons, Inc.

Lescouarc, 'H Y, (1982). Les pieds de poteaux articules en acier. St.Rémy-lés-Chevreuse: *CTICM*, 163-170.

Lescouarc', H Y. (1988). Les pieds de poteaux encastres en acier. St.Rémy-lés-Chevreuse : *CTICM*, 220-228.

Murray, T. M. (1983). Design of lightly loaded steel column base plates. *Engineering Journal AISC*, 20: 143-152.

Nakashima, S. (1994). Steel column bases in Japan: Experimental research and design practice. In *Proceedings of the Second State of the Art Workshop COST C1* (ed. Wald F.). Brussels: 247 - 258.
Niyogi, S. K. (1973). Effect of side slopes on concrete bearing strength. *Journal of the Structural Division, ASCE* 104: 599-604.
Pertold, J.(1997). *Embedded Column Base*. Doctoral Thesis. VUT. Praha
Pertold, J., Wald, F., and Xiao, R. (1999). Embedded column base. *Journal of Structural Steel Research*. London (in print).
Pertold, J. (1999). *Embedded Column Base. Introduction of Structural Steel Eurocodes into the Czech and Austrian Praxes*. Praha: 59-71.
Salmon, C. G., Schenker, L., and Johnston, G. J. (1957). Moment - rotation characteristics of column anchorages. *Transaction ASCE*, 122: 132-154.
SBR (1973). *Mortelvoegen in montagebouw*. Rapport no. 34. Stichting Bouwreseach: Samson Alphen aan de Rijn.
Schleich, J. B. and Wiry, A. (1999): Steel Columns Embedded in Concrete Foundations, *EUR paper* No: 16057, Brussels.
SCI (1995). *Joints in Simple Construction. Practical Application*. BCSA, No.207/95:192.
Shelson, W. (1957). Bearing capacity of concrete. *Journal of the American Concrete Institute*, 29: 405-414.
Sokol, Z. and Wald, F. (1997). *Experiments with T-stubs in Tension and Compression*. Research Report. Praha: ČVUT.
Steenhuis, M. (1994). Pre - design of semi-rigid joint in steel frame. In *Proceeding of the Second State of the Art Workshop COST C1*. Praha: 131 - 140.
Stockwell, F. W. Jr. (1975). Preliminary base plate selection. *Engineering Journal AISC* 21: 92-99.
Treiberg, T. (1987). *Base plates* (in Swedish). Pub. 101, Oct. Stockholm: Staalbyggnadsinstitutet, 1-99.
Tschemmernegg, F. and Huber, G. (1994). Shear region on the panel zone of a composite Joint. *Technical paper T 3/4*. TU Insbruck.
Wald, F. and Seifert, J. (1991). The column Base stiffness classification. In *Nordic Steel Colloquium*. Odensee: 413-421.
Wald, F. (1993). *Column Base Connections. A Comprehensive State of the Art Review*. Praha: ČVUT.
Wald, F., Obata, M., Matsuura, S., and Goto, Y. (1993). Flexible base plates behaviour using FE strip analysis. *Acta Polytechnica* 33. Praha: ČVUT, 83-98.
Wald, F., Obata, M., Matsuura, S., and Goto Y. (1993). *Prying of Anchor Bolts*. Nagoya University Report: 241-249.
Wald, F. (1995). *Patky sloupů, Column Base*. Praha:ČVUT, 1-137.
Wald, F., Bauduffe, N., Sokol, Z., and Muzeau, J. P. (1996). Pre-design model of column base stiffness. In *Semi-Rigid Structural Connections IABSE. II.*:11.
Wald, F. and Sokol, Z. (1997). Column base stiffness classification. In *SDSS 97 - Colloquium Stability and Ductility of Steel Structures*. Nagoya: 675 - 682.
Wald, F. and Jaspart, J. P. (1998). Stiffness design of column bases. In *2nd World Conference on Steel on Construction*. San Sebastian: paper No. 135.
Zoetemeijer, P. (1985). *Summary of the Researches on Bolted Beam-to-Column Connections*. Report 6-85-7. Delft University of Technology.

# PART V

## NUMERICAL SIMULATION
### PRINCIPLES, METHODS AND MODELS

**C.C. Baniotopoulos**
**Aristotle University of Thessaloniki, Thessaloniki, Greece**

**Abstract:** Aim of the present chapter is to present certain basic principles, numerical techniques and algorithmic models applied to the simulation of the structural response of steel semi-rigid connections. For the numerical treatment of this problem, the Finite Element Method is applied. Nonlinearities introduced by unilateral contact and friction over the connection interfaces, as well as material nonlinearities including yielding have been taken into consideration in all of the herein presented models. Within this framework, a variational inequality and a quadratic programming approach have been applied to the unilateral contact problem in steel connections and the corresponding sensitivity analysis has been performed. Based on this methodology, an effective 2-D numerical model for steel connections including contact and plasticity is first presented and, in the sequel, a parametric analysis of the response of steel base plate connections is carried out. Mesh size, choice of finite elements and other parameters affecting the accuracy and the numerical behaviour of the models are also discussed. In the last paragraphs of this chapter a parametric analysis of the response of hollow section joints by means of an effective 3-D model has also been performed giving rise to interesting numerical results that describe with accuracy any possible failure mode of the joint under investigation.

## 1 Introduction

The simulation of semi-rigid connections by the Finite Element Method helps to understand the phenomena that determine the response of the connection. Parametric simulations give us the possibility to produce behaviour curves, which can be applied for the analysis of steel frames within a prismatic-members structures framework. Obviously, such phenomenological numerical simulation efforts always have to be supported by experimental testing, so that the numerical results may be compared to laboratory data for the calibration of the respective numerical technique.

The variety of structural configurations and quality in steel connections combined with the two major nonlinearities, i.e. unilateral contact (with or without friction) and plasticity, significantly complicate the numerical simulation process. As a matter of fact, during this effort a lot of difficulties arise and have to be tackled with utmost care so that the numerical response of the semi-rigid joints under investigation results close enough to their response as predicted by the theory, the existing formulas and any available laboratory tests. Such difficulties are, among others, the choice of the most effective analysis method, the choice of the finite elements for the various parts of the joint (cf. e.g. plates, bolts, welds), the preference of a dense or a sparse finite element mesh, the modelling of yielding, hardening and possibly softening in ductile and even cracking in brittle steel, the question of using 1-D, 2-D and 3-D models, the unilateral contact, friction and the gaps in the connection interfaces. In any case, as computational effort becomes continuously less expensive due to availability of advanced computational tools, i.e. modern software, large memory and innovative techniques,

Finite Element modelling and numerical experimentation constitute a reliable and less expensive alternative to laboratory testing.

The essence of modeling is to focus on the predominant structural function of the connection. When the latter is clear, the simulation can be simple enough to be carried out with prismatic elements (1-D formulation). In more complicated cases, folded shell elements have to be used. Although nowadays a plethora of convenient tools to tackle the simulation problem are available, experience with the software used is always sure to produce better results.

Concerning the types of finite element used in the numerical modelling of steel connections, shell and solid elements are the most common ones. In particular, although shell elements are almost always applied to produce the folded plate model for joints, there are also cases where the use of solid elements is inevitable (cf. e.g. modelling of bolts). The most common types of these solid elements are the 8-noded brick element having the advantage to easily simulate thickness variations in the various parts of a connection and the 10- and 20-noded brick element that both easily lead to nice results.

The material law for steel in the respective structural connections can obviously be assumed either within the elastic region or to include yielding. The nonlinear material behaviour that includes yielding and isotropic strain hardening can be seen as classic for structural steel used in connections and covers successfully most cases. As the connection is a compact structure where individual members are far from being slender, the usual stress situation is triaxial and therefore, the full 3-D material law has to be implemented.

In the investigated models, specific attention has been drawn to the modelling of two more factors that may significantly affect the numerical results: the boundary conditions of the connection at hand and the application of loading into the model. For instance, in the numerical experimentation a commonly used technique is to substitute the sequence of the applied loads by a sequence of imposed statically equivalent displacements so that a more stable algorithm to be applied, or more complicated mathematical formulations (cf. e.g. nonconvex energy optimization problems) to be avoided.

In the next paragraphs the basic principles, numerical techniques and algorithmic models applied to the simulation of the structural response of steel connections are presented. The paramount role of the phenomena that significantly affect the joint response, i.e. unilateral contact and friction, as well as material nonlinearities including yielding with isotropic strain hardening have been taken into account in all the herein presented models.

Within this finite element modelling framework, a variational inequality and a quadratic programming approach have been applied to the unilateral contact problem in steel connections and the corresponding sensitivity analysis has been performed. Based on this methodology, an effective 2-D numerical model for steel connections including contact and plasticity is first presented and, in the sequel, a parametric analysis of the response of steel base plate connections is carried out. Mesh size, choice of finite elements and other parameters significantly affecting the accuracy and the numerical behaviour of the models are also discussed. A parametric analysis of the response of hollow section joints by means of an effective 3-D model has also been performed. As a conclusion, it could be noticed that in the present chapter some modern methods and innovative techniques for the simulation of the structural behaviour of semi-rigid joints are presented. Comments on the correct way to interpret the obtained results from the numerical analysis complement the present chapter.

## 2 A Variational Inequality and a Quadratic Programming Approach to the Unilateral Contact Problem over Steel Connection Interfaces

A method is herein proposed for the investigation of the separation problem over bolted plates in steel connections. Applying the classic unilateral contact law, the continuous problem is first formulated as a variational inequality or as a quadratic optimization problem and then, the discrete problem is formulated as a quadratic optimization problem with inequality constraints. Two methods from the Theory of Optimization, in particular the Theil-van de Panne and the Hildreth-d'Esopo method are proposed for the numerical treatment of the formulated problem. The applicability and the effectiveness of the proposed method are illustrated by means of a numerical example.

### 2.1 Introduction

Bolted plates are used in steel structures to transmit internal forces between structural elements. Such connections applied for instance as column-base, column-to-column and beam-to-column connections are extensively used in any possible combination and variation in structural steelwork. It is therefore, obvious that any improvement on the analysis methods for such splices leads to improved principles concerning both design and safety criteria dictated by modern steel construction codes.

According to the classic methods used for the analysis of the response of bolted plates in steel connections, different deformability of connecting plates is in most cases disregarded, whereas complete contact between plates is assumed under any type of loading. The consequence of the latter hypothesis is that compression forces between contact fronts are absorbed by the plates, whereas tension forces are transmitted through the bolts (CECM (1978) and AISC (1981)). However, the latter assumption does not cover a plethora of cases where the flexibility of the steel connection plates combined to specific loading leads to the appearance of regions of separation between the plates. The evidence of both laboratory testing and construction practice certifies that steel bolted connection plates tend to separate under certain loading conditions and thus, nonactive contact regions between them are developed. The problem of explaining and analyzing the structural behavior of steel bolted or riveted joints has recently attracted the interest of numerous researchers who applied theoretical, numerical and experimental methods to investigate it (see Ballio and Mazzolani (1977), Bufler (1968), Bufler and Müßigmann (1977), Kato and McGuire (1973), Bogdan et al. (1977), Chen and Patel (1981), Chen and Lui (1986), Paker and Morris (1977), Hyer and Klang (1985), and Bortman and Szabó (1992)). Results of considerable importance on the problem under consideration have been recently obtained by means of the numerical simulation of the structural response of bolted plates in steel connections by means of the use of the Finite Element and the Boundary Element Method. These studies led to the ascertainment of the appearance of regions of separation between the connecting plates (see Raffa and Strona (1984), Thomopoulos (1985) and (1986), Thambiratnam and Krishnamurthy (1989), and Abdalla (1988)).

The separation phenomenon is of geometrically nonlinear nature, whereas regions of active and nonactive contact are not *a priori* known.Therefore, classic Structural Analysis methods for the investigation of this problem may lead to erroneous results and must therefore, be used with care. However, the methods of Nonsmooth Mechanics, a recently developed branch of Mechanics, can be successfully applied to the simulation of the separation problem of bolted steel splice plates because they permit the accurate mathematical and numerical description of the separation process and consequently lead to the better estimation of the structural behaviour of these connections. Indeed Nonsmooth Mechanics can account for the exact nature of the unilateral contact non-linearity which, as will be seen in the sequel possesses some traits of its own (Panagiotopoulos (1985), Moreau et al. (1988), and Moreau and Panagiotopoulos (1988)). Such a treatment of the problem exhibits among others the advantage of the exact determination of the active contact and separation zones between the steel plates, as well of the exact evaluation of the loss of strength of the bolted steel connection due to the development of the detachment phenomenon.

The mathematical description of the separation phenomenon of the steel plates is obtained by appropriately applying the unilateral contact law of Signorini first investigated by Fichera (1972). By means of this theory, the continuous problem constituting a Boundary Value Problem (B.V.P.), is first formulated as a variational inequality problem expressing the principle of virtual work of the structure in inequality form at the state of equilibrium (see Panagiotopoulos (1976)). The latter formulation permits the proof of the theorem of minimum potential energy at the state of equilibrium, thus leading to the formulation of an equivalent quadratic optimization problem that involves a quadratic energy function with inequality kinematical constraints. Applying an appropriate finite element discretization, the continuous problem is equivalently transformed into a discrete quadratic optimization problem; this formulation seems to be very promising since numerous quadratic programming algorithms are nowadays available for its effective numerical treatment. It is worth noting that a dual approach can also be employed: the variational inequality problem with respect to stresses expresses, from the standpoint of Mechanics, the principle of complementary virtual work and leads, equivalently, to the formulation of a quadratic optimization problem with inequality constraints which constitutes the mathematical expression of the theorem of minimum complementary energy at the state of equilibrium (cf. e.g. Panagiotopoulos (1976) and (1985)).

Concerning the numerical treatment of the aforementioned quadratic optimization problems, two methods from the theory of Quadratic Optimization and in particular, the Theil-van de Panne algorithm and the Hildreth-d'Esopo algorithm that is a Mangasarian type method are proposed to be employed (see Künzi and Krelle (1962)).

The first algorithm transforms the initial quadratic problem into an equivalent sequence of linear problems, which in turn can be solved by any available linear system solver. This method is not by any means a trial-and-error method since it provides criteria to the search for the optimal solution of the problem, as well a criterion for the final test of a seemingly correct solution concerning its optimality (for details, see Panagiotopoulos and Talaslidis (1980), Talaslidis and Panagiotopoulos (1982), Panagiotopoulos et al. (1984), Baniotopoulos and Abdalla (1993), and Baniotopoulos et al. (1992)).

The second algorithm is a typical iterative procedure that in many cases seems to be much more easily programmable and computationally efficient for the numerical treatment of such problems (see Abdalla and Stavroulakis (1989), and Abdalla and Baniotopoulos (1991)). The previous method and the range of its applicability and effectiveness are illustrated by means of a numerical application.

## 2.2 The Continuous Problem

Let us consider in an orthogonal Cartesian system $Ox_1x_2x_3$ an elastic body $\Omega$ with boundary $\Gamma$ made up of three non-overlapping parts $\Gamma_U$, $\Gamma_F$ and $\Gamma_S$. On $\Gamma_U$ (respectively $\Gamma_F$) the displacements (respectively the surface forces) have given values $U_i$ (respectively $F_i$), whereas on the part $\Gamma_S$, unilateral contact boundary conditions hold (Fig.1a). Assuming that the unilateral contact is frictionless and defining as positive normal direction this one directed inwards to the boundary, the unilateral contact conditions with respect to an inelastic support can be expressed in the following form

$$\text{if } u_N < 0 \quad \text{then} \quad S_N = 0 \tag{1}$$

$$\text{if } u_N \geq 0 \quad \text{then} \quad S_N + k(u_N) = 0 \tag{2}$$

where $u_N$ (respectively $S_N$) denotes the normal (with respect to the boundary) displacements (respectively reaction forces) on $\Gamma_S$ and k(.) is a non-decreasing function. The above relations are illustrated in Fig.1b, whereas Fig.1c and Fig.1d correspond respectively to unilateral contact with a linearly elastic and rigid support (for details on the mathematical nature of these diagrams see e.g. Fichera (1972), Panagiotopoulos (1976) and (1985), Baniotopoulos et al. (1992), and Baniotopoulos (1994)).

Under the assumption of small strains and displacements, the Boundary Value Problem under consideration consists of the equation of equilibrium, the compatibility relations, the constitutive

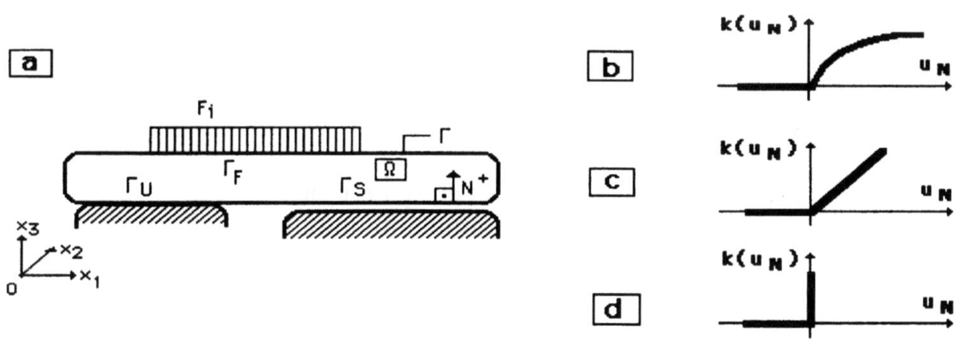


**Figure 1.** On the continuous problem and the forms of function k(.).

relations and the boundary conditions holding on $\Gamma$. We define a field $X^*$ of strains and displacements $\varepsilon_{ij}^*$, $u_i^*$ to be kinematically admissible if it satisfies the compatibility relations, the kinematical boundary conditions on $\Gamma_U$ and the kinematical conditions imposed by the boundary conditions holding on $\Gamma_S$. Volume forces are denoted by $p_i$, whereas actual strains and displacements at the position of equilibrium are denoted by $\varepsilon_{ij}$ and $u_i$ respectively, and thus, the differences $(\varepsilon_{ij}^* - \varepsilon_{ij})$ and $(u_i^* - u_i)$ represent the kinematically admissible variations of strains and displacements.

Moreover, $\sigma_{ij}^*$ denotes the stress field obtained from $\varepsilon_{ij}^*$ by means of the elasticity law. By splitting now $u_N^*$ into its positive and negative parts $u_{N+}^*$ and $u_{N-}^*$ defined by the forms

$$u_{N+}^* = ( u_N^* + |u_N^*| ) / 2 \geq 0 \tag{3}$$

and

$$u_{N-}^* = ( - u_N^* + |u_N^*| ) / 2 \geq 0 \tag{4}$$

the identity

$$\int_\Omega \sigma_{ij}^* (\varepsilon_{ij}^* - \varepsilon_{ij})\, d\Omega = \int_\Omega p_i (u_i^* - u_i)\, d\Omega + \int_{\Gamma_S} S_{Ni}(u_{Ni}^* - u_{Ni})\, d\Gamma + \int_{\Gamma_F} F_i (u_i^* - u_i)\, d\Gamma \qquad \forall\, \varepsilon_{ij}^*, u_i^* \in X^* \tag{5}$$

combined to the inequality

$$\int_{\Gamma_S} [S_{Ni} (u_{Ni}^* - u_{Ni}) + k(u_{N+}) (u_{N+}^* - u_{N+}) ]d\Gamma \geq 0 \qquad \forall u_{Ni}^* \in X^* \tag{6}$$

holding on $\Gamma_S$ yields by means of the boundary conditions on $\Gamma_F$ the variational inequality

$$\int_\Omega \sigma_{ij}^* (\varepsilon_{ij}^* - \varepsilon_{ij})\, d\Omega - \int_\Omega p_i (u_i^* - u_i)\, d\Omega + \int_{\Gamma_S} k(u_{N+})(u_{N+}^* - u_{N+})\, d\Gamma - \int_{\Gamma_F} F_i (u_i^* - u_i)\, d\Gamma \geq 0 \qquad \forall\, \varepsilon_{ij}^*, u_i^* \in X^* \tag{7}$$

By using the method of special variations it is proved that variational inequality (7) yields the equation of equilibrium and the boundary conditions on $\Gamma_S$ and $\Gamma_F$, and thus it characterizes completely the position of equilibrium.

Inequality (7) expresses from the standpoint of Mechanics the principle of virtual work in its inequality form for the body under consideration. It has been proved that any solution of the variational inequality problem (7) minimizes over $X^*$ at the position of equilibrium the potential energy of the structure

$$\Pi = {}^1/_2 \int_\Omega \sigma_{ij}\, \varepsilon_{ij}\, d\Omega - \int_\Omega p_i\, u_i\, d\Omega + \int_{\Gamma_S} K(u_{N+})\, d\Gamma - \int_{\Gamma_F} F_i\, u_i\, d\Gamma \tag{8}$$

where K(.) is a convex - due to the monotonicity of k(.) - function defined by the following integral

$$K(\xi) = \int_0^\xi k(\xi)\, d\xi. \tag{9}$$

Conversely, it has been proved that any solution of the previous minimization problem (8) satisfies the variational inequality problem (7) (cf. e.g. Panagiotopoulos (1976) and (1985)). Note also that a dual approach with respect to stresses and forces can be also equivalently applied. In the latter case, a variational inequality problem expressing the principle of complementary virtual work in inequality form is formulated equivalently giving rise to a minimization problem of the complementary energy of the body $\Omega$.

## 2.3 The Discrete Problem

The present paragraph deals with the mathematical description of the separation problem of bolted steel plates in discrete form by applying the previously presented theory of unilateral contact problems.

To start with, a typical steel column-to-column connection formed by bolted steel endplates is first considered (Fig.2). In this connection, bolts are located in single lines parallel to the columns sides. Due to the symmetry of the splice, bolts are equally stressed in each direction under single axial loading, whereas under certain external loading conditions a reduction of the contact pressure on certain regions of the splice plates combined to additional forces depending on the stiffness of the splice may appear on the plate tips (see, Abdalla and Stavroulakis (1989), Baniotopoulos (1994) and (1995), and Abdalla and Baniotopoulos (1993)).

In the case of rather flexible flanges, prying forces appear on the splice plates, reducing in this sense the contact forces on certain regions between the splice plates (cf. e.g. Fisher and Struik (1978)). The development of the latter phenomenon greatly affects the overall strength of the connection under consideration. It is therefore, obvious, that in order to better estimate the response of this connection, the regions of possible detachment have to be with accuracy defined

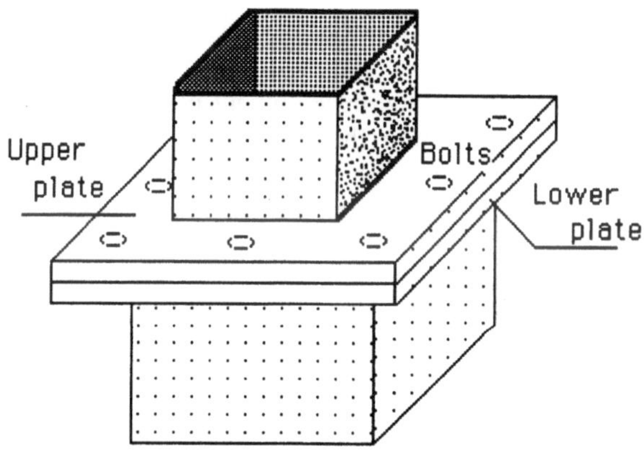


**Figure 2.** A typical bolted steel plate connection.

taking into account that on the active contact regions of the plates, contact forces appear, whereas on the noncontact regions, reaction forces are equal to zero.

The steel connection at hand is now discretized by means of an appropriately chosen finite element scheme. In particular, plate elements are used to simulate the behaviour of the splice plates, whereas the detachment conditions holding on the adjacent fronts of the splice plates are realized by means of one-dimensional elastic couplers (fictitious springs of length tending to zero) connecting the adjacent nodes of the two discretized splice plates. The mechanical behaviour of these couplers simulating the possibility of partial separation of the adjacent nodes of the plates can be mathematically described for instance for the i-th element by means of the following law

$$\text{if } [u_z(i)]>0 \text{ then } R_z(i)=0 \tag{10}$$

$$\text{if } [u_z(i)]=0 \text{ then } R_z(i)\geq 0 \tag{11}$$

where $[u_z(i)]$ denotes the relative displacement along the z-axis between the two endplates and $R_z(i)$ the respective reaction force. The first relation expresses that if separation occurs, no reaction forces appear on the plates, whereas the second one certifies that on the contact regions reaction forces do appear (frictionless type of contact). Assuming now that one of the plates can be rigidified having displacements along the z-axis equal to zero, the above relations (10)-(11) can be written in the following form

$$\text{if } u_z(i)>0 \text{ then } R_z(i)=0 \tag{12}$$

$$\text{if } u_z(i)=0 \text{ then } R_z(i)\geq 0 \tag{13}$$

where $u_z(i)$ denotes the displacements of the flexible splice plate along the z-axis in the neighbourhood of the i-th coupler. Assembling the above relations in matrix terms for all the n imposed couplers, the following Linear Complementarity Problem (L.C.P.) is formulated

$$\mathbf{u}_z\geq\mathbf{0},\quad \mathbf{R}_z\geq\mathbf{0}, \tag{14}$$

$$\mathbf{u}_z^T\,\mathbf{R}_z=0 \tag{15}$$

where $\mathbf{0}$ denotes a vector with zero all its entries and superscript $^T$ denotes transpose vectors and matrices. The previously formulated problem (14)-(15) completely describes mathematically in a quasi-static way the development of the separation phenomena over the steel plates. Taking into consideration the experience gained by laboratory testing data certifying that deformations are small on the splice plates (see Vogt (1947)), the classic stiffness method is next applied giving rise to the following matrix equation

$$\mathbf{K}\,\mathbf{u}=\mathbf{P} \tag{16}$$

where $\mathbf{K}$ is the stiffness matrix of the structure, $\mathbf{u}$ the displacement vector for the whole discretized structure including vector $\mathbf{u}_z$ and $\mathbf{P}$ the external load vector. The previous problem can be equivalently put in the following quadratic optimization form

$$\Pi(\mathbf{u})=\min\{{}^1/_2\mathbf{u}^T\mathbf{K}\,\mathbf{u}-\mathbf{P}^T\mathbf{u} \mid \mathbf{A}\,\mathbf{u}\leq\mathbf{b}\} \tag{17}$$

where $\mathbf{A}$ is an appropriately chosen transformation matrix and $\mathbf{b}$ a vector which defines the unilateral contact conditions (14)-(15) holding on the adjacent fronts of the splice plates.

The latter quadratic optimization problem expresses from an engineering point of view the principle of minimum potential energy of the steel connection at the state of equilibrium. The actual displacements of the steel plates caused by the combined external loading, as well as the active contact and separation regions over them can be with accuracy defined by solving the previously formulated problem (17). As has been previously pointed out for the continuous problem, a dual approach can be as well employed for the discrete problem giving rise to a quadratic programming problem of the same type where in that case, stresses are the unknown variables appearing in the quadratic term instead displacements and the constraints concern the equilibrium equation and the reaction forces appearing on the plates, i.e.

$$\Pi^c(\mathbf{s})=\min\{{}^1/_2\,\mathbf{s}^T\mathbf{F}_0\,\mathbf{s}-\mathbf{s}^T\mathbf{e}_0 \mid \mathbf{G}\,\mathbf{s}=\mathbf{P},\ \mathbf{R}_z\geq\mathbf{0}\} \tag{18}$$

where $\mathbf{G}$ (respectively $\mathbf{F}_0$) is the equilibrium (respectively flexibility) matrix of the steel connection and $\mathbf{s}$ (resp. $\mathbf{e}_0$) is the stress (resp. initial strain) vector. The latter formulation (problem (18)) expresses from the standpoint of engineering the principle of minimum complementary energy for the steel connection at the state of equilibrium.

## 2.4 On the Algorithmic Treatment of the Analysis Problem

In the following paragraphs two methods are proposed for the numerical treatment of the previously formulated problems (17) or (18).

The first method is known as Theil-van de Panne method and is based on three theorems from the theory of Quadratic Optimization: it transforms the initial quadratic programming problem into an equivalent sequence of linear problems which in turn, can be solved by any available linear system solver. This method provides criteria to the search for the optimal solution of the problem and a criterion for the final optimality test of a seemingly correct solution.

The second applied method is the Hildreth-d'Esopo algorithm, which is a typical iterative procedure and exhibits the advantage to be easily programmable and computationally efficient for the numerical treatment of the formulated problems.

**The Theil - van de Panne Algorithm.** A solution strategy to the search for the solution of problem (17) will be employed. The term "active contact region" of the steel plates denotes a set of the previously mentioned couplers with zero displacements along the vertical z-axis, whereas the term "nonactive contact regions" or "regions of separation" correspond to those couplers where displacements along the z-axis are greater than zero.

The following propositions proposed by Theil-van de Panne (see Künzi and Krelle (1962), Panagiotopoulos and Talaslidis (1980), and Talaslidis and Panagiotopoulos (1982)) are used as basis for the solution procedure:

<u>Proposition 1</u>

Let $V_1(\hat{\mathbf{u}})$ denotes the set of those inequality restrictions that correspond to the active contact regions, i.e.

$$\hat{u}(i)=0 \text{ for } i\in V_1(\hat{\mathbf{u}}) \tag{19}$$

$$\hat{u}(i)<0 \text{ for } i\notin V_1(\hat{\mathbf{u}})\,. \tag{20}$$

Then it holds that $\hat{\mathbf{u}}=\mathbf{u}^{V_1(\hat{u})}$.

If the solution $\mathbf{u}^0$ of the unrestricted problem violates some of the inequality constraints of the problem, then the actual (optimal) solution $\hat{\mathbf{u}}$ must satisfy at least one of these inequality restrictions in equality form.

<u>Proposition 2</u>

If $\hat{\mathbf{u}}=\mathbf{u}^{V_1(\hat{u})}$ and $V_1(\hat{u})\neq\varnothing$, then it holds that for all subsets $V\subset V_1(\hat{u})$ (including $V=\varnothing$) that $h\in W(\mathbf{u}^V)$ for at least one $h\in V_1(\hat{\mathbf{u}})\text{-}V$. Here W denotes the set of constraints satisfied either as greater or less than zero.

If a solution $\mathbf{u}^V$ of an intermediate problem violates some inequality constraints then, at least one of these restrictions must also be satisfied as equality in order to obtain the actual solution.

Proposition 3

If $W(\mathbf{u}^V)=\varnothing$ and $h\in W(\mathbf{u}^{V-\{h\}})$ for all $h\in V$, then $\mathbf{u}^V=\hat{\mathbf{u}}$.

Here $\mathbf{u}^{V-\{h\}}$ denotes the solution of the problem where superscript $(V-\{h\})$ corresponds to those restrictions that are satisfied in equality form. Proposition 3 can be interpreted as follows: A solution $\mathbf{u}^V$ satisfying all restrictions of the initial problem coincides with the actual solution $\hat{\mathbf{u}}$, if and only if, for every $h\in V$, the subsequent solutions $\mathbf{u}^{V-\{h\}}$ violate the restriction h omitted in the solution $\mathbf{u}^V$.

The solution method which is based on the previously presented propositions, is now sketched. The unrestricted solution $\mathbf{u}^0$ of problem (17) is first calculated. If $\mathbf{u}^0$ violates certain inequality constraints, then at least one of these restrictions has to be satisfied in equality form in the actual (optimal) solution $\hat{\mathbf{u}}$. In the sequel, if the solution of problem (17) violates certain inequality constraints, then, in order to obtain the actual solution of the problem at hand, we must consider in addition to the holding equality constraints, at least one more of the constraints that violates the intermediate solution $\mathbf{u}^V$. The solution of such a problem that fulfills the inequality constraints, coincides with the actual solution $\hat{\mathbf{u}}$ of the problem, if and only if, the solution of each problem $\Pi(\mathbf{u}^{V-\{h\}})$ violates the h-th inequality constraint for every h that belongs to V. The employment of the previous propositions leads to the solution of the quadratic programming problem by solving iteratively a sequence of linear problems, i.e. by appropriately combining the three previously presented theorems, the quadratic programming problem is linearized in the sense that within each solution step a linear problem has to be solved by means of the classic methods of Structural Analysis. Note that the iterative steps are not arbitrarily chosen, but they are controlled by the prescribed criterion defined by proposition 2. A final optimality criterion defined by proposition 3 also exists indicating whether an intermediate seemingly correct solution is the actual (optimal) solution of the problem, ensuring in this sense the convergence into a small number of iterative steps. The present method cannot by any means be characterized as a trial-and-error method, due to the existence of the latter optimality criterion (for details on the application of this method see Panagiotopoulos and Talaslidis (1980), Talaslidis and Panagiotopoulos (1982), and Panagiotopoulos et al. (1984)).

**The Hildreth-d'Esopo algorithm.** The application of the Hildreth-d'Esopo algorithm is in the sequel sketched for the solution of the previously formulated problem (17). As is well known, the Kuhn-Tucker optimality conditions for the quadratic programming problem (17) can be written in the following form

$$\mathbf{A}\,\mathbf{u}+\mathbf{y}=\mathbf{b}\ , \tag{21}$$

$$\mathbf{K}\,\mathbf{u}+\mathbf{A}^T\boldsymbol{\lambda}=\mathbf{P}\ , \tag{22}$$

$$\mathbf{y}\geq\mathbf{0}\ ,\ \boldsymbol{\lambda}\geq\mathbf{0}\ , \tag{23}$$

$$\mathbf{y}^T\boldsymbol{\lambda}=\mathbf{0} \tag{24}$$

where $\mathbf{y}$ is a vector corresponding to the unilateral constraints of the problem and $\boldsymbol{\lambda}$ the vector of reactions on the same constraints. Solving equation (22) with respect to $\mathbf{u}$, we obtain

$$\mathbf{u} = -\,\mathbf{K}^{-1}(\mathbf{A}^T\boldsymbol{\lambda} - \mathbf{P}) \quad (25)$$

and then by putting

$$\mathbf{h} = -\,\mathbf{A}\,\mathbf{K}^{-1}\mathbf{P} + \mathbf{b} \quad (26)$$

and

$$\mathbf{F} = {}^1/_2\,\mathbf{A}\,\mathbf{K}^{-1}\,\mathbf{A}^T \quad (27)$$

relations (21) - (24) are put in the form

$$2\,\mathbf{F}\,\boldsymbol{\lambda} - \mathbf{y} = -\,\mathbf{h}\ , \quad (28)$$

$$\mathbf{y} \geq \mathbf{0}\ ,\ \boldsymbol{\lambda} \geq \mathbf{0}\ , \quad (29)$$

$$\mathbf{y}^T\,\boldsymbol{\lambda} = \mathbf{0}\ . \quad (30)$$

The latter relations constitute the Kuhn-Tucker optimality conditions for the following quadratic optimization problem

$$\Pi(\boldsymbol{\lambda}) = \min\ \{{}^1/_2\,\boldsymbol{\lambda}^T\,\mathbf{F}\,\boldsymbol{\lambda} + \mathbf{h}^T\,\boldsymbol{\lambda}\ \mid\ \boldsymbol{\lambda} \geq \mathbf{0}\} \quad (31)$$

where matrix $\mathbf{F}$ (cf. relation (27) ) is a flexibility matrix relating contact forces to the corresponding unilateral contact displacements. When the solution of problem (17) exists, then problem (31) does also have a unique solution (Künzi and Krelle (1962)). Problem (31) can be numerically treated by means of the Gauß-Seidel iteration method by taking the values

$$\lambda_i^{p+1} = \max\ \{\ 0\ ,\ \omega_i^{p+1}\ \} \quad (32)$$

where

$$\omega_i^{p+1} = -\,{}^1/_{g_{ii}}\,\Big(\sum_{j=1}^{i-1} g_{ii}\,\lambda_j^{p+1} + \sum_{j=j+1}^{m} g_{ii}\,\lambda_j^{p+1}\Big) \qquad i=1,2,\dots m \quad (33)$$

where m is the number of linear constraints of the problem.

Iterations stop as son as computed contact reactions pass the imposed accuracy, i.e.

$$\|\ \lambda_j^{p} - \lambda_j^{p+1}\ \| < \varepsilon \quad (34)$$

where symbol $\|\ .\ \|$ denotes an appropriately chosen norm.

## 2.5 Numerical Application

The steel column-to-column splice of Fig.3 subjected to tension has been investigated. The dimensions of the upper and lower column are 120x120x5.6 (mm), whereas the dimensions of the upper and lower splice plate are 300x300xt and 300x300xt (mm) with t=12,15,18 (mm).

The discretized model obviously depends on the effectiveness and the facilities of the finite element package that was used. A revised edition of the program SAP IV-B has been here employed. The use of a powerful pre- and post-processor makes simpler the input data loading and the checking of the output data through graphics. Plate elements and couplers have been used to simulate the response of the steel connection. Due to the double symmetry of the splice, the solution procedure was limited to the one fourth of the whole splice region, whereas the same discretization scheme was adopted for both the upper and the lower endplate. The one fourth of each plate was discretized by means of 196 plate elements where appropriate condensation (mesh refinement) was made where needed (cf. e.g. neighbourhood of columns and bolts). Note that in order to have a good accuracy, the factor between plate sides must not be greater than 2 and not less than 0.5. Having assumed that the contact is frictionless, the respective degrees of freedom of the plate elements have been put equal to zero, so that the element behaves completely as a plate and movements along the plate level are avoided. In par-

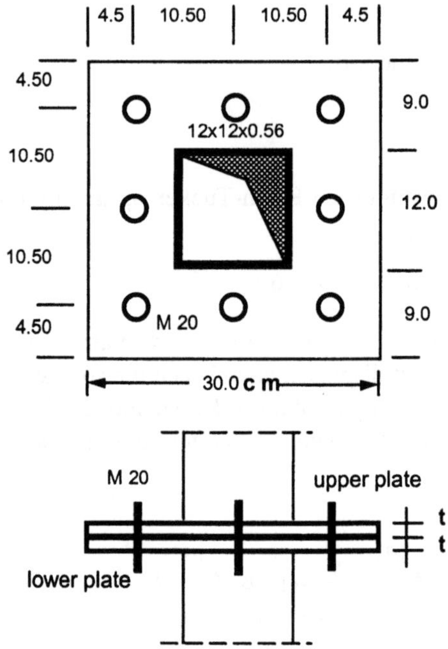


**Figure 3.** On the numerical application.

ticular, the plate element 6 of SAP IV having 6 degrees of freedom has been employed: 2 degrees of freedom have been restricted (movement along the plate level and rotation around z-axis). Each node of the upper plate was connected to the respective node of the lower plate by a coupler having length of 1 mm and of cross-section the $^1/_6$ of the bolts. Couplers corresponding to the bolts have the elasticity characteristics of steel, whereas couplers simulating the possibility of separation, 30 times greater.

The steel plate connection at hand is subjected to external tension force. The one fourth of the discretized steel connection is composed of 392 plate elements and 225 couplers, which correspond to 1259 unknown variables. The solution procedure starts with the assembling of the stiffness matrix of the discretized steel splice. Then, the potential energy of the structure in quadratic form is formulated. To begin with, the bilateral problem, i.e. the problem with no inequality constraints, is first solved. The violated restrictions are imposed in the next step by equality constraints with respect to vertical displacements of the plates. The same procedure is continued till no inequality restrictions are violated. Then, a seemingly correct intermediate solution of the problem has been obtained.

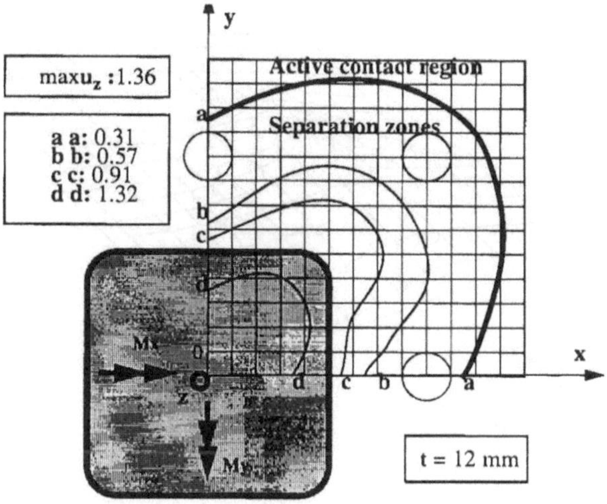


**Figure 4.** The displacement contours of the upper plate for plate thickness t=12 mm.

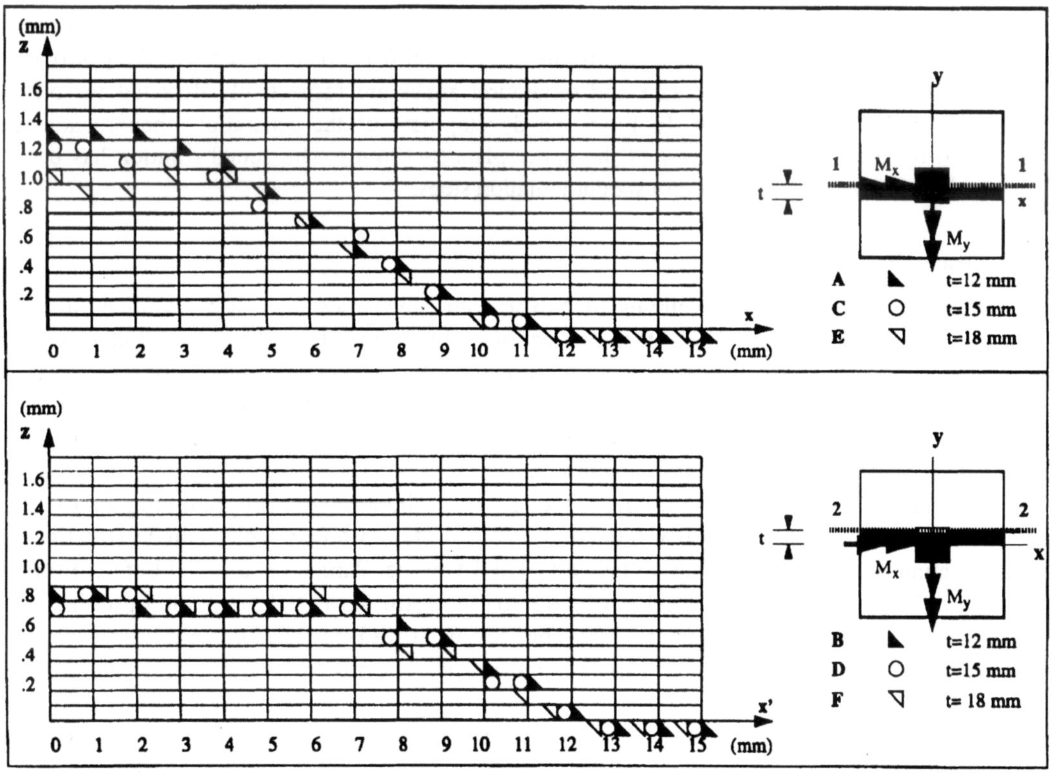


**Figure 5.** Deformation graphs for plate thickness t=12, 15,18 mm along sections 1-1 and 2-2.

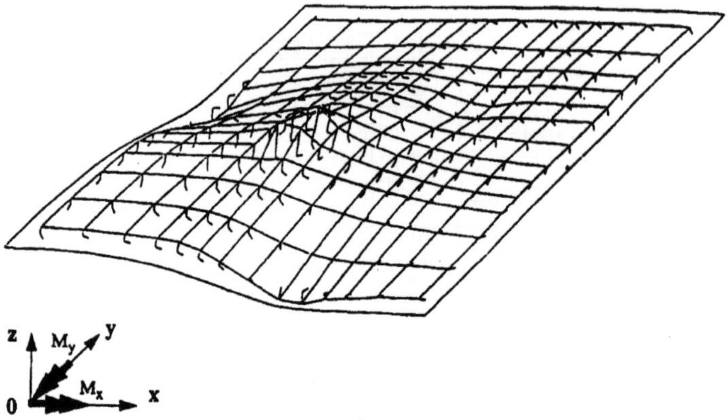


**Figure 6.** Deformed shape of the upper base plate for thickness t=12 mm.

This solution has to be checked to fulfill the optimality criterion of the Theil and van de Panne method (proposition 3). The constraints imposed as equalities in the latter solution must be eliminated one-by-one, and then, the respective structure must be solved. The respective results must violate at least the same restriction that had been eliminated in the same solution step. If this holds, then the actual (optimal) solution of the problem has been obtained. In Fig.4 the upper plate displacement contours in the actual (optimal) solution $\hat{\mathbf{u}}$ for plate thickness t=12 (mm) are depicted. In Fig.5 the active contact and separation zones are respectively depicted for plate thickness t=12, 15 and 18 mm. In Fig.6 the development of the separation phenomenon is illustrated by means of a 3-D view of the upper plate (t=12mm). The actual solution has been obtained in all investigated cases into six computation steps.

## 2.6 Discussion

In the above paragraphs the methods of Nonsmooth Mechanics have been applied for the analysis and calculation of bolted steel connections where separation phenomena between the steel connection endplates appear. Such a treatment of the problem exhibits among others the advantage of the exact determination of the active contact and separation regions between the connection plates, as well of the exact evaluation of the loss of strength of the connection due to the development of the separation phenomenon.

By means of the presented variational inequality and the quadratic programming approach, the problem at hand is formulated as a quadratic optimization problem which can be effectively treated numerically by applying either the Theil-van de Panne or the Hildreth-d'Esopo algorithm. The former has the great advantage to transform the initial quadratic optimization method into a sequence of linear problems that can be solved by any available linear system solver, whereas the latter - being easily programmable - seems to be more computationally efficient.

As a conclusive remark it can be stated that the presented theory and the numerical results obtained thereof which exhibit conformance to experience obtained through laboratory testing, can be considered as a contribution, on the rational side, to the set of ideas aiming to the improvement of modern structural steelwork codes.

# 3 Sensitivity Analysis of the Contact Problem over Steel Bolted Plates

The present section deals with the sensitivity analysis of bolted plates in steel connections by taking into account unilateral contact effects on the plates. The methods of Nonsmooth Mechanics combined to the Finite Element Method are herein applied in order to obtain a more realistic simulation of the structural behaviour of bolted plates in steel structures. As previously mentioned, the exact calculation of this type of connections can be obtained by minimizing a quadratic programming problem expressing from the standpoint of Mechanics the potential energy of the steel connection. In order to investigate the variation of the structural response of this type of steel connections due to variation of critical design parameters, the sensitivity analysis problem of the previous quadratic programming problem is formulated and studied. The latter problem is as well a quadratic programming one, where design parameters appear only in the quadratic term. The numerical treatment of the sensitivity analysis problem is obtained by applying some appropriately chosen quadratic optimization algorithm as is e.g. the Hildreth-d'Esopo algorithm. Sensitivity analysis results for the design of two steel column splices illustrating the previous theoretical results are presented in the last part of this section.

## 3.1 Introduction

The variation of the mechanical behaviour of steel bolted plates due to variation of critical design parameters is herein investigated. Such design parameters are, for instance, plate thickness, geometry and material of the steel connections. The great significance of the sensitivity analysis results of the present problem arises from the fact that column splices are extensively applied in any possible combination in structural steelwork. As is therefore obvious, since local failure phenomena on such connections due to undesirable variation of design parameters may cause a total destruction of the respective steel structure, it is important, such a behaviour to be with accuracy predicted through a sensitivity analysis of the respective design. In this sense, such investigations lead to an amelioration of the design principles of bolted column connections and to a refinement of the steel construction standards.

As previously stated, the analysis problem of bolted plates in steel connections taking into consideration the possibility of separation between them, is formulated as a quadratic programming problem (cf. e.g. problem (17) or (18)). In order to obtain sensitivity analysis results of the structural response of these joints, following the method proposed by Bendsoe et al. (1985) directional derivatives with respect to variations of the variables appearing in the quadratic term have to be defined. It worths here to note that, according to the method proposed by Bendsoe and Sokolowki (1988), this problem is similar to the sensitivity elastoplastic problem with the only difference that in the latter, design parameters enter in both the quadratic and the linear term.

The present study deals with the sensitivity analysis of the previously mentioned quadratic problem that describes the structural behaviour of bolted connections taking into account as well the possibility of detachment between the connecting plates. By means of such an analysis, the influence of small variations of the design parameters (as is e.g. the plate thickness) to the overall structural response of the steel connections is investigated. An appropriate algorithm for the effective numerical treatment of the present sensitivity analysis problem is also discussed, whereas in the last part, numerical examples concerning sensitivity analysis of two different bolted connections, illustrate the previously presented topics.

### 3.2 Formulation of the Sensitivity Analysis Problem

Let us consider the steel connection of Fig.2 formed by bolted endplates. Let us also consider the primal quadratic programming problem (17) that describes the structural behaviour of the steel connection at hand, written in the following convenient form

$$\min \{ {}^1/_2\, k_{ij}^r u_i u_j - P_i^r u_i \;\mid\; \alpha_{\kappa i} u_i \leq b_\kappa \quad (i=1,....n;\ \kappa=1,....m) \} \ . \tag{35}$$

Here the summation convention holds for the repeated matrices and $\{k_{ij}^r\}$ is a set of symmetric matrices that are uniformly positive definite with respect to the design parameter r that belongs to the interval $[\,0\,, r_F\,)$ with $r_F > 0$. Loading $P_i^r$ and quantities $\alpha_{\kappa i}$, $b_\kappa$ defining the unilateral contact restrictions on the connection endplates take on each node a given value for $i=1,....n;\ \kappa=1,....m$. We assume next that the unilateral constraint set is nonempty and denote by $\mathbf{u}^r$ the unique solution of problem (35). Then, the Kuhn-Tucker conditions, i.e. the necessary conditions for $\mathbf{u}^r$ to be the unique solution of problem (35), take the following form

$$k_{ij}^r u_j - P_i^r + \lambda_\kappa^r \alpha_{\kappa i} = 0 \qquad (i=1,....n;\ \kappa=1,....m)\,, \tag{36}$$

$$\lambda_\kappa^r = 0 \quad \text{if} \quad \alpha_{\kappa i} u_i^r < b_\kappa \qquad (i=1,....n;\ \kappa=1,....m) \quad \text{and} \tag{37}$$

$$\lambda_\kappa^r \geq 0 \quad \text{if} \quad \alpha_{\kappa i} u_i^r = b_\kappa \qquad (i=1,....n;\ \kappa=1,....m)\,, \tag{38}$$

where $\lambda_\kappa^r$ are the respective Lagrange multipliers which express from an engineering point of view, the reaction distribution on the contact regions of the steel bolted plates having as consequence the appearance of separation phenomena of a certain part of the endplates interface.

Let define now the set $V(\mathbf{u}^0)$ of active constraints of problem (35) (i.e. the set of nodes being in contact ), and the sets $V_0(\mathbf{u}^0)$ and $V_1(\mathbf{u}^0)$ denoting respectively the sets of active constraints with zero and nonzero Lagrange multipliers

$$V(\mathbf{u}^0) = \{ \kappa \mid \alpha_{\kappa i} u_i^0 = b_\kappa \}, \tag{39}$$

$$V_0(\mathbf{u}^0) = \{ \kappa \text{ that belongs to } V(\mathbf{u}^0) \mid \lambda_\kappa^r = 0 \} \qquad \text{and} \tag{40}$$

$$V_1(\mathbf{u}^0) = \{ \kappa \text{ that belongs to } V(\mathbf{u}^0) \mid \lambda_\kappa^r \neq 0 \}. \tag{41}$$

Defining now the sets

$$C = \{ \mathbf{v} \mid \alpha_{\kappa i} v_i^r \leq \mathbf{0} \text{ for } \kappa \text{ belonging to } V(\mathbf{u}^0) \} \qquad \text{and} \tag{42}$$

$$H = \{ \mathbf{v} \mid k_{ij}^0 u_i v_j - P_i^0 v_i = 0 \} \tag{43}$$

and taking into account relation (36) multiplied by $v_i$, i.e.

$$k_{ij}^0 u_j^0 v_i - P_i^0 v_i + \lambda_\kappa^0 \alpha_{\kappa i} v_i = 0 \qquad (i=1,....n;\ \kappa=1,....m)\,, \tag{44}$$

where superscripts 0 denote vectors and matrices corresponding to the basic solution of problem (35) for $r=r_0$, we obtain the set

$$\Lambda = \{ \mathbf{v} \mid \alpha_{\kappa i} v_i \leq 0 \text{ for } \kappa \text{ belonging to } V_0(\mathbf{u^0})$$

$$\text{and } \alpha_{\kappa i} v_i = 0 \text{ for } \kappa \text{ belonging to } V_1(\mathbf{u^0}) \}. \quad (45)$$

Assume that $k_{ij}' = \lim ( k_{ij}^r - k_{ij}^0 ) / r$ where r tends to 0 and $P_i' = \lim ( P_i^r - P_i^0 ) / r$ with r tending to zero. Then, for r positive and small enough

$$\mathbf{u^r} = \mathbf{u^0} + r\,\mathbf{u'} + \mathbf{O}(r) \quad (46)$$

with $\|\mathbf{O}(r)\| / r$ tending to 0 (for r tending to 0) and $\mathbf{u'}$ the solution of the following quadratic programming problem

$$\min\{ {}^1/_2\, k_{ij}^0 v_i v_j - P_i' v_i + k_{ij}' u_i^0 v_j \mid \alpha_{\kappa i} v_i \leq b_\kappa \text{ for } \kappa \text{ that belongs to } V_0(\mathbf{u^0}) \text{ and } \alpha_{\kappa i} v_i = b_\kappa \text{ for } \kappa \text{ that belongs to } V_1(\mathbf{u^0}) \}. \quad (47)$$

Since problem (47) involves the same stiffness matrix with the initial problem (17), sensitivities can be computed by applying the same solution method applied to the solution of the initial quadratic programming problem (17). As is obvious, in the general case that vectors $\alpha_{\kappa i}$ are linearly independent, the Lagrange multipliers $\lambda_\kappa' = \lim (\lambda_\kappa^r - \lambda_\kappa^0) / \rho$ (for $\rho$ tending to zero) of the problem (47) coincide to the reaction forces on the activated contact nodes of the endplates interface and are uniquely determined.

The previous sensitivity analysis problem has been formulated as a quadratic optimization problem which involves the same quadratic term with the initial analysis problem. Therefore, stiffness matrix is the same in both the analysis and the sensitivity problem and has to be only once assembled. The latter exhibits a significant advantage in numerical modeling. Obviously, the same solution method is applied to the numerical treatment for both problems (17) and (47).

### 3.3 Numerical Applications

The previously presented method is now applied in order to obtain sensitivity analysis results with respect to endplate thickness for the design problem of two steel column-to-column splices. In both numerical examples, modulus of elasticity and Poisson's ratio for the material of the connections under investigation are respectively $E=2.1*10^6$ Kg/cm$^2$ and $\nu=0.30$.

The first example considers a steel connection of two columns of dimensions 60x60x5 mm (upper) and 150x150x5 mm (lower). The dimensions of upper and lower endplates are 300x300xd mm where d is the thickness of the endplates and is assume to be the critical design parameter. By considering as basic endplate thickness d=12 mm, deflections for both upper and lower endplates are calculated by solving problem (17) by applying the previously presented Hildreth and d'Esopo algorithm (Fig.7). Note here the separation phenomena that appear along some regions on the endplates. Then, by formulating problem (47) and numerically treating it by the same algorithm, sensitivity analysis results with respect to endplate thickness for both upper and lower plates are obtained (Fig.8).

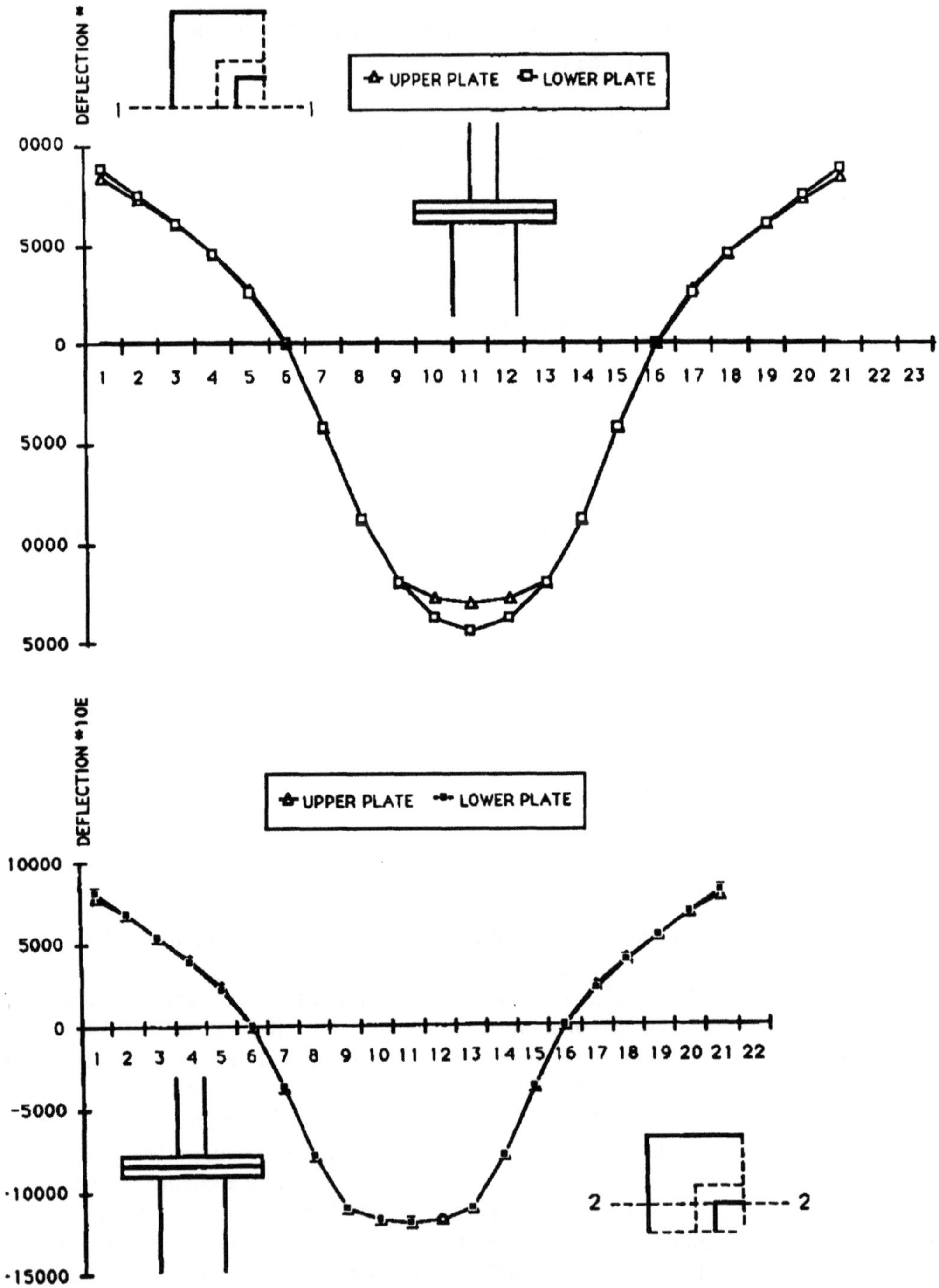


**Figure 7.** Basic solution: Deformation of the upper and lower plate (t=12mm) along the sections 1-1 and 2-2 respectively (square column intersection-square endplates).

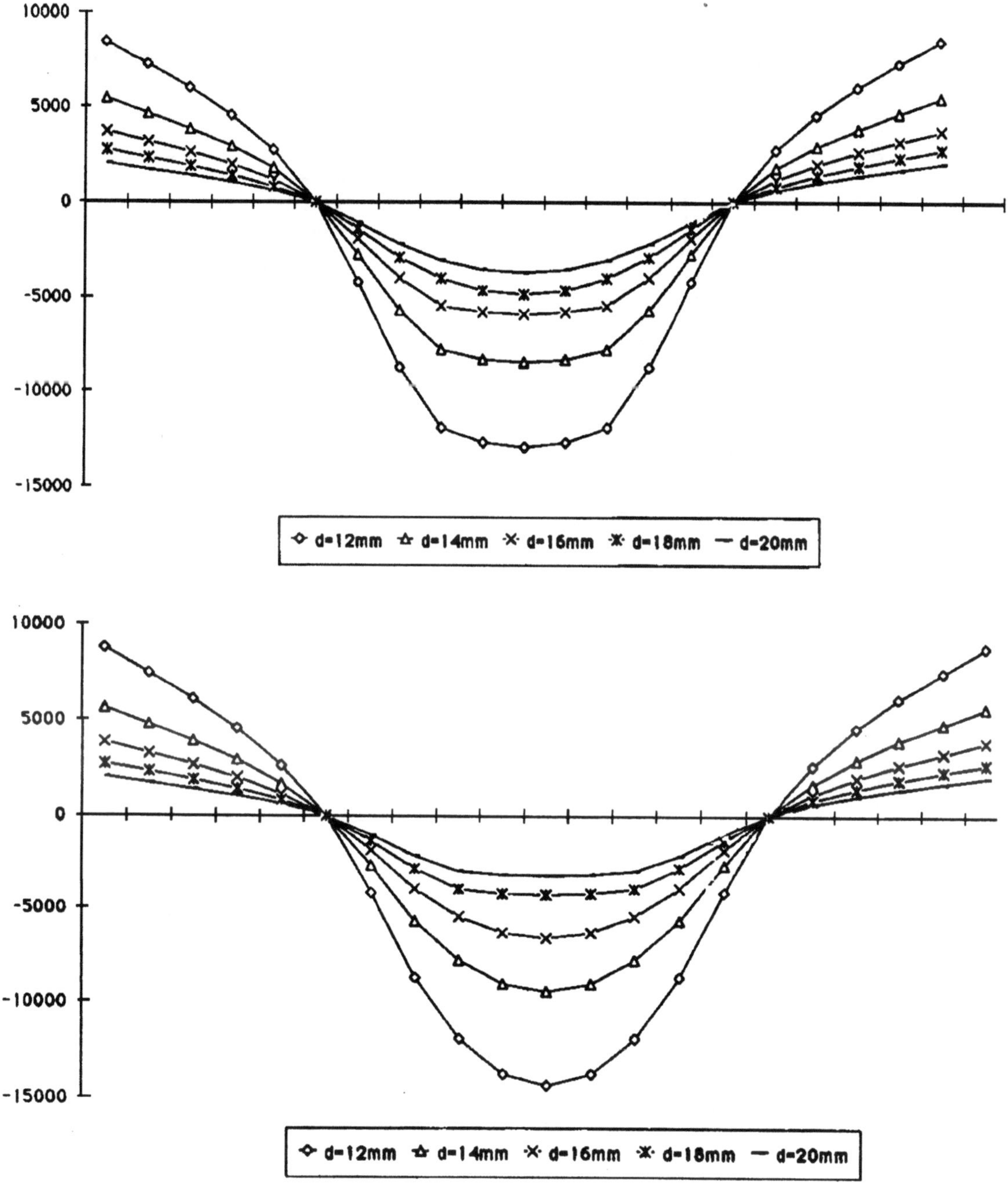


***Figure 8.*** Sensitivity analysis results with respect to endplate thickness for upper and lower plate.

Similarly, the column-to-column steel splice design problem for orthogonal columns (250x150x6.3 / 450x150x10 mm) and orthogonal endplates of dimensions 450x700xd mm is investigated.

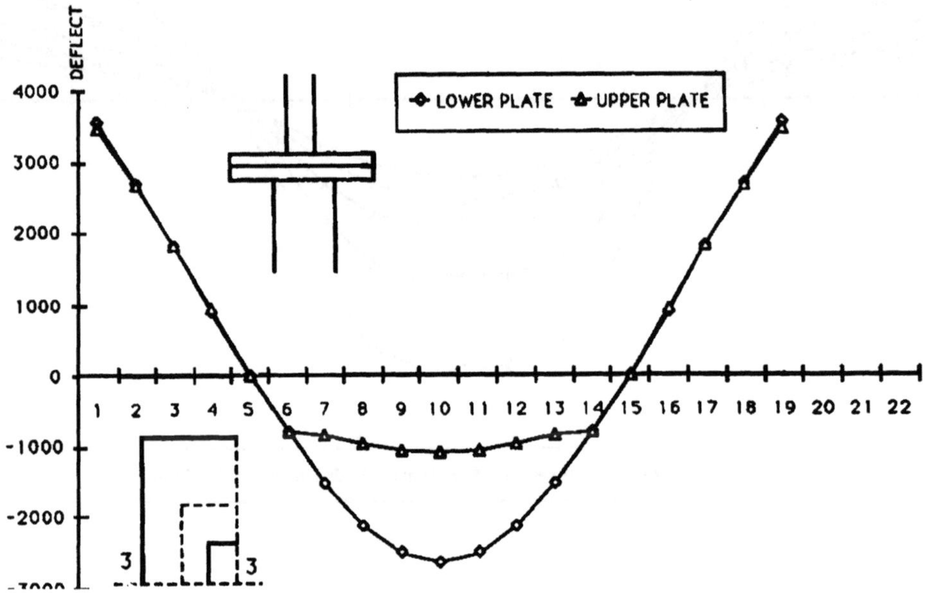


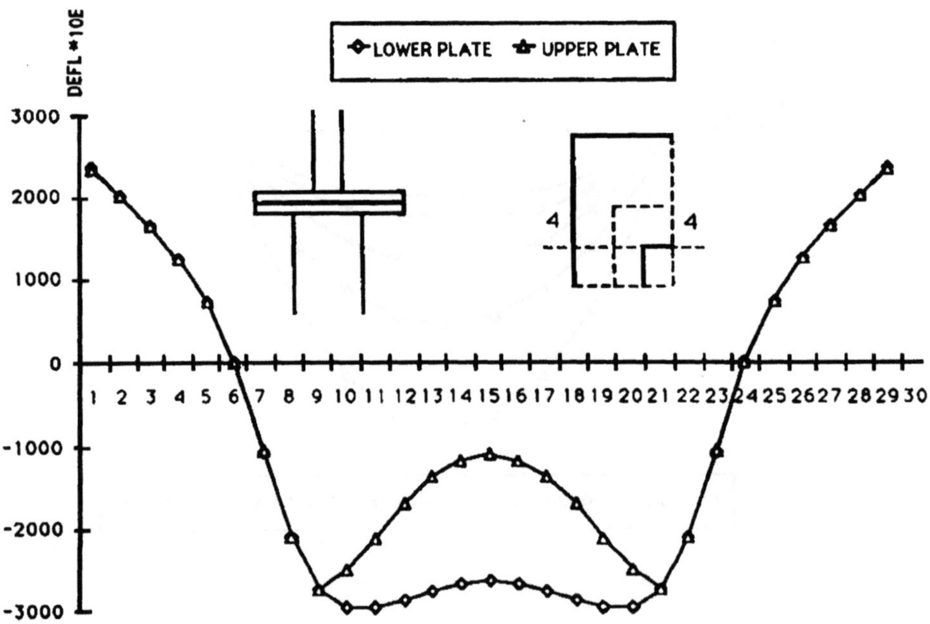


**Figure 9.** Basic solution: Deformation of upper and lower orthogonal endplates (t=14mm).

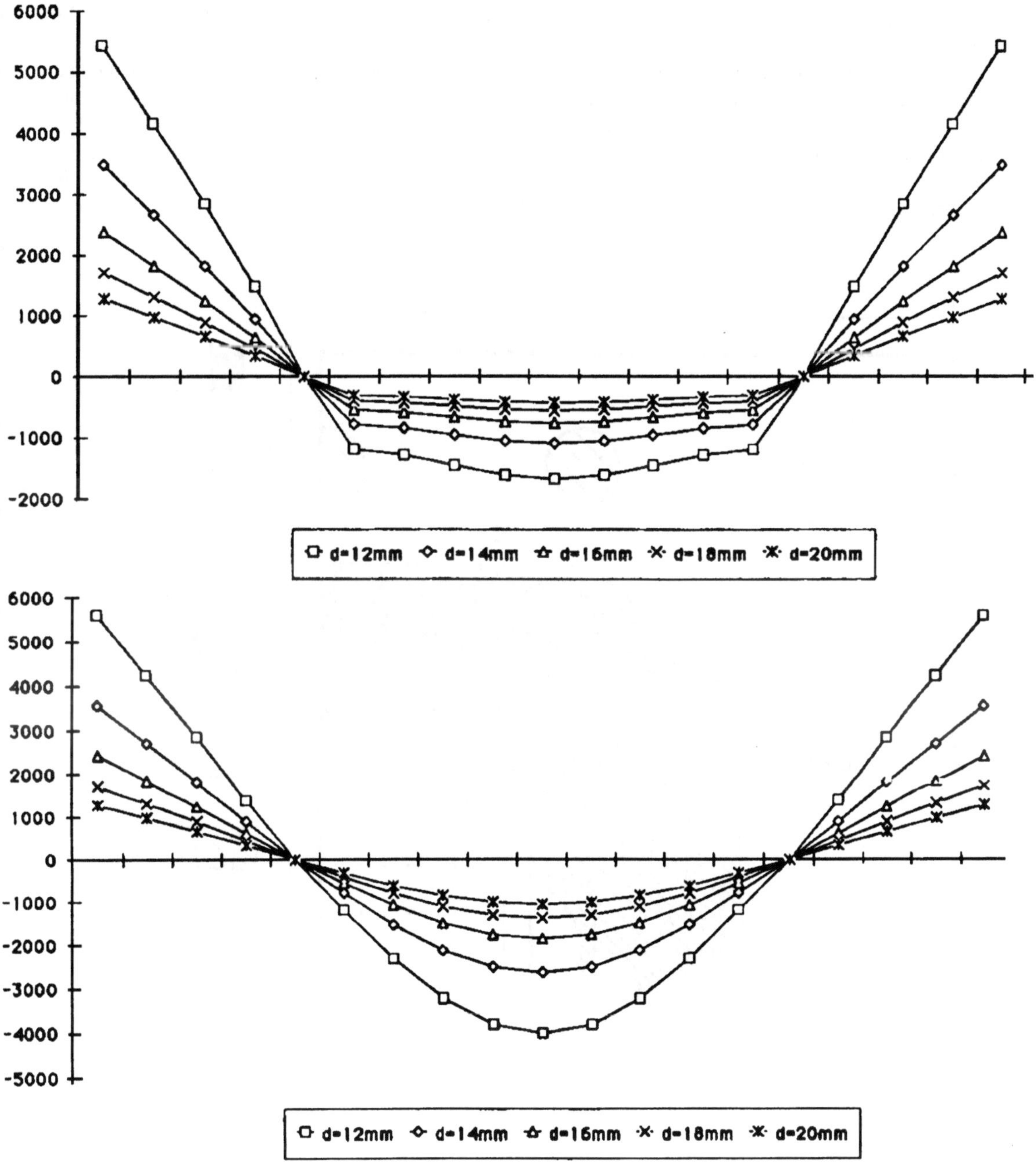


**Figure 10.** Sensitivity analysis results with respect to plate thickness for the endplates deformations: (a) upper endplate (b) lower endplate.

Basic solution for endplate thickness d=14 mm is depicted in Fig.9 where separation and complete contact regions on the endplate interface can be also recognized. Sensitivity analysis results with respect to plate thickness for deflections of the column steel splice under consideration are given in Fig.10 for both the upper and the lower endplate. An obvious consequence of the previous numerical investigations is that the active contact area along the connection interface under investigation is that this zone changes with respect to the variation of certain parameters, as is e.g. the plate thickness. The previously obtained results are schematically depicted in the two following Figs.11 and 12, where the variation of the active contact area (dashed area) with respect to the variation of the plate thickness is respectively shown for the square and the orthogonal plate studied.

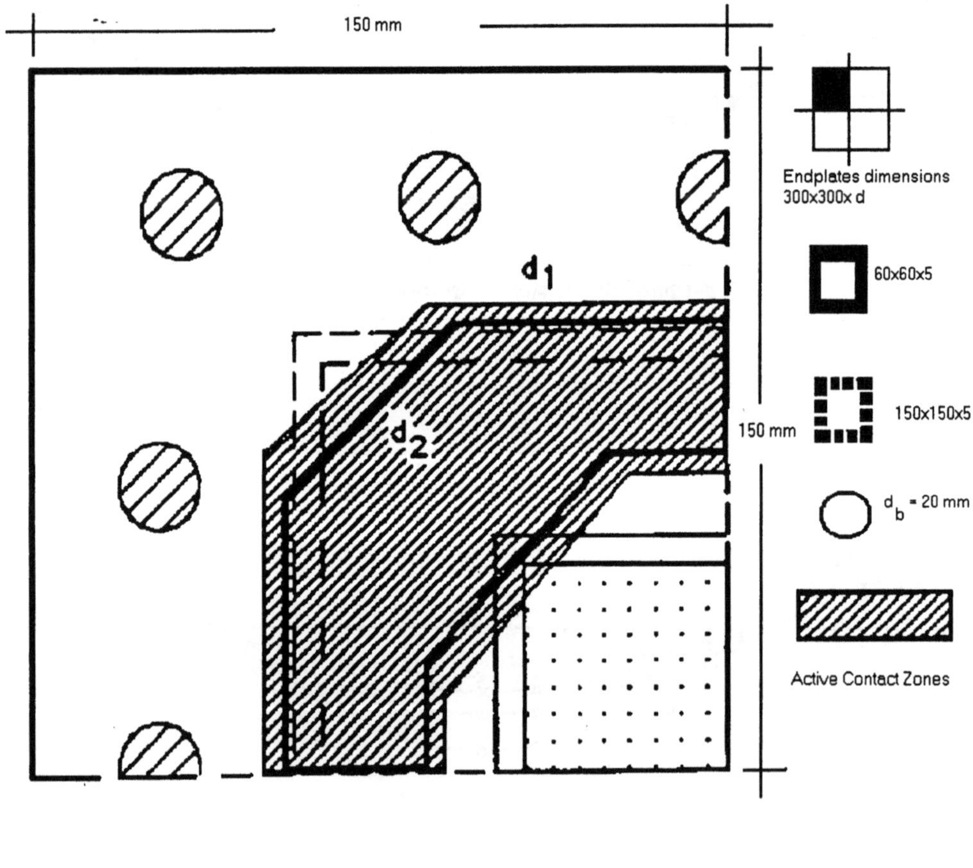


**Figure 11.** Variation of the contact and separation zones over the square endplates interface due to thickness variation.

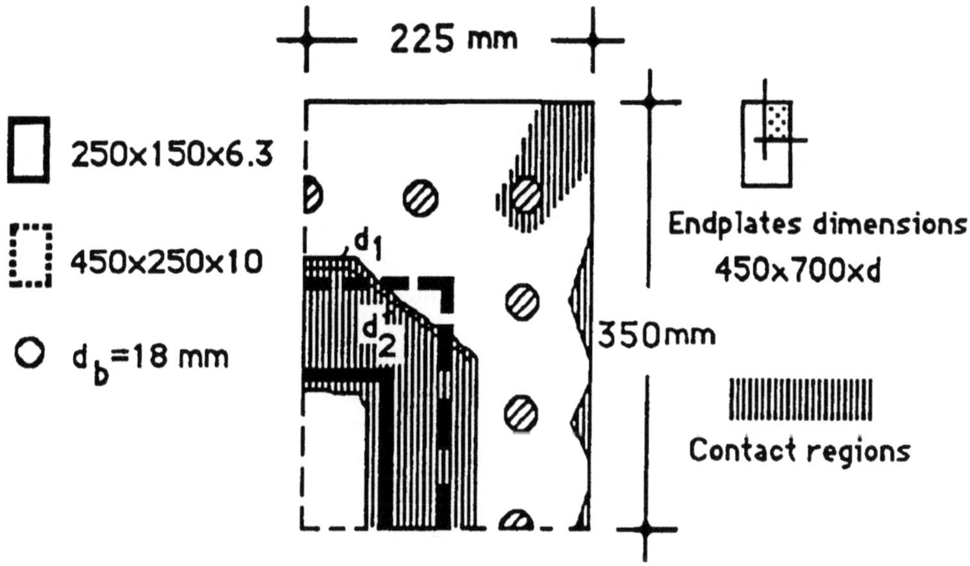


**Figure 12.** Variation of the contact and separation zones on the orthogonal endplates interface due to thickness variation.

## 4 Effective 2-D Numerical Models for Steel Connections in the Presence of Unilateral Contact and Plasticity Phenomena

In this section a method is presented, which has been proved to be a particularly appropriate method for the analysis of steel plate connections under static loading. This method that has been proposed by Mistakidis et al. (1998) concerns the simulation of the structural behaviour of steel connections in the case that the development of zones of plastification, as well as unilateral contact and friction effects between connection members and bolts are taken into account. Within this framework, effective two-dimensional finite element models capable to describe plasticity, unilateral contact and friction effects are constructed; the latter constitute easy-to-use and accurate numerical models for the analysis of steel connections subjected to static loading. These models are simplifications of the respective three-dimensional ones and aim to reduce in a reliable way the extensive computational effort required for the analysis of three-dimensional fine meshes of discretized steel connections.

### 4.1 Introduction

The method proposed in the present section aims to contribute to the study of the structural behaviour of steel connections by proposing an effective two-dimensional geometrical simplification of the respective three-dimensional one that takes into account both plasticity and unilateral contact effects. The respective two-dimensional model leads to accurate numerical results provided that the connection geometry and the loading conditions lead to a more or less two-dimensional deformed configuration. Its main advantage is that this model does not require so much computational effort as any three-dimensional models does. The proposed finite element model is this way constructed that the interaction at any interface of the steel connection at hand is taken into account by means of unilateral contact boundary conditions as proposed by Thomopoulos (1985), Baniotopoulos and Abdalla (1993), and Baniotopoulos (1994) and (1995). In addition, the development of plastification zones is also taken into account in the analysis. The method seems to be a reliable tool for the numerical simulation of the structural behaviour of most of the types of steel connections, because, from one side, the response of the different parts (flanges, bolts, washers etc.) of the modelled steel connections are taken into account in an interactive way, whereas on the other side the actual thickness of the various parts of the two-dimensional model are defined by applying an easy-to-use and efficient technique proposed by Mistakidis et al. (1996a). It is obvious that the results of the proposed two-dimensional finite element model can also be applied for quick benchmark tests in validating commercial three-dimensional finite element codes for the analysis of steel connections.

As a matter of fact, the numerical modelling of the structural behaviour of steel joints up to the ultimate limit state is in general characterized by certain difficulties due to the arising of a plethora of highly nonlinear effects (e.g. material yielding, contact interface slip and interface interaction) that have to be in a realistic way and with utmost care incorporated into any numerical model. The first attempts for two- and three-dimensional modelling of steel connections based on several simplified assumptions are dated back to the seventies; these efforts are till nowadays continued by deducting one by one the simplifying hypotheses of the initially proposed models, thus producing more and more interesting and realistic results (cf. e.g. the papers by Krishnamurthy and Graddy (1996), Kato and McGuire (1973), Paker and Morris (1977), Kukreti et al. (1987), Chen and Patel (1981), Chen and Lui (1986), Bortman and Szabo (1992), Thambiratnan and Krishnanurthy (1989), and Baniotopoulos et al. (1992)). In particular, among them certain two-dimensional finite element models were proposed because the question for minimization of the computational cost for an accurate simulation of steel connections response was always an important criterion and often an inevitable choice. However, only these models could be characterized as appropriate, if they contain all the essential features of the problem at hand.

Applying an appropriately developed numerical model, the stress flow between the various components of a steel structure has to be followed up, whereas the main deformation and stress distribution patterns must be present in the model and directly recognizable and interpretable. As a matter of fact, the three-dimensional models having been appropriately formulated and computed, usually contain the right stress distribution patterns, but in a form that requires too much computation/evaluation effort. In the engineering problem at hand, the stress and deformation patterns are mainly influenced by the unilateral contact effects between the interfaces, the variation of stress along the thickness of the flanges, and the stress concentration effects which usually arise in regions with high three-dimensional stress gradients.

Such a three-dimensional model can describe all the aforementioned phenomena that can often be hardly interpretable, especially those referred to bending. This is due to the fact that a plate/shell model has usually difficulties related to the contact stress distribution and to the stress concentration. Since the connected flanges of steel connections are of the order of other problem dimensions, i.e. comparatively rather thick, the application of a plate model is in general questionable because the field variation effects along the thickness have to be taken into account. Much more simpler one-dimensional (Bernoulli beam) models with the bolts modelled as springs have been recently proposed for the analysis of such connections, but in these models only primary bending action has been incorporated.

The herein proposed two-dimensional plane stress model encompasses all the thickness effects, primary bending/membrane effects and the contact stress distribution on the connection interfaces between the several parts of the connection. One of the main advantages of this model is that stress concentration zones can be easily identified although the analysis remains incomplete. Secondary bending effects are not present in the model, but they cannot drastically influence the whole picture of the stress fields caused by the static loading. The model takes into consideration the dimensions of the connection also along the third direction by assigning different thickness values to the various regions of the F.E.M. mesh achieving in this way a realistic character of the overall response of the model. In the regions where the thickness cannot be directly prescribed, as is the neighbourhood of the bolts, the washers and the zones with holes, several different assumptions have been made and the respective results have been compared to accurate numerical models which take into account the exact thickness value of each finite element. Although the later model gives the most accurate numerical results, the task of assigning the thickness of each region element by element is a very time-consuming work. Thus, comparison of the results of the various versions of the proposed two-dimensional model is unavoidable, in order to estimate the errors introduced by rougher assignments of the thickness of each region, which requires less effort during the modelling process

The numerical method applied does not use any *a priori* assumption on the flexibility of the several parts of the steel connection, thus obtaining first its deformed shape where contact (i.e. compressive reaction) and separation (i.e. zero reaction) zones are developed, and next the actual stress distribution on the connection members by taking into account friction effects on the interfaces (cf. e.g. the PANA-method proposed by Panagiotopoulos (1975) and (1985)). By means of this numerical method, friction forces are also taken into account along all the interfaces of the steel connection applying a double-step iterative numerical procedure. In particular, solving iteratively first a unilateral contact problem for the steel connection in the presence of friction forces and next a friction problem in the presence of normal reaction forces computed in the previous solution step, the deformed shape of the connection (that includes all the contact, noncontact and plasticity zones) and the actual stress distribution are accurately defined. Then, using the existing standards and norms, the design of the connection at hand can be easily completed.

With respect to the two-dimensional finite element model used to simulate the structural behavior of the steel connection, it is worthy to note that the classical finite element models do need some appropriate modification in order to accurately take into account the development of separation (noncontact) zones between the connection interfaces. The latter phenomenon that is of highly nonlinear character, has to be with care handled because it possesses certain difficulties of its own. Here again, applying the previously mentioned methods of Nonsmooth Mechanics, promising results for the numerical simulation of steel connections have been obtained. This is due to the fact that the exact numerical study of the local separation process on the steel interfaces can be easily performed as proposed by Panagiotopoulos (1988),

Panagiotopoulos and Strang (1988), Moreau and Panagiotopoulos (1988). Following this method, local separation zones between the interfaces of the steel connection are computed, whereas the deformed shape of the steel connection is with accuracy evaluated. Introducing the unilateral contact law with friction to simulate the boundary conditions on the steel connection interface, the separation process can be mathematically described giving rise either to a primal quadratic programming problem with respect to the connection displacements or equivalently, to its dual formulation with respect to the stresses. As it has been proved by Panagiotopoulos (1975), and Baniotopoulos and Abdalla (1993), the primal problem expresses, from the standpoint of Mechanics, the principle of minimum of the potential energy of the connection under consideration at the state of equilibrium, whereas the dual one, the principle of the minimum complementary energy. As has been proved by Panagiotopoulos (1975), solving iteratively these two problems and using the output of the former as input for the latter, the method converges after a few steps to the actual solution of the problem (as proved by means of a fixed point algorithm by Necas et al. (1980). The transformation of the initial structural steel connection problem into a sequence of constrained quadratic programming problems makes things, from a computational point of view, much more simple and allows for the application of a wealth of theoretical work and software. Note that the numerical results calculated by applying the proposed two-dimensional model qualitatively conform well to those obtained by experiments (cf. e.g. Mistakidis et al. (1996b), Jaspart (1994) and (1997)).

Within such a theoretical framework, the separation, the active contact, as well as the plastification zones are with accuracy calculated, leading thus, to the computation of the exact stress state conditions holding on the steel connection under investigation. The next paragraphs of this section is devoted to the presentation in details of two-dimensional modelling of the structural behaviour of a T-stub steel connection and a column base plate steel connection under static loading.

### 4.2 Description of the Method

A steel connection composed by several structural elements such as plates, bolts, etc. is first considered. The material of the parts of the connection that may be different for the various parts (i.e. high strength steel for the bolts, mild steel for the plates, etc.) exhibits an elastic-plastic behaviour with or without hardening. The interfaces formed between the various components of the joint are denoted by $\tilde{A}_s$. The forces applied on the connection are transmitted by the interfacial forces developed on those parts of the connection that come in contact. The latter are the contact forces that develop in the normal to the interfaces direction, and the frictional forces that develop in the tangential to the interfaces direction, and only between those parts that come in contact.

From a mathematical point of view, the unilateral contact conditions on the interfaces are expressed by the formulas (10) and (11) (or in a simpler way by relations (12) and (13)) that hold for each one of the k nodes of the interfaces $\tilde{A}_s$ belonging to the discrete model. Furthermore, the friction boundary conditions holding on the active contact regions (defined by the right-hand part of relation (10) or (12)) are written in the form (48), (49) where absolute values denoted by the symbol | . | are present

$$\text{if } [u_T] = 0 \text{ that } |S_T| < \mu |S_N| \tag{48}$$

and

$$\text{if } |[u_T]| > 0 \text{ then } |S_T| = \mu |S_N| \tag{49}$$

where $\mu$ is the Coulomb's friction coefficient and $[u_T]$ is the symbol that corresponds to the tangential to the interface relative displacements.

Following the method proposed by Maier (1984), the static behaviour of the discrete model at hand can be described by

(i) the constitutive equations

$$\mathbf{e} = \mathbf{F}_0\,\mathbf{s}, \tag{50}$$

$$\mathbf{e} = \mathbf{e}_0 + \mathbf{e}_E + \mathbf{e}_p \tag{51}$$

$$\mathbf{e}_p = \mathbf{N}\,\boldsymbol{\lambda} \tag{52}$$

$$\boldsymbol{\phi} = \mathbf{N}^T\mathbf{s} - \mathbf{H}\,\boldsymbol{\lambda} - \mathbf{k} \tag{53}$$

$$\boldsymbol{\lambda} \geq \mathbf{0}, \quad \boldsymbol{\phi} \leq 0, \quad \boldsymbol{\lambda}^T \boldsymbol{\phi} = 0 \tag{54}$$

where $\mathbf{F_0}$ is the natural flexibility matrix of the structure, $\mathbf{e}$ the respective strain vector consisting of three parts, i.ε. the initial strain vector $\mathbf{e}_0$, the elastic strain vector $\mathbf{e}_E$ and the plastic strain vector $\mathbf{e}_p$, $\lambda$ are the plastic multipliers vector, $\phi$ is the yield functions, $\mathbf{H}$ is the work-hardening matrix, $\mathbf{N}$ is the matrix of the gradients of the yield functions with respect to the stresses and $\mathbf{k}$ is a vector of positive constants,

(ii) the equilibrium equations

$$\mathbf{G}\,\mathbf{s} = \mathbf{P} \tag{55}$$

where $\mathbf{G}$ is the equilibrium matrix and $\mathbf{P}$ the external load vector,

(iii) the compatibility relation

$$\mathbf{e} = \mathbf{G}^T\,\mathbf{u} \tag{56}$$

where $\mathbf{u}$ is the displacement vector, $\mathbf{G}^T$ is the transpose matrix of the equilibrium matrix and in addition

(iv) the unilateral contact conditions and the friction boundary conditions holding on the unilateral contact interfaces as described by (12), (13) and (48), (49).

A direct solution of the above problem is not possible due to the fact that it is not *a priori* known which parts of the connections are in contact under the application of the static loading (i.e. which are the exact values of the contact forces) and subsequently, for those parts that come in contact, it is not known which of them are in sticking contact or in slipping contact, i.e. for which of them relation (48) or (49) hold. For details on the mathematical difficulties arising in the previous problem and the approaches to overcome them, see the book by Panagiotopoulos (1985), and the paper by Necas et al. (1980).

An additional problem is the interaction between the normal contact forces and the frictional ones due to the fact that the latter depend on the value of the contact force, whereas the values of the frictional forces may lead to the change of the active contact area or the exact value of the contact forces.

In order to avoid the previous difficulties, a method first proposed by Panagiotopoulos (1975) for the solution of the unilateral contact problems with friction and mathematically investigated concerning convergence by means of a fixed point theorem by Necas et al. (1980) is followed. By means of this method the problem is transformed into two sub-problems. The first sub-problem corresponds to the problem formulated with respect to unknowns that are the normal to the interfaces displacements, whereas the second one corresponds to the problem formulated with respect to unknowns which are the tangential to the interface displacements. Then, the solution of the problem is achieved by applying the following iterative scheme:

**Contact sub-problem.** The unilateral contact problem is first considered assuming that the friction forces are constant and known. The solution of this problem is obtained by minimizing the potential energy of the discretized structure with subsidiary constraint the non-penetration condition holding on the contact interface (relations (12), (13))

$$\Pi(\mathbf{u})=\min\{{}^1/_2\mathbf{u}^T\mathbf{K}\mathbf{u} + {}^1/_2\,\boldsymbol{\lambda}^T\mathbf{H}\boldsymbol{\lambda} - \mathbf{u}^T\mathbf{G}\mathbf{K}_0\,\mathbf{N}\boldsymbol{\lambda} + {}^1/_2\,\boldsymbol{\lambda}^T\mathbf{N}^T\,\mathbf{K}_0\mathbf{N}\boldsymbol{\lambda} +$$

$$+\,\mathbf{e}_0{}^T\mathbf{K}_0(\mathbf{N}\boldsymbol{\lambda}-\mathbf{G}^T\mathbf{u}) - \mathbf{P}_1{}^T\mathbf{u} - \mathbf{k}\,\boldsymbol{\lambda} \mid \boldsymbol{\lambda}\geq \mathbf{0},\ \mathbf{u}_k \geq \mathbf{0}\,\} \tag{57}$$

where vector $\mathbf{P}_1$ includes both the external loading and the friction forces assumed known, $\mathbf{K}$ is the stiffness matrix of the discrete model and $\mathbf{K}_0$ is the inverse of $\mathbf{F}_0$.

The solution of the above minimization problem leads to the computation of the contact and non-contact areas, the values of the contact forces and the respective stress and stain states of the connection.

**Friction sub-problem.** Assuming next that the contact forces are constant and known (equal to the ones obtained in the previous step), we solve in turn an analysis problem taking into account that the Coulomb's friction boundary conditions have to be also satisfied. This problem is now formulated as a quadratic programming problem with respect to the stresses, having as subsidiary conditions the relations (48) and (49), i.e.

$$\Pi^C(\mathbf{s})=\min\{\,{}^1/_2\mathbf{s}^T\mathbf{F}_0\mathbf{s} + {}^1/_2\boldsymbol{\lambda}^T\mathbf{H}\boldsymbol{\lambda} + \mathbf{s}^T\mathbf{e} \mid \mathbf{N}^T\mathbf{s}-\mathbf{H}\boldsymbol{\lambda}-\mathbf{k} \leq \boldsymbol{\phi}\,,\ \mathbf{G}\mathbf{s} = \mathbf{P}_2,\ |S_T| \leq \mu|S_N|\,\} \tag{58}$$

where $\mathbf{P}_2$ includes both the external loading and the contact forces assumed as known. The latter expresses the principle of minimum complementary energy for the considered connection at the state of equilibrium taking into account the frictional conditions on the contact interface. Notice that the previous problem (58) is formulated with respect to stresses and not with respect to displacements so that the appearance of non-differentiable terms due to the absolute values to be avoided.

The two previously described sub-problems constitute the first double step iteration of the proposed technique that is in the sequel repeated. In the second double step, the unilateral static problem is solved by taking into account as nonactive contact and sliding regions, those ones obtained in the first double step by also assuming as known the tangential loading calculated from the previous step. In this sense, a new external loading is applied to the connection which is in turn solved again, i.e. first minimizing the potential energy of the structure and then, minimizing its complementary energy. Obviously, the equilibrium configuration of the connection may be different between two successive solution steps. This means that the activated contact and the sliding regions computed in the second step are probably different from those calculated in the first step. The present solution rocedure must be continued by iteratively solving quadratic programming problems that ke p the same basic mathematical characteristics, but take as input data the output of the previous step. This iterative procedure is ended as soon as the difference of two successive stress states of the connection is less than or equal to a desirable accuracy $||s(n+1) - s(n)|| < \varepsilon$, where $n$ is the number of the solution step.

Concerning the numerical treatment of the formulated quadratic programming problems, any effective algorithm from the Theory of Optimization can be applied (cf. e.g. Fletcher (1987)). Concerning the first step of the solution procedure that corresponds to the computation of the active contact and separation regions between the various interfaces (problem (57)), the optimization algorithm explicitly determines the nodes coming into contact with the respective contact forces. In turn, the optimization algorithm is applied in a similar way within the second step of the method. The post-processing part of the algorithm concerning the computation of the new stress distribution is done in a similar way of this one in the usual finite element analysis. It is therefore, obvious that the only real difference of the necessary software from classical finite element codes will be the substitution of the solver of equations by a quadratic programming routine and the treatment of the additional data that introduce the inequality constraints.

### 4.3 Analysis of a T-stub

The previously presented method was applied by Mistakidis et al. (1998) for the analysis of a T-stub reference connection subjected to tension (Fig.13). The two T-stubs are extracted from hot-rolled IPE300 profiles and the connecting bolts are M12 10.9 ones (bolts and washers according to DIN 6914-6915-6916).

In order to analyze the connection by applying the previous method, the two-dimensional model of Fig.13 (right) was constructed. Taking advantage of the two symmetry axes, 1159 nodes and 975 plain stress quadrilateral elements were used to represent one quarter of the structure. The structure is loaded by applied displacements introduced as a sequence of 50 increments, at each increment a displacement of 0.2 mm was applied into the structure.

The structure under investigation consists of three different parts and in particular, the bolt, the washer and the T-stub. The thickness of the plain stress elements was properly adjusted in order to take into account the three-dimensional properties of the steel joint. Thus, a thickness of 80 mm was given to the T-stub except the region of the holes, where a reduced thickness of 61.6 mm was given. A mean thickness of 18.8 mm was used for the shank of the bolt, a mean value of 34.6 mm was used for the head of the bolt and for the washer a mean value of 25.48 for the washer. In the zones around the holes, the two 2-D components were overlapping.

The interaction between these two parts is taken into consideration by assuming unilateral contact conditions to hold along the interfaces. Unilateral contact conditions are also assumed to hold on the x-symmetry axis because it is not *a priori* known which parts of the structure is in contact with the other symmetric part. With respect to the quality of the steel of the T-stub connection, the stress-strain law of Fig.14a has been assumed to hold, whereas for the material of the bolt the diagram of Fig.14b was taken into account. For the ultimate strength of the steel $f_{u,st}$ the nominal value was used (360 N/mm$^2$) and for the bolts $f_{u,b}$=1000 N/mm$^2$.

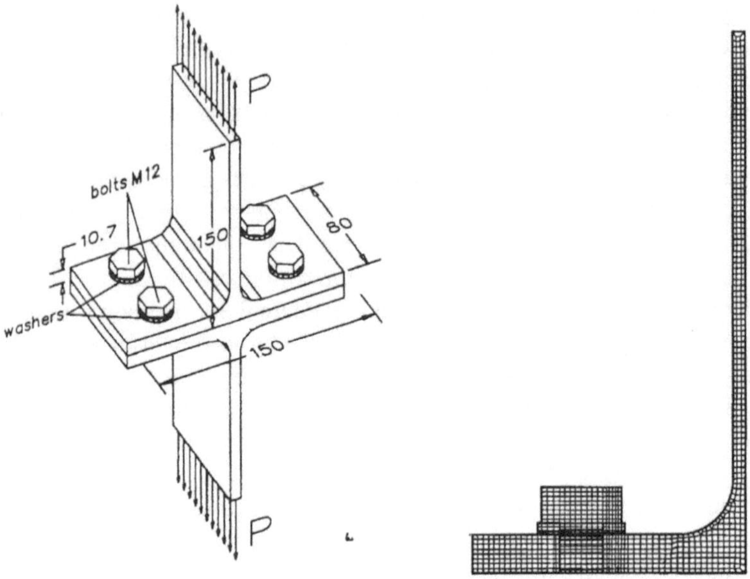


**Figure 13.** A typical steel bolted T-stub connection and the proposed 2-D F.E. model.

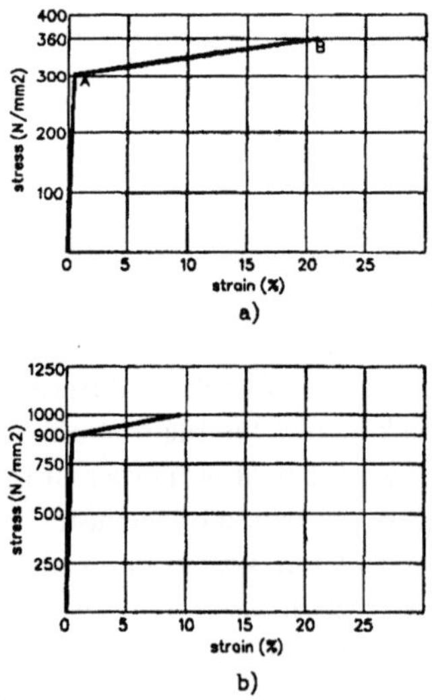


**Figure 14.** Stress-strain diagrams for (a) T-stub (b) bolts.

The options in the analysis of the connection at hand included contact, plasticity and large displacements. Fig.15 presents the successive deformed shapes of the T-stub at the increments 10, 20, 30, 40, 50. In increment 10, the structure just rises from the x-symmetry axis. In the 30$^{th}$ increment, the shank of the bolt comes in contact with the left part of the T-stub. In increment 40 significant deformations are developed at the region where the bolt comes in contact with the right part of the T-stub. The situation is the same in increment 50, but the deformation of the bolt is now much more considerable.

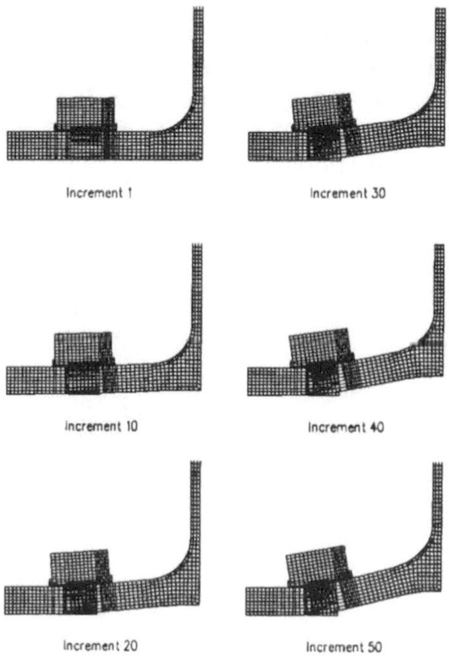


**Figure 15.** Successive deformed configurations of the T-stub.

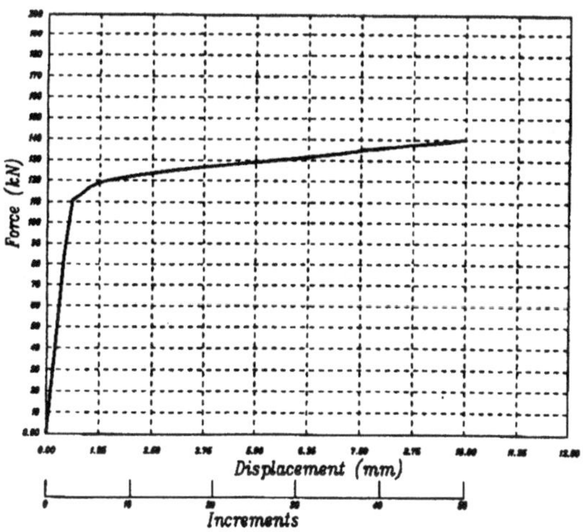


**Figure 16.** Load-displacement diagram.

The study of the obtained stress fields and of the force-displacement diagram of Fig.16 reveals that the strongest non-linearity is due to plasticity. As the loading increases, two plastic hinges are formed. The first appearance of a plastic hinge takes place in the left part of the flange, near the web. The second plastic hinge is formed close to the connection of the flange with the bolt. In addition, the stress in the shank of the bolt is considerable and increases in an analogous way to the loading. The results of an extensive numerical study by means of the present model including the effect of the pre-stressing of the bolts into the joint structural response has been recently presented by Mistakidis et al. (1997).

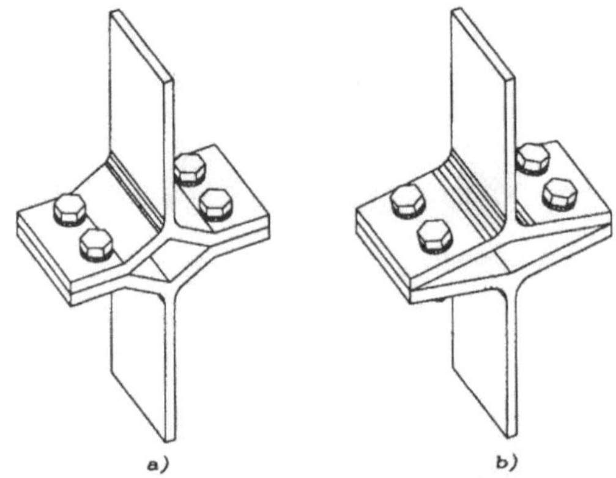


**Figure 17.** Certain typical T-stub collapse modes (a) I and (b) II.

As a matter of fact, the proposed model seems to be a reliable tool for a two-dimensional analysis. Indeed, the structure collapses in mode I (complete flange yielding), which means that two plastic hinges are developed as shown in Fig.17a. As a consequence, the whole structure comes in a more or less two-dimensional deformed configuration something that justifies the validity of the assumptions that led to the present model. Note that if other collapse modes are activated due to a different combination of bolt diameter and flange thickness, such as mode II (bolt failure with flange yielding (Fig.17b)) or mode III (bolt failure without flange yielding), then the structure seems again to come in a two-dimensional deformed configuration.

The case of pre-tensioning can be easily incorporated into the proposed model by using a variety of techniques. For instance, pre-tensioning is introduced as an appropriately defined additional force. However, this technique often leads to erroneous results. Another approach leading to much more accurate results is to introduce pre-tension as an initial strain of the bolts that corresponds to the pre-tensioning force. If the F.E.M. code used does not support the introduction of initial strains in the model, then an equivalent temperature change yielding the equivalent initial strain can be introduced.

### 4.4 Analysis of a Steel Bolted Column Base Plate Connection

The same approach has been applied by Baniotopoulos (1994) and (1995), and Mistakidis et al. (1998) to the simulation of the structural behaviour of the base plate connection shown in Fig.18. The steel column section is a RHS 120/200/10.0 and is connected to a concrete block by means of a 20mm-thick steel plate. Four M20-10.9 bolts acting as anchors embedded to the concrete block ensure a reliable structural joint response. The connection is idealized by means of a two-dimensional finite element mesh consisting of 2122 nodes and 1914 plain stress quadrilateral elements shown in the right part of Fig.18.

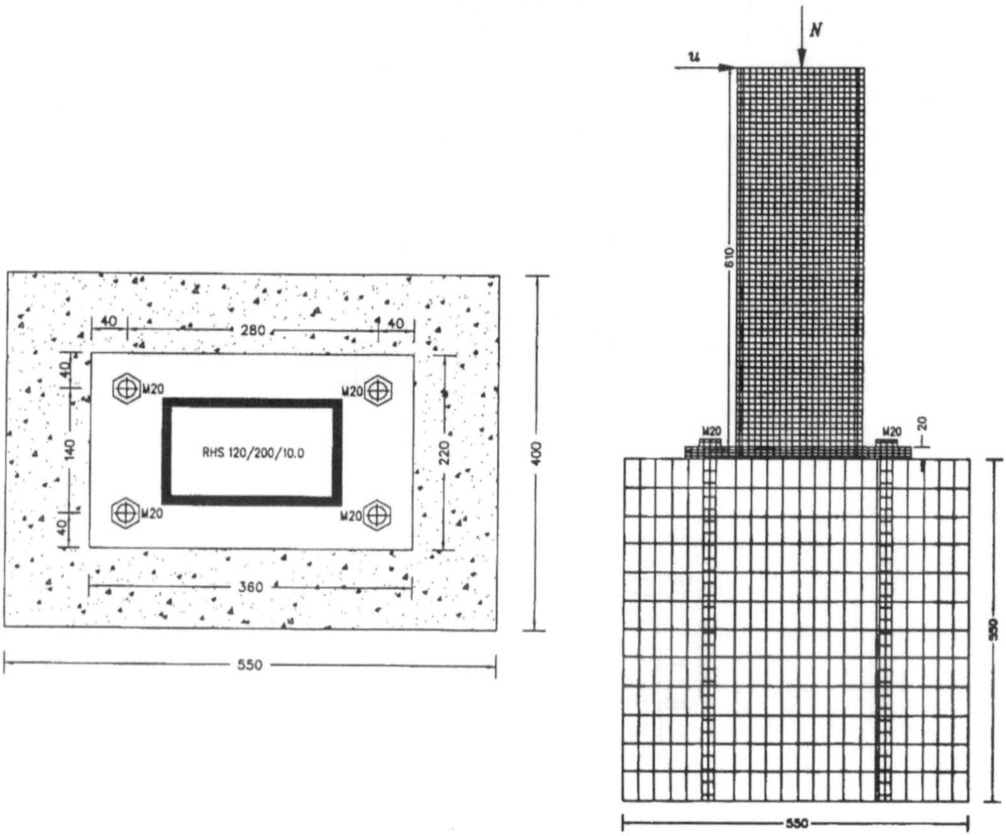


**Figure 18.** Steel bolted column base plate connection: plan view and F.E. model.

The discrete model consists of four different parts, i.e. the steel profile together with the plate, the two bolts and the concrete block. The thickness of the plain stress elements is properly adjusted in order to take into account the three-dimensional properties of the connection under investigation as shown in Table 1.

| Connection part | Equivalent thickness in mm |
|---|---|
| Concrete block | 400.0 |
| Base plate | 220.0 |
| Steel profile (weak area) | 20.0 |
| Steel profile (strong area) | 120.0 |
| Bolt | 31.4 |

**Table 1.** Equivalent thickness for the various parts of the connection.

As already discussed in the previous numerical model, in the region of the holes of the base-plate, the two parts representing the plate and the bolt are overlapping. The interaction between the two bodies is taken into account by considering unilateral contact conditions between them. Unilateral contact conditions are also assumed to hold between the plate and the concrete block. For the material of the steel profile and the plate, we assume that the stress-strain law of Fig. 19a holds. For the material of the bolt, a similar diagram is considered with the only difference that the ultimate strain is considerably smaller as shown in Fig. 19b. Finally, the material of the concrete block is considered to be linear with modulus of elasticity $E_C$=29 Gpa. The friction coefficient between all the connected parts of the joint was taken equal to $\mu$=0.3.

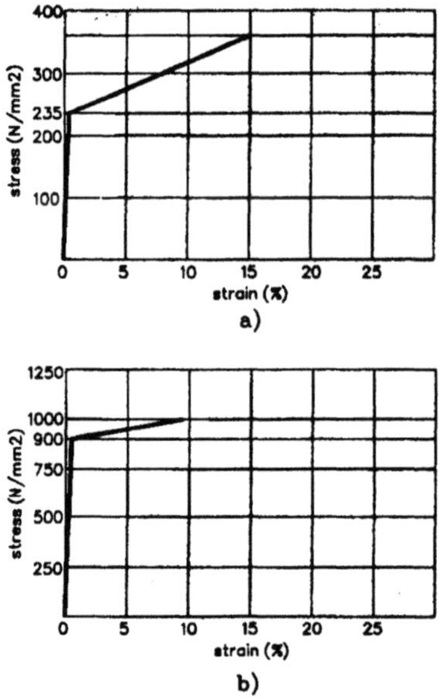


**Figure 19.** Stress-strain diagrams for (a) the steel profile (b) the bolts.

The structure is subjected to applied displacements introduced as a sequence of 50 increments on the top edge of the column; at each increment a displacement of 1 mm is applied into the structure. Moreover, for the axial loading of the connection, the following three cases are considered: 0kN, 250kN and 500kN.

In Fig.20 certain successive deformed configurations of the connection subjected to moment and to three cases of the axial loading, at the increments 10,20,30,40,50 are depicted where the displacements are multiplied by a magnification factor of 2. In this Figure, the left column presents the deformations of the connection for the case of zero axial force. We notice that very quickly a contact zone is established under the right end of the base plate. The increase of the horizontal loading does not change the contact area (the right part of the plate remains in contact with the concrete base and the rest part rotates having as center of rotation the right point where the column meets the plate). After the $30^{th}$ increment, plasticity phenomena appear

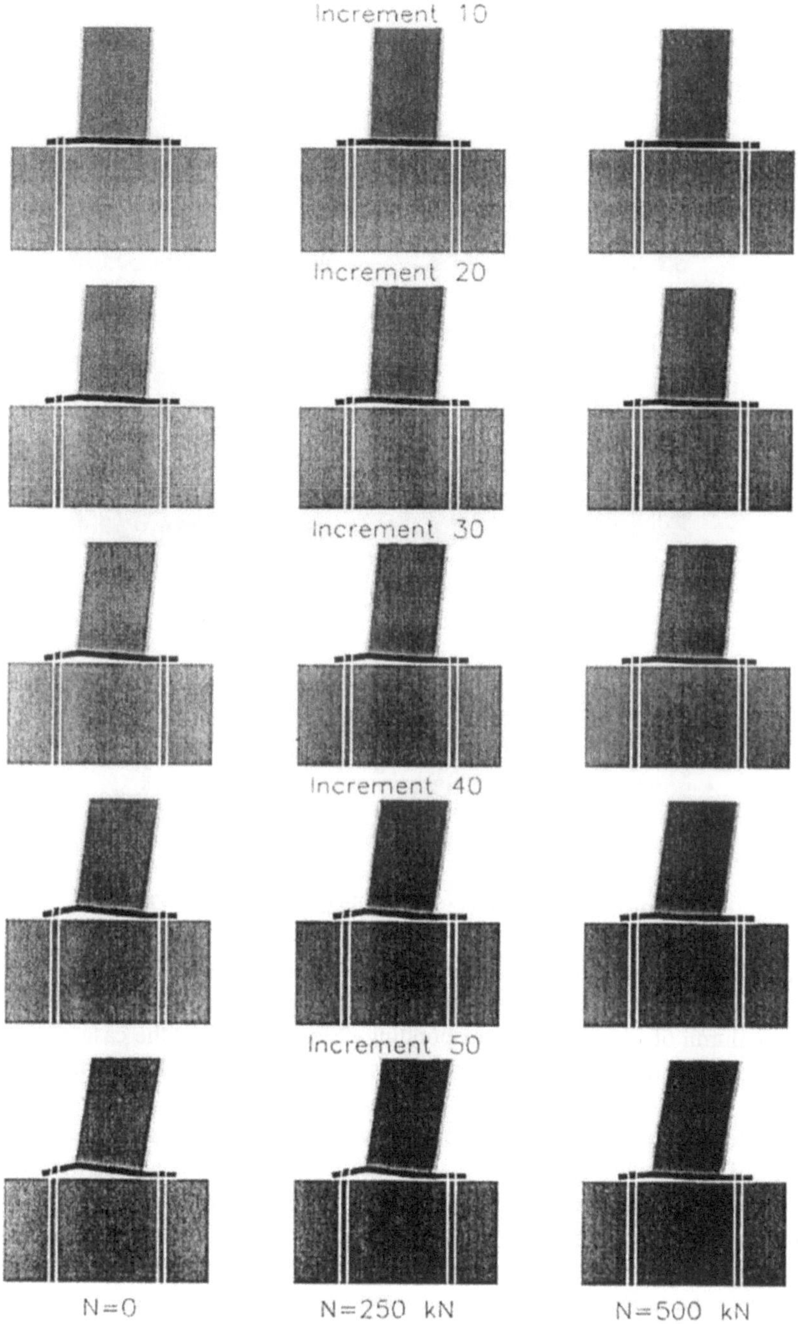


**Figure 20.** Successive deformed configurations of the steel base plate connection subjected to moment M for different axial loading N.

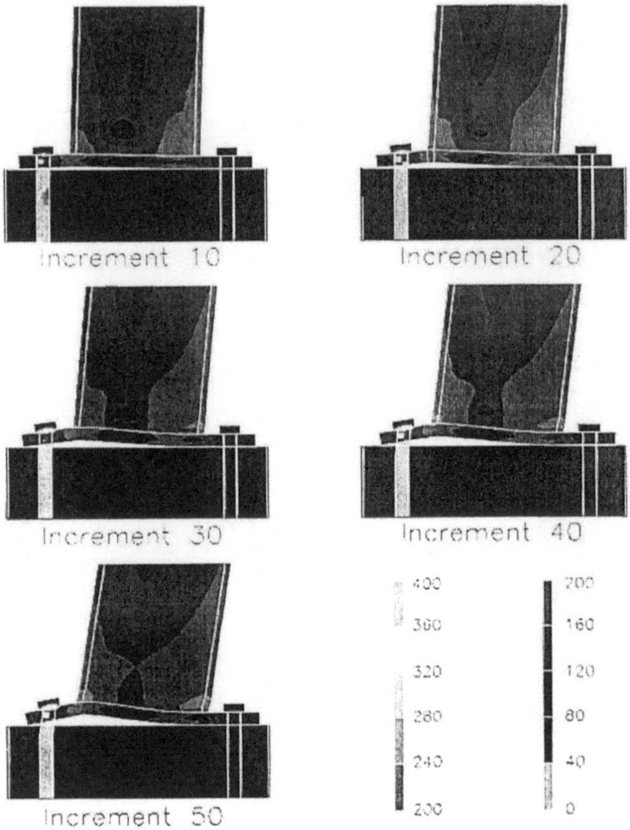


**Figure 21.** Von Mises stress distribution for N=250 kN.

along certain zones of the plate leading to considerably large. It is clear that in this case the collapse of the connection occurs due to the plastification of the left part of the base plate. In the middle column of Fig. 20 the corresponding deformations for the case of the axial loading equal to 250kN are presented. The deformations in this case are considerably smaller due to the effect of the axial force. In this case the contact area changes as the horizontal loading increases. After the $10^{th}$ increment, almost every point of the base plate comes in contact with the concrete. Then, the left part of the plate rises following a similar pattern with the one of the previous case. Finally, the left part of the plate again yields and the connection collapses. The same conclusions can also be drawn from Fig.21 where the von Mises stress distribution for various increments for the case of axial loading N=250kN is presented. The right column of Fig.20 presents the deformations of the connection for axial loading equal to 500kN. In this case the effect of the axial force is much more sound. The larger part of the plate is always in contact with the concrete. It is however, clear that a plastic hinge at the lower part of the steel column is formed. This is obvious after the $40^{th}$ increment where we can see that the column is not perpendicular to the base plate. Thus, the collapse of the structure occurs due to the plastification of the steel column and therefore, the strength of the other parts of the connection appear to be only a secondary factor in the overall strength of the joint.

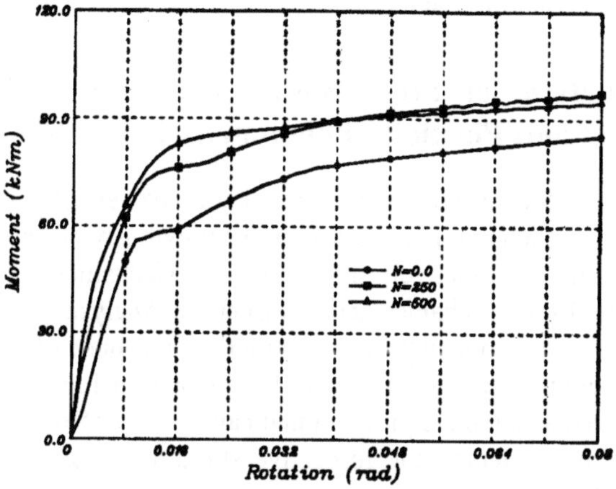


**Figure 22.** Moment –rotation diagrams for the three cases of axial loading studied.

The previous remarks can also be verified from the moment-rotation curves of Fig.22. As the axial loading increases, the moment capacity for the same rotation increases as well. It is also interesting here to define this part of the moment-rotation curve that is mainly influenced by the change of the contact area: in the case of zero axial force, it is noticed between the $5^{th}$ and the $10^{th}$ increment, in the second case it is noticed between the $10^{th}$ and the $20^{th}$ increment and finally, in the last case it is not present at all. It is also clear from the moment-rotation curve corresponding to the third case that the connection behaves like a plastic hinge having a very limited capacity to undertake additional moment as the rotation increases.

### 4.5 Conclusive Remarks

In the previous paragraphs a numerical method leading to certain effective discrete models have been studied with respect to their applicability and effectiveness for the analysis of certain types of steel connections. The applied analysis method which is mainly based to the ideas of one of the methods for the effective treatment of frictional contact problems, leads to the exact determination of the active contact and separation regions on the interfaces of the connection, of the sliding and sticking regions and the plastification zones. The analysis problem formulated as a quadratic programming one, can be numerically treated by fast and effective algorithms leading to numerical results that conform well to alternative numerical simulations of the same problem and to experimental testing results.

In the proposed two-dimensional model all the thickness effects, primary bending/membrane effects and the contact stress distribution on the connection interfaces between the several parts of the connection have been successfully incorporated leading to deformed configurations where stress concentration zones can be easily. Obviously, secondary bending effects are not present in the model, but - as previously mentioned - they cannot drastically influence the whole picture of the stress fields caused by the static loading.

Although the dimensions of the steel joint along the third direction are in a realistic way taken into account, the analysis still remains incomplete, being a little bit different from a complete 3-D analysis (cf. e.g. Sakellariadou et al. (1996)), but simultaneously exhibiting the great advantage of a restricted computational effort.

## 5 Parametric Analysis of Steel Bolted Base Plate Connections in the Presence of Contact and Plasticity Phenomena Using 2-D F. E. Models

The present section aims to contribute to the parametric analysis of semi-rigid steel base plate connections. The herein applied method is based on the previously presented theoretical results of Nonsmooth Mechanics that is a relatively new branch of Mechanics initiated three decades ago and deals with problems from Mechanics and/or Structural Analysis that involve generalizations of the gradient. The numerical modelling of the structural response of a steel column base plate under static loading where unilateral contact with friction and yielding phenomena are taken into account can be considered as a typical problem that can be very effectively treated within such a theoretical framework. In particular, the respective stress states of the steel connection under static loading are calculated by taking into account the development of plastification zones and the unilateral contact and friction effects along the interfaces between connection members. The previously presented 2-D finite element model being capable to describe the previously mentioned phenomena is again taken into account. The application of this model into the parametric analysis of the steel base plates reduces in a reliable way the huge computational effort that is often required for the analysis of 3-D fine meshes of discretized steel connections. A parametric analysis of the steel bolted base plate response where two critical parameters are altered (the base plate thickness and the axial load on the model) demonstrates the effectiveness and the applicability of the proposed method.

### 5.1 Introduction

During the last decades, the assumption of complete contact between the connection interfaces has been widely used for the study of steel column base plate connections because it generally leads to results on the safe side. However, in the case that a loading process leads to the development of moment, shear and tension, this simplification fails to describe the actual response of the base steel plate interface. Laboratory experimentation and construction practice do prove that separation is often caused on the contact surface between the deformable column base plate and the rigid concrete foundation. This phenomenon has already attracted during the last years the interest of numerous researchers who applied analytical, experimental and numerical approaches to investigate it. Of significant importance are the analytical approaches that consider along the column base plate effective contact surface, where various normal reaction distributions appear in the presence of concentric and/or light loading (cf. e.g. the works by Fling (1970), Stockwell (1975), and Murray(1983)). Extensive experimental investigations, aiming to the accurate description and evaluation of the response of column base connections, have been conducted for concentric and eccentric loads (cf. e.g. Dewolfe (1978), Khrishnamurthy (1978), Dewolfe and Sansley (1983), Thambiratnam and Paramasivam (1986), Cook and Klinger (1990), and Wald (1993) and (1994)).

In the present section the previous 2-D F.E.M. plane stress model is constructed for the parametric analysis of the structural behaviour of the same column base plate connection. The model contains all the essential features that characterize the separation problem, whereas material yielding, contact interface slip and interface interaction are also taken into account. Secondary bending effects due to the static loading are not present. Note that the third dimension of the connection is also considered by assigning different thickness values to the various regions of the F.E.M. mesh.

Following the method presented in the previous paragraphs, local separation zones between the interfaces of the steel connection are computed and the deformed shape of the steel connection is with accuracy evaluated. Introducing the unilateral contact law with friction to simulate the boundary conditions along the steel connection interfaces, the separation process is mathematically described giving rise either to a primal quadratic programming problem with respect to the displacements, or equivalently to its dual formulation with respect to the stresses. As it has been proved by Panagiotopoulos (1975), the primal problem expresses from the standpoint of Mechanics, the principle of minimum of the potential energy of the connection under consideration at the state of equilibrium, whereas the dual one the principle of the minimum complementary energy. Solving iteratively these two problems and using the output of the former as input for the latter, the method converges after a few steps to the actual solution of the problem (see the work by Necas et al. (1980)). The transformation of the initial structural steel connection problem, into a sequence of constrained quadratic programming problems makes things from the computational point of view, much more simple and allows for the application of a wealth of numerical techniques and software. Within this theoretical framework the separation zones, the active contact, as well as the plastification ones are with accuracy calculated, leading thus to the computation of the exact stress state conditions holding along the several parts of the steel connection under investigation.

The numerical formulation of such problems in the form of quadratic programming problems permits the investigation of the appearance of prying action. In the case of the column base plate connection, the prying action phenomenon is directly connected with the flexibility of the connections. In flexible connections, e.g. in splice plates, such forces do not appear. Exactly opposite is the reaction in the case of steel column base plates with underlying concrete foundation. The difference is due to the fact, that the one part of the connection (concrete foundation) is not deformable and this leads this way the deformable base plate to be locally separated from the concrete surface. As is obvious, the thickness of the base plate is one of the most significant parameters that affect the response of such steel connections. Therefore, considering the thickness as a critical parameter for the analysis of the connection interface under various axial external forces and moment rotation, such a parametric analysis should be considered as a contribution, to the research on the structural response of steel column base plate connections. The parametric study of a numerical application is presented in the last paragraphs taking into account the base plate thickness as a parameter for analysis of the above described connections. The obtained results are of great interest because they show the different response of each connection of different base plate thickness under different loading conditions.

### 5.2 Brief Description of the Method

A steel connection composed of several structural elements such as plates, bolts, etc. is first considered (Fig.18). We assume that the material of the parts of the connection (which may be different for the various parts, e.g. high strength steel for the bolts, mild steel for the plates, etc.) exhibits an elastic-plastic behaviour with or without hardening. The interfaces that are formed between the various parts of the connection are in the following denoted by $\Gamma_s$. The forces applied on the connection are transmitted by the interfacial forces developed on the parts of the connection that come in contact. The latter are the contact forces that develop in the normal to the interfaces direction, and the frictional ones that develop along the tangential to the interfaces direction, and only between those parts that come in contact.

A direct solution of the above problem described by relations (48)-(56) is not possible due to the fact that it is not *a priori* known which parts of the connection come in contact under the application of the static loading and which are the exact values of the contact forces, whereas for those parts that come in contact, it is not known which of them are in sticking contact or in slipping contact. Moreover, there exists an interaction between the contact and the frictional forces, i.e. the frictional forces depend on the value of the contact force and the values of the frictional forces may alter the active contact area or the values of the contact forces.

For the above reasons the problem at hand is split into two sub-problems. The first sub-problem corresponds to the problem formulated with respect to the unknowns at the normal to the interface direction (problem (57)), whereas the second one corresponds to the problem formulated with respect to the unknowns at the tangential to the interface direction (problem (58)). The solution of the problem at hand is achieved by applying the PANA-iterative scheme. The solution of the minimization problem (57) gives the contact and non-contact areas, the values of the contact forces $S_N$ and the respective stress and strain states of the connection. The second minimization problem (58) gives the friction forces distribution along the active contact interfaces. The formulation of the second sub-problem has to be done with respect to stresses and not with respect to displacements because otherwise non-differentiable terms (due to the absolute value in friction and normal contact force) do appear. The iterative solution procedure continues until the difference of two successive stress states of the connection is less than a desirable accuracy.

Concerning the numerical treatment of the formulated quadratic programming problems, effective algorithms from the theory of Optimization can be applied (see e.g. Fletcher (1987)). As for the first step of the solution procedure (i.e. computation of the active contact and separation regions between the various interfaces), the optimization algorithm explicitly determines the nodes coming into contact with the respective contact forces. Then, the optimization algorithm is applied in a similar way within the second step of the method. The post-processing part of the algorithm concerning the computation of the new stress distribution is done as in the usual finite element analysis. It is therefore obvious, that the only real difference of the necessary software from classical finite element codes is the substitution of the solver of equations by a quadratic programming routine and the treatment of the additional data concerning the inequality constraints.

### 5.3 Numerical Application

The previously described algorithm is applied in order to simulate the behavior of the base plate connection shown in Fig.18. The connection consists of an RHS 120/200/10 steel column, which is connected to a concrete block through a steel plate. The thickness of the base plate is considered as a critical parameter, which varies taking the values 20mm, 25mm and 30mm. Also six M20-5.6 bolts are used. The connection is idealized through a two dimensional finite element mesh (Fig.18), consisting of 4044 nodes and 2829 plane stress quadrilateral elements. The presented finite element mesh is the final one of a series of F. E. meshes with different densities that were studied by Kontoleon et al. (1999). Attempts to further refinement of the mesh gave results very close to the ones presented here. Therefore, the results could be considered satisfactory for the analysis and on the other hand, the computational effort is reasonable. The thickness of the plane stress elements are properly adjusted in order to take into account the three-dimensional properties of the structure and are given in Table 1, except the bolt thickness that was taken equal to 47.1. In the region around the holes of the plate, the plate and the bolt are overlapping. The interaction between the two bodies is taken into account by considering unilateral contact conditions between them. Unilateral contact conditions are also assumed to hold between the plate and the concrete block. The elastic-plastic stress-strain law of the steel profile and the plate are given in Fig.19a. A similar diagram is used to describe the material of the bolt (Fig.19b). Finally, the material of the concrete block is considered as linear with modulus of elasticity $E_c$ = 29GPa. The friction coefficient between the base plate and the concrete is taken equal to $\mu$=0.3. The structure is loaded by applied displacements introduced as a sequence of 50 increments on the top of the edge of the column. At each increment a displacement of 2mm is applied into the structure. We distinguish three sets of solutions, each one corresponding to the three base plates with different plate thickness. In each set, the axial compressive loading of the connection consists of the following six cases: 0, 100kN, 200kN, 300kN, 400kN, 500kN.

From the first set of results for base plate thickness t=20mm, we notice that a contact zone is established under the right end of the base plate. The resting part of the plate starts separating from the concrete foundation, tensioning the left bolt. The maximum detachment of a selected base plate node near the left bolt is obtained for each increment and shown in Fig.23a, for the six cases of axial load. The deformation of the base plate, as well as the plastic strains decrease as axial load increases. The opposite situation occurs for the column that develops greater plastic strains with the increasing of the axial force. This fact is verified from the plastic strains of the base plate for the cases of axial force: 0kN, 100kN, 200kN where the plastic strains in the base plate are significantly larger than in the column. In these cases the structure collapses due to the plastification of the base plate as well as the failure of the left bolt. The left bolt develops tension forces near 400N/mm$^2$ at the 50$^{th}$ increment due to the fact that as the plate rises, it causes tension at the left bolt, creating an additional critical member failure. For the applied axial force 300kN, the plastic strains are similar in the column and base plate and finally, for the load cases of 400kN and 500kN the plastic strains of the column are larger leading the RHS column to fail first, developing plastic hinges at the lower parts of the edges of the steel column. Fig.24a gives the deformed shapes of the connection for the six cases of axial loading at increments 10, 20, 30, 40, 50. From the deformed shapes it is observed that the larger part of the steel base plate during the first 20 increments is always in contact with the concrete. For the

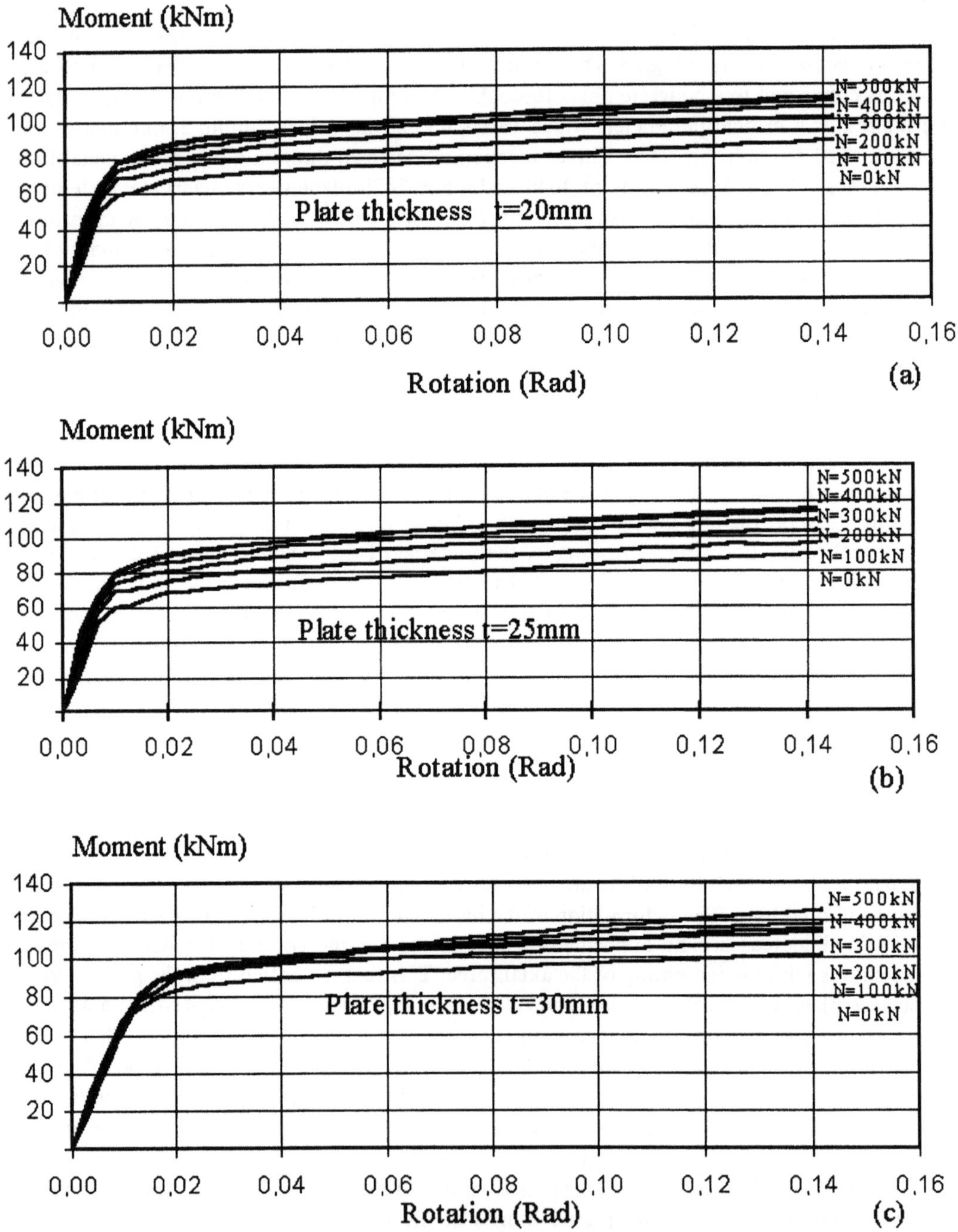


**Figure 23.** Moment-rotation diagrams.

following increments, as the base plate uplifts, only the left edge node and the base plate nodes near the right bolt remain in contact with the concrete foundation. This contact area increases along with the axial loading from 0kN to 500kN.

Concerning the stress condition of the connection, although the right region of the column and the base plate are naturally the first expected failure areas, similar stresses are developed in the left region after the 14th-18th increment. This phenomenon occurs because of the prying forces that developed in this area, when the left edge of the base plate comes in contact with the concrete base. From the moment-rotation curves (Fig.23) it becomes obvious that the moment capacity of the connection increases along with the increasing of the axial force. For axial force 0kN the moment capacity is near 90kNm. In the case of 500kN the moment capacity reaches 116kNm.

The stress condition as well as the plastic strains for the second set of results (base plate thickness t=25mm) change because prying forces are not present in this case. Comparing these results with the model with base plate thickness t=20mm, the highest uplifting is slightly smaller, as well as the separation length. These results are natural, since the stiffness of the base plate increases for thickness t=25mm, permitting smaller deformability and reducing its final plastic strains. This fact is also observed from the maximum node detachment of the base plate (Fig.25b, at the left edge of the base plate). The steel column for each case of axial force, fails around the area of its right foot near the 40th increment, with stresses that exceed its ultimate strength. The stresses appearing in the left region of the base plate are beneath its ultimate strength. This occurs because the prying forces which were developed in the previous analysis for plate thickness t=20mm do not appear. Through the deformed shapes of the base plate for the six cases of axial loading at increments 10, 20, 30, 40, 50 (Fig.24b), we notice that the bending of the base plate is reduced in comparison with the connection of plate thickness t=20mm. This proves that the increasing of the stiffness of the base plate, significantly affects its response under the applied axial loading and bending moment combination.

For axial force 0kN and 100kN the base plate and the left bolt exceed their yield strength and reach first their ultimate plastic strains in a similar way with the case for the t=20mm base plate. For the remaining loads from 200kN-500kN the column starts to yield creating plastification areas, which in comparison with the plate with thickness 20mm are slightly smaller. The other parts of the connection are not critical and still preserve strength capacity. From the moment-rotation curves (Fig.23b), for axial force 0kN the moment capacity is near 100kNm and in the case of 500kN, the moment capacity reaches 120kNm.

From the third set of results for base plate thickness t=30mm, the stress condition of the plate is similar with the case of plate thickness t=25mm, because the prying forces are not present. Increasing the base plate thickness, the stiffness it possesses permits limited deformation that is slightly visible only for the last increments. As a result, the base plate does not fail for any case of axial load. Significant stresses are developed mainly at the area of the right foot of the column, which fails first exceeding its ultimate plastic strain. Figure 25c provides us a full picture of the maximum detachment of the node at the left edge of the base plate. In comparison with the previous plate thickness 20mm, 25mm the maximum height the plate rises is smaller, especially in the case of axial force 500kN. Figures 24c presents the deformation of the connection for six cases of axial loading at increments 10,20,30,40,50. Through all the cases of loading, the column fails after the 40th increment. The developed plastic areas create a plastic hinge at the lower right

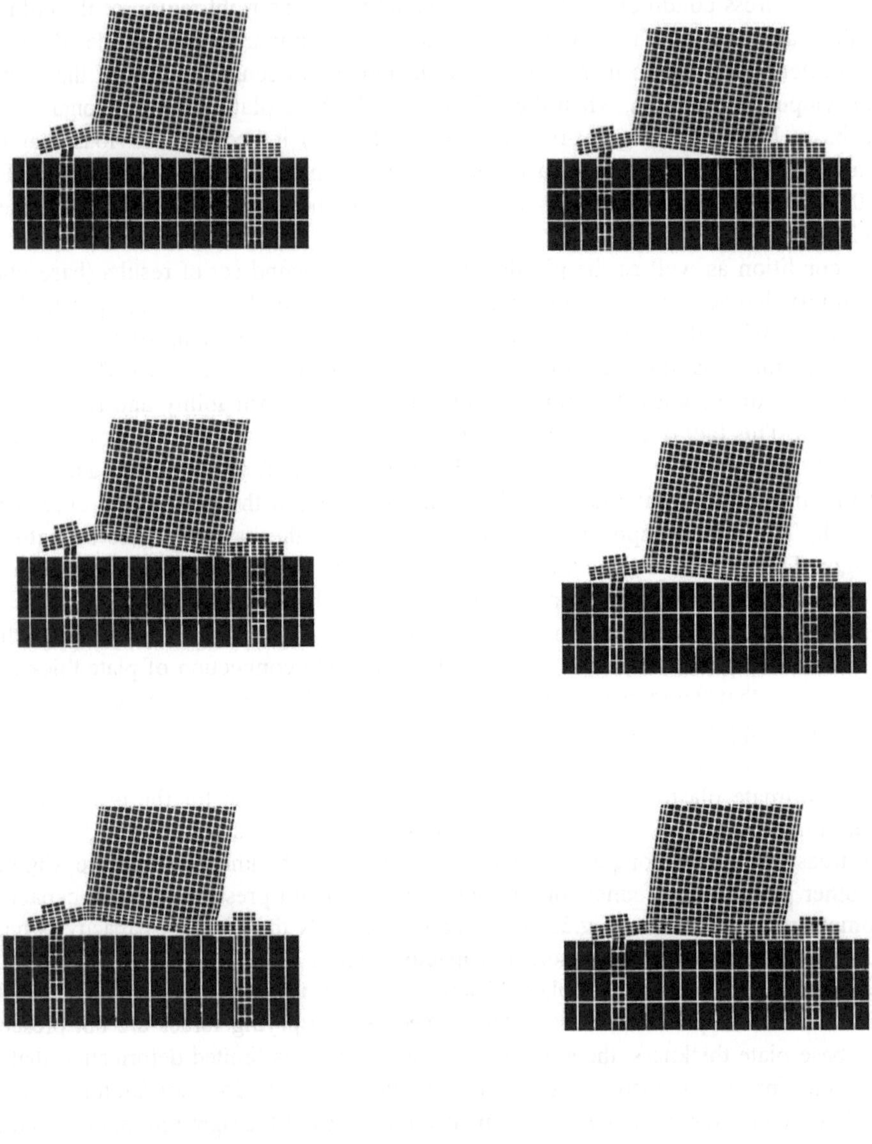

**Figure 24a**. Deformed shapes of the column base-plate with plate thickness t=20 mm for axial loading N=0, 100, 200, 300, 400, 500 kN (see the columns downwards).

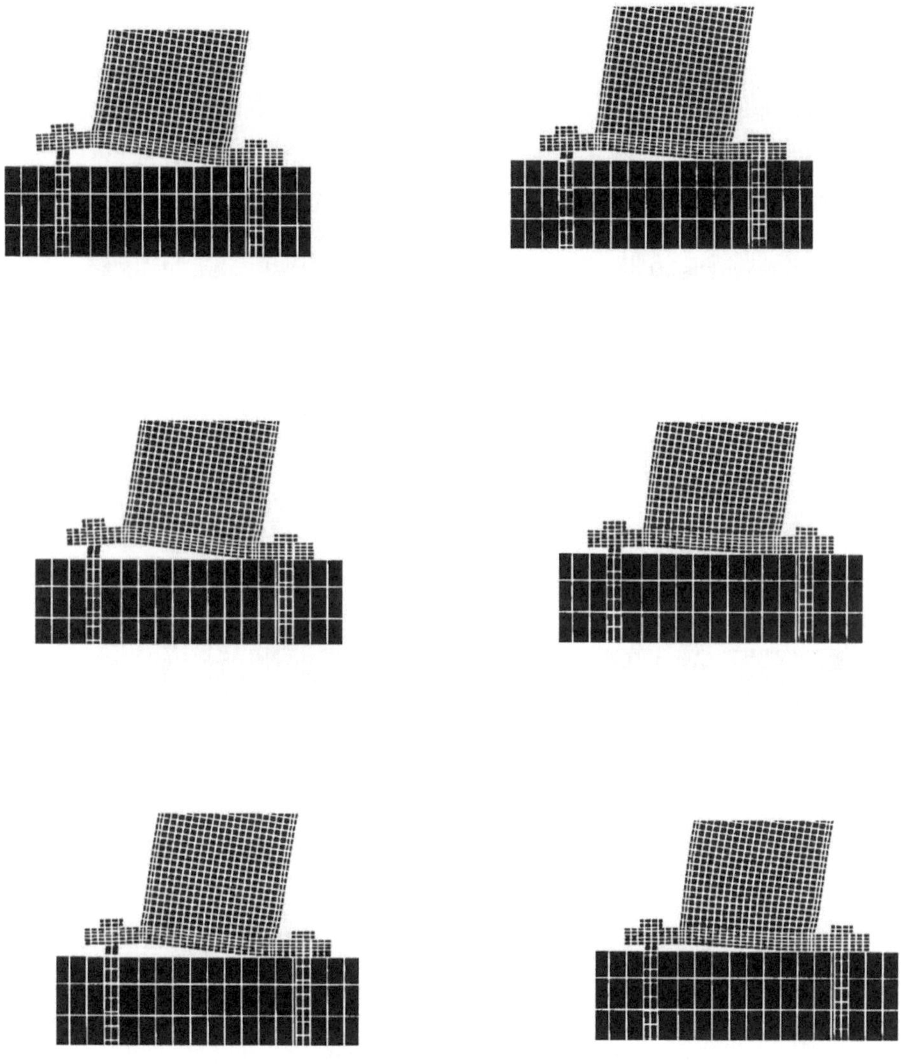

**Figure 24b**. Deformed shapes of the column base-plate with plate thickness t=25 mm for axial loading N=0, 100, 200, 300, 400, 500 kN (see the columns downwards).

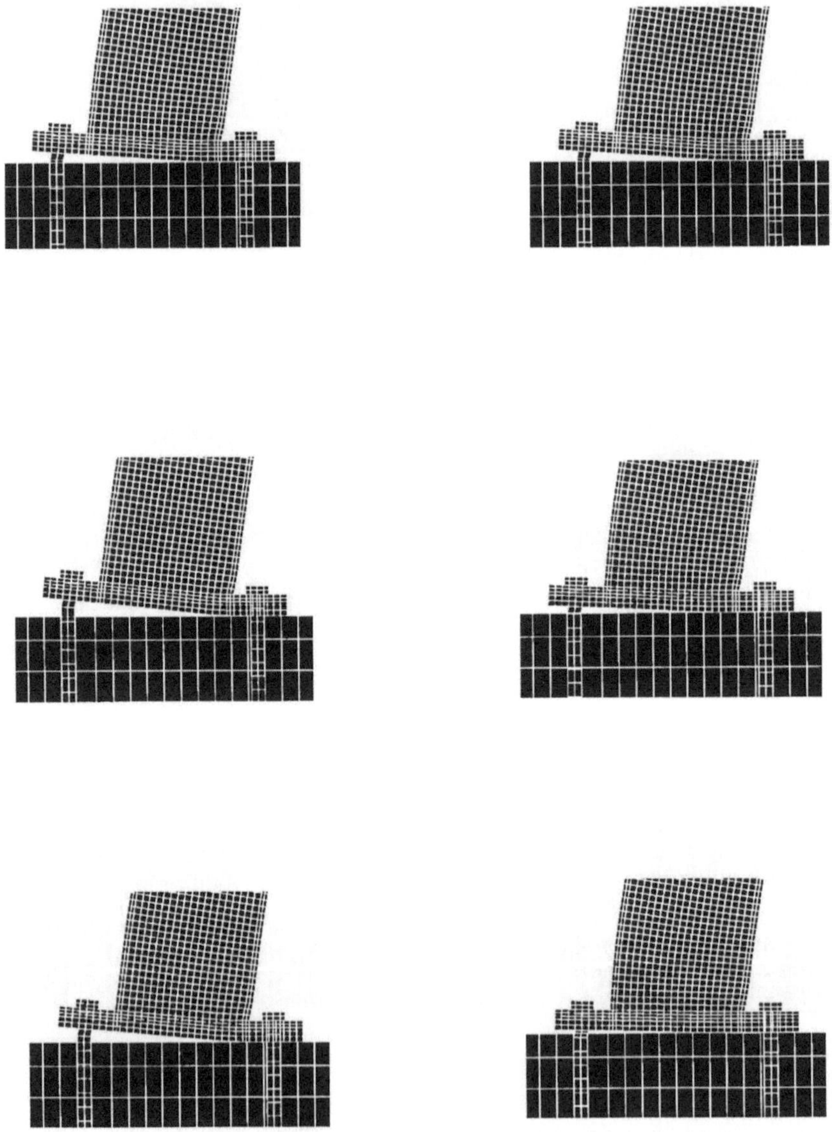

**Figure 24c.** Deformed shapes of the column base-plate with plate thickness t=30 mm for axial loading N=0, 100, 200, 300, 400, 500 kN (see the columns downwards).

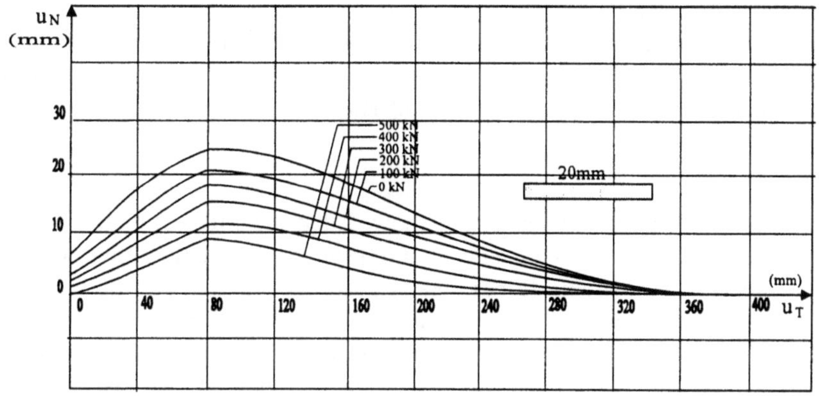


(a)

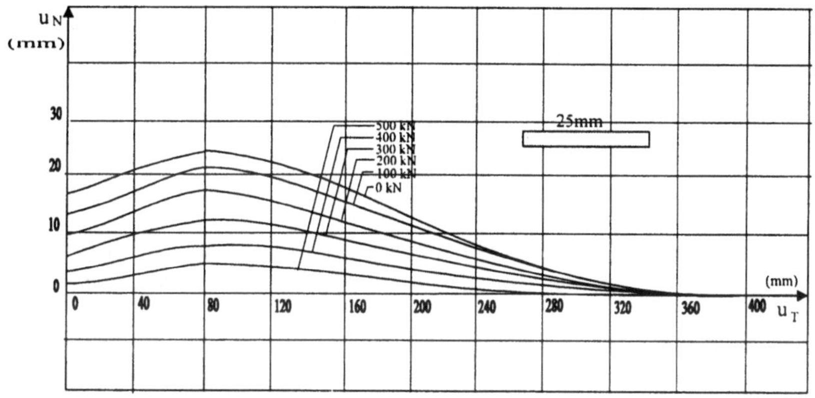


(b)

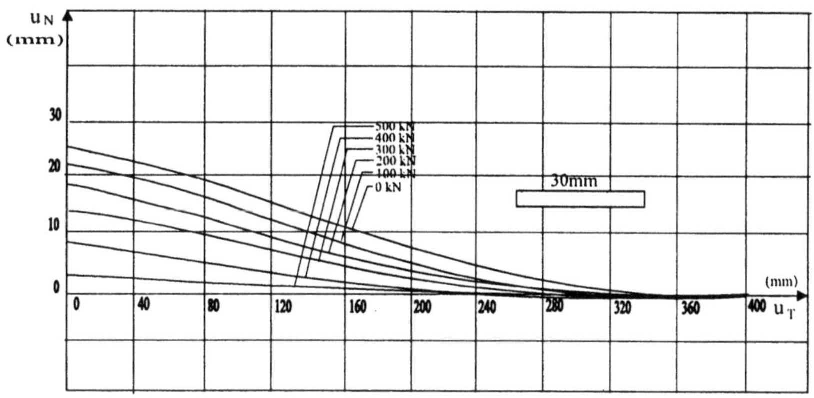


**Figure 25.** Deformed shapes of the base plates .

part of the steel column. Thus, the collapse of the structure occurs due to the plastification of the steel column and the other parts of the connection are not critical. The obtained moment-rotation curves (Fig.23c) are similar to the previous cases. For axial force 0kN, the moment capacity is near 100kNm. In the case of 500kN, the moment capacity reaches 125kNm.

### 5.4 Conclusion

The present parametric analysis of the model showed that the stiffness of the base plate is a significant parameter that affects the development of prying action at the active contact areas of the plate. The appearance of prying forces creates plastification zones at the interfaces of the connections in areas that could not be considered using classical design and calculation methods. The reader should have in mind that a 2-D model is herein analysed instead a more accurate 3-D one; this 2-D model obviously is not suitable to predict secondary processes, as is e.g. a cone mechanism, which however, in most cases do not have any significance into the overall strength of the joint. The present 2-D analysis encompasses all the essential characteristics and dominant plastification mechanisms of the problem under investigation. Indeed, the connection at hand under axial force and moment loading comes in a more or less 2-D deformed configuration, something that justifies the validity of the assumptions that led to the herein proposed numerical model.

## 6 Analysis of Tubular K-joints by Means of 3-D Numerical Modelling

The subject discussed in this paragraph is the behaviour of steel square hollow section (SHS) gap K-joints, as studied with the use of the F.E.M. The analysis was achieved by the development of a 3-D model of gap K-joints and the influence of geometrical parameters to the strength and to the overall stability of the joint was investigated. The geometry of the joints was in agreement with the general requirements of the EC3 and therefore, the results obtained could be directly compared to those obtained by the analytical formulas provided by the EC3-Annex K.

### 6.1 Introduction

The strength of structural steel hollow section joints has been widely studied during the last decades by many researchers (see e.g. Korol and Packer (1992), Koskimaki and Niemi (1990), Paker et al. (1992), Soh and Soh (1994) and Wardenier (1984)). The present study concerns a further investigation of both strength and deformation aspects by means of a finite element analysis, aiming at highlighting the geometric parameters that contribute to a credible design. The analysis simulated a planar gap K-joint loaded on the brace members anti-symmetrically. The load distribution was in respect with the usual static load conditions (service load conditions) and was gradually (step-by-step) applied to the numerical model.

The mesh of the generated 3-D model consisted of more than 2000 thick shell elements, designed in agreement with the EC3-Annex K requirements. As parameters of this analysis the chord wall thickness and the chord width have been considered, while the brace sections, the angle and the gap between them have been considered as constant.

## 6.2 On the 3-D F.E.M. Model

The numerical model shown in Fig.26 was composed by the two brace members and the chord member in such way that symmetry was assured: the brace sections were exactly the same and were located symmetrically (same angle (θ) between them and the chord member). The distance between them (gap) was also standard for all of the models.

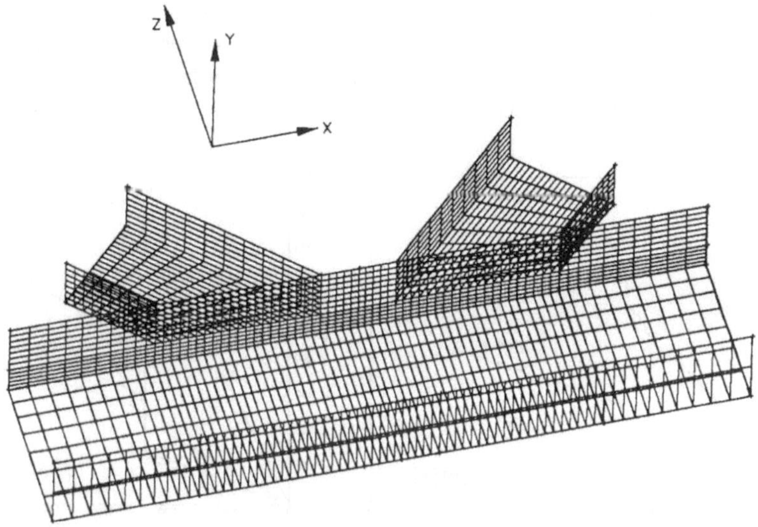


**Figure 26**. Mesh of the 3-D model.

The variables of the analysis were the chord wall width and the chord wall thickness. Each model was provided with a different combination of chord width ($b_0$) and chord thickness ($t_0$). There were four groups of $b_0$ with four different $t_0$ (sixteen models in total) (Table 2).

| Model | $b_o$ (mm) | $t_o$ (mm) | $\beta = b_i / b_o$ | Model | $b_o$ (mm) | $t_o$ (mm) | $\beta = b_i / b_o$ |
|---|---|---|---|---|---|---|---|
| 1 | 200 | 6,3 | 0,500 | 9 | 150 | 5 | 0,667 |
| 2 | 200 | 8 | 0,500 | 10 | 150 | 6,3 | 0,667 |
| 3 | 200 | 10 | 0,500 | 11 | 150 | 8 | 0,667 |
| 4 | 200 | 12,5 | 0,500 | 12 | 150 | 10 | 0,667 |
| 5 | 180 | 6,3 | 0,555 | 13 | 120 | 5 | 0,833 |
| 6 | 180 | 8 | 0,555 | 14 | 120 | 6,3 | 0,833 |
| 7 | 180 | 10 | 0,555 | 15 | 120 | 8 | 0,833 |
| 8 | 180 | 12,5 | 0,555 | 16 | 120 | 10 | 0,833 |

**Table 2.** Geometry of the 3-D F.E.M. models.

In an attempt to reduce the computer time, it was decided to take advantage of the symmetry on the x-z level. Therefore, only the half of the connection was finally simulated in respect with the required boundary conditions.

All the elements have been provided with isotropic and plastic properties that comply to the von Mises criterion. Two groups of material properties were defined; they were characterized by a different element thickness : the group characterized by the double thickness represents the welds and their contribution to the stiffness around the connection area, as well as the influence of the rounded edges of the brace section to the even transition of the stresses. The thick shell four-nodded elements that have been used have 6 degrees of freedom per node. The results that are presented and discussed in this study refer to the two outer layers and to the mid-layer through the shell thickness of the elements.

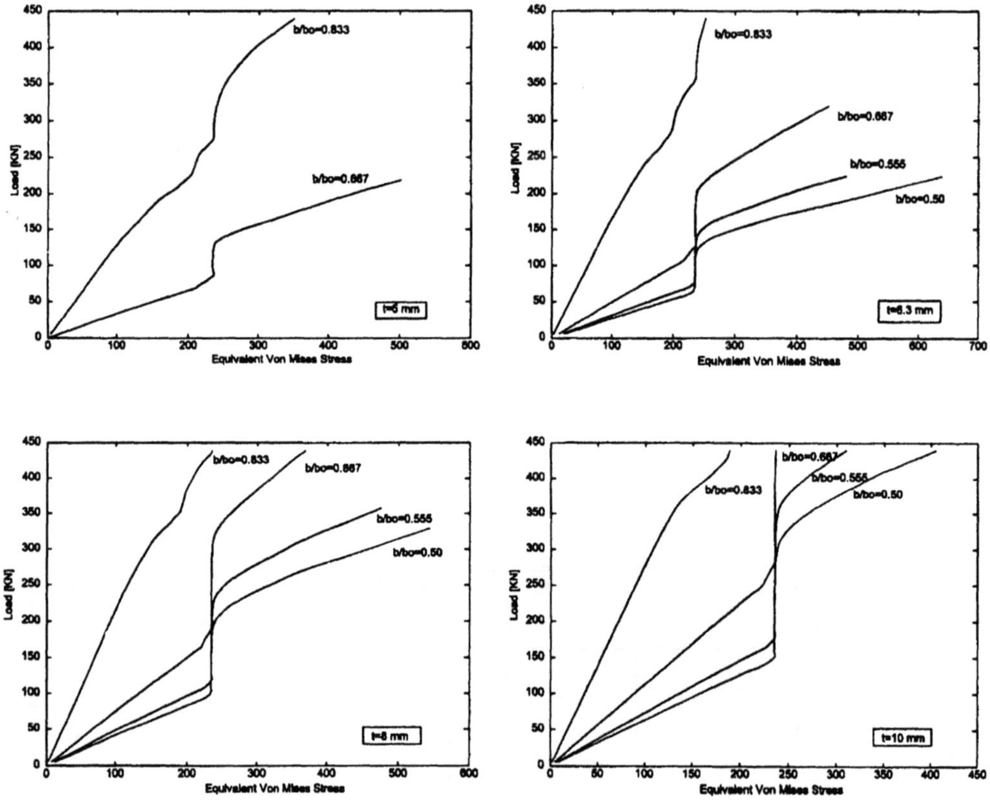


**Figure 27**. Load-von Mises stress curves for constant chord wall thickness and variable $b/b_0$ ratio.

## 6.3 Numerical Results

The dominating issue is how the stress and strain values develop throughout the step load application procedure, from the first step to the defined failure state. The definition of failure differs respectively to the stress and strain paths, dependent on the excess of a specific deformation and stress concentration. The criteria that have been chosen in the present analysis are the excess of 20% for the plastic strain and the 360 MPa for the von Mises stress.

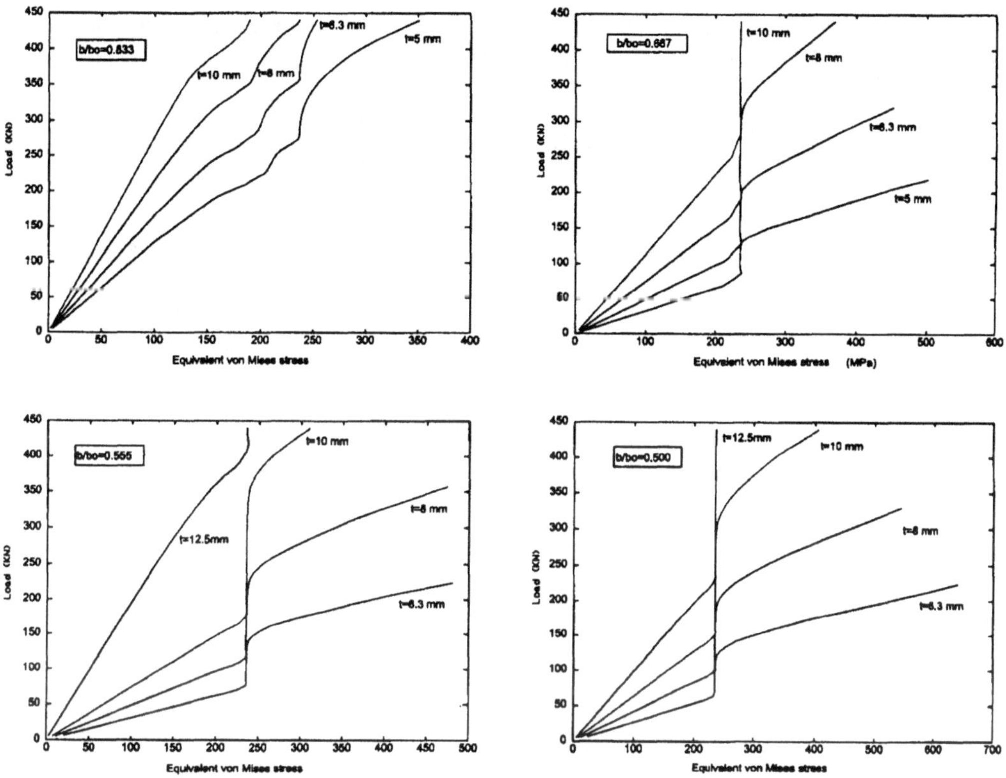


**Figure 28**. Load-von Mises stress curves for constant $b/b_0$ ratio and variable chord wall thickness.

Four types of response have been observed :

a. Chord face failure -It took place in the majority of the models.

b. Brace failure with reduced effective width: It was observed only for two of the joints, and they both had a high $\beta$ ratio.

c. Combination of the chord face and brace failure modes: The joint with the smallest and thinnest chord developed the characteristics of both of the aforementioned failure modes.

d. No Failure under the total applied load: There were some connections that did not fail according to the defined criteria of failure. They had all a quite thick chord.

Comparing the types of response (on the parameters $\beta = b_0/t_0$ and $t_0$), the following remarks are made:

1. For the same chord wall thickness $t_0$, the strength of the joint increases for higher $\beta$ ratios.
2. However, for very high $\beta$ ratios it is possible that the brace failure takes place first, as the function of the joint is more likely a Y-type function rather than a K-type, where the wall thickness of the brace is highly important. In that case, the behaviour of the connection can be improved by selecting a thicker brace member.

3. For the same $\beta$ ratio and a different $t_0$, the strength of the connection is higher proportionally to the increase of the $t_0$. Thus, a stiffer chord contributes to the improvement of the joint behaviour.

### 6.4 Comparison of the Analytical and Numerical Results

The calculation formulas included in EC3-Annex K were used in order to obtain comparable strength values and therefore, to inspect the credibility of the numerical results. The adopted

| | Analytical formulas (EC3) | | 3-D F.E.M. | | Div. |
|---|---|---|---|---|---|
| Model | Strength (KN) | ($N_{i,rd}/N_{it,rd}$) | Strength (KN) | ($N_{i,rd}/N_{it,rd}$) | (%) |
| 1 | 233,86 | 0,53 | 166,40 | 0,37 | 28,8 |
| 2 | 334,64 | 0,75 | 275,00 | 0,62 | 17,8 |
| 3 | 467,67 | >1 | 418,00 | 0,94 | 10,6 |
| 4 | 653,59 | >1 | >440 | >1 | - |
| 5 | 246,51 | 0,56 | 192,00 | 0,43 | 22,1 |
| 6 | 352,74 | 0,79 | 313,50 | 0,71 | 11,1 |
| 7 | 492,97 | >1 | >440 | >1 | - |
| 8 | 688,95 | >! | >440 | >1 | - |
| 9 | 190,93 | 0,43 | 175,00 | 0,39 | 8,3 |
| 10 | 270,04 | 0,61 | 275,20 | 0,62 | -1,9 |
| 11 | 386,41 | 0,87 | 412,50 | 0,93 | -6,8 |
| 12 | 540,02 | >1 | >440 | >1 | - |
| 13 | 309,42 | 0,70 | 363,00 | 0,82 | -17,3 |
| 14 | 366,95 | 0,83 | 390,05 | 0,88 | -6,4 |
| 15 | 462,17 | >1 | >440 | >1 | - |
| 16 | 603,17 | >1 | >440 | >1 | - |

**Table 3.** Comparison of the joint strength values.

assumptions and the techniques used to facilitate the modelling led to a slight deviation of the numerical (FEM) from the calculated (EC3) strength values (Table 3), but generally the comparison was satisfactory (Fig.30). Nevertheless, the deformed shape was the expected one for all the models and in this aspect the simulation was fully acceptable.

### 6.5 Conclusion

The results of the 3-D model presented in this section have shown that the geometric parameters ($t_0$, $b/b_0$) are important factors for the safety of the joints. The design of the connection is a very important stage, as a potential failure of the connection might lead to the modification in the geometry of the whole structure.

The choice of brace and chord members must lead to high $\beta = b_i/b_0$ ratios, as well to a thick walled chord. It is also important to avoid a great size difference (dimensions) between the

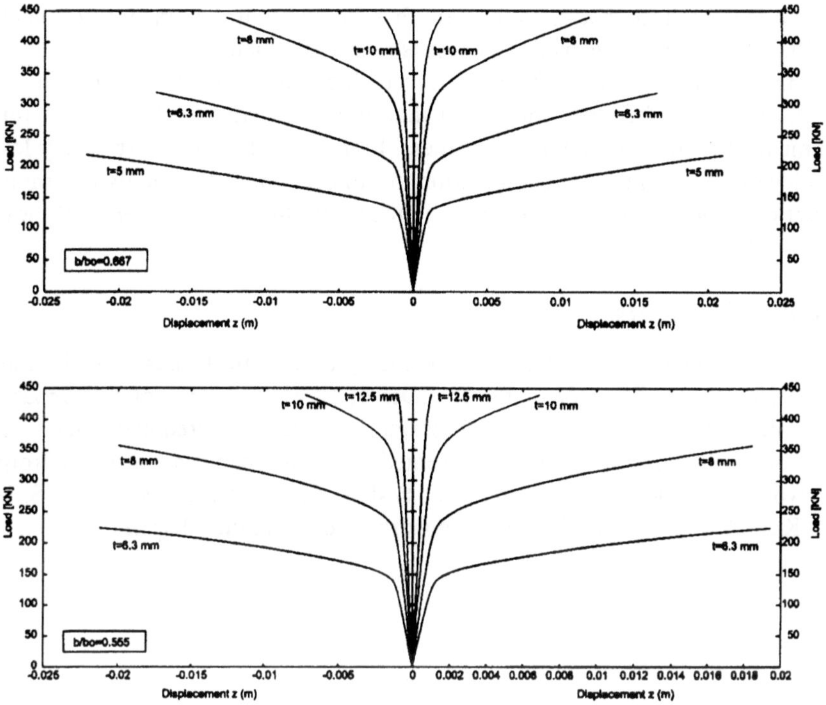


**Figure 29**. Load-displacement curves at a certain node of the brace members base on the symmetry axis for $b/b_o$=0.667 and $b/b_o$=0.555.

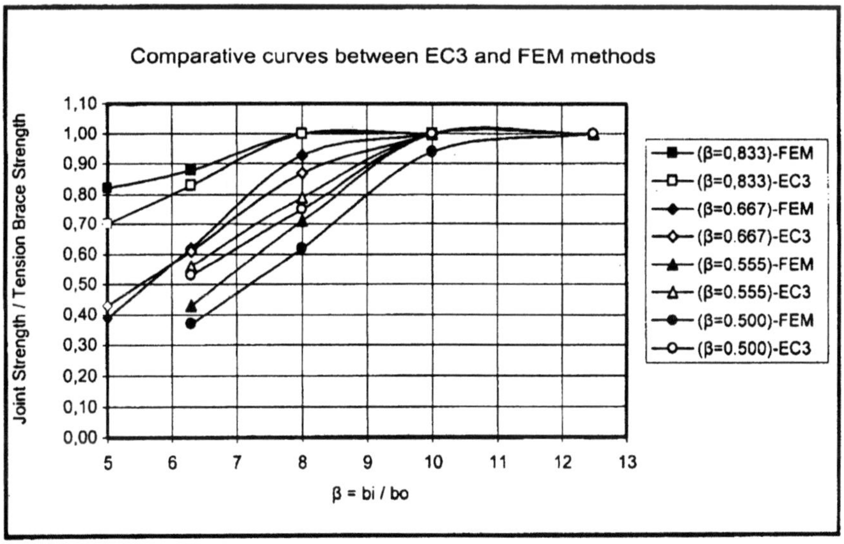


**Figure 30**. Joint strength/tension brace strength with respect to variation of $b/b_o$ according to the EC3 and the F.E.M. results.

brace and the chord members, taking care that the width of the brace is not much smaller than this of the chord, even if its wall thickness is relatively small.

In the case that the $\beta = b_i/b_0$ ratio is very high (0.9-1.0), the brace failure with reduced effective width can become crucial. In this case, a greater wall thickness for the brace is recommended to increase the brace stiffness. In the case that joints carry very low brace loads, the choice of lower $\beta$ ratios in combination to a thicker chord is strongly recommended because for great chord wall thickness the gap K-joints rarely fail, even for relatively low $\beta$ ratios (but at least $\beta \geq 0.4$).

## Acknowledgements

The present chapter is dedicated to the memory of the late Professor P. D. Panagiotopoulos (1950-1998) who as Director of the Institute of Steel Structures of the Department of Civil Engineering, Aristotle University of Thessaloniki, Greece inspired and guided all of us for two decades along the lines of the herein presented research. Thanks are also due from the author to his co-workers Associate Professor Dr. C. Bisbos, Assistant Professor Dr. E.S. Mistakidis and Dr. M. Kontoleon who contributed to the work reported in this chapter.

# References


Abdalla, K. M. (1988). *Application of the Theory of Unilateral Problems to the Calculation of Connections of Steel Structures*. Ph.D. Dissertation, Aristotle University, Thessaloniki.

Abdalla, K. M., and Baniotopoulos, C. C. (1991). Design sensitivity investigations of column splices in steel structures. In *Proceedings of the First Greek Conference on Steel Structures*. Athens: M.S.R.S., 120-129.

Abdalla, K. M., and Stavroulakis, G. E. (1989). Zur rationalen Berechnung des "Prying-Actions"-Phänomens in Schraubenverbindungen. *Stahlbau* 58:233-238.

AISC (1981). *Manual of Steel Structures*. Chicago: AISC, 8th Edition.

Ballio, G. and Mazzolani, F. M. (1977): *Theory and Design of Steel Structures*. Bristol: J. W. Arrowsmith.

Baniotopoulos, C. C. (1994). On the numerical assessment of the separation zones in semi-rigid column base plate connections. *Structural Engineering and Mecanics* 2:1-15.

Baniotopoulos, C. C., (1995): On the separation process in bolted steel splice plates. *Journal of Constructional Steel Research* 32:15-35.

Baniotopoulos, C. C., Karoumbas, G., and Panagiotopoulos, P. D. (1992). A contribution to the analysis of steel connections by means of quadratic programming techniques. In *Proceedings of the First European Conference on Numerical Methods in Engineering*. Amsterdam, London, New York, Tokyo: Elsevier, 519-525.

Baniotopoulos, C. C., and Abdalla, K. M. (1993). Steel column-to-column connections under combined load: A quadratic programming approach. *Computers and Structures* 46:13-20.

Baniotopoulos, C. C., Sokol, Z., and Wald, F. (1999). Column base connections. In COST C1 Report of WG6-Numerical simulation. Brussels-Luxemburg: European Commission, 32-47.

Bendsoe, M. P., Olhoff, N., and Sokolowski, J. (1985). Sensitivity analysis of problems of Elasticity with unilateral constraints. *Journal of Structural Mechanics* 13:201-222.

Bendsoe, M. P. and Sokolowski, J. (1988). Design sensitivity analysis for elastic-plastic analysis problems. *Mechanics, Structures and Machines* 16:81-102.

Bogdan, H., Kuzmanovic, O., and Willems, N. (1977). *Steel Design of Structural Engineers*. New Jersey: Prentice Hall.

Bortman, J., and Szabó, B.A. (1992). Nonlinear models for fastened structural connections. *Computers and Structures* 43:909-923.

Bufler, H. (1968): Zur Theorie diskontinuierlichen und kontinuierlichen Verbindungen. *Ingenieur Archiv* 37:176-188.

Bufler, H., and Müßigmann, F. (1977). Kraftübertragung bei diskontinuierlichen Scher-verbindungen. *Stahlbau* 46:12-20.

CECM (1978). *European Recommendation for Steel Construction*. Brussels: ECCS.

Chen, W. F., and Patel, K. V. (1981). Static behavior of beam-to-column moment. *ASCE Journal of Structural Division* 107:1815-1838.

Chen, W. F., and Lui, E. M. (1986). Steel beam-to-column moment connections. Part I: Flange moment connections. *Solid Mechanics Archives* 11:257-316.

Cook, R. A., and Klinger, R. E.(1990). Ductile multiple-anchor steel-to-concrete connections. *ASCE Journal of Structural Division* 118:1645-1665.

Dewolfe, J. T. (1978). Axially loaded column base plates. *ASCE Journal of Structural Division* 104:781-794.

Dewolfe, J. T., and Sansley, E. F. (1983). Column base plates with axial loads and moments. *ASCE Journal of Structural Division* 106:2167-2184.

Eurocode 3- Part 1, Annex K (normative) (1994). *Hollow section lattice girder connections,* ENV 1993-1-1 : 1992/A1, Brussels.

Fichera, G. (1972). Boundary value problems in Elasticity with unilateral constraints. In *Encyclopedia of Physics*. Berlin: Springer, VIa/2:391-424.

Fisher, J. W., and Struik, J. H. A. (1978). *Guide to Design Criteria for Bolted and Riveted Joints*. New York: John Wiley and Sons.

Fletcher, R. (1987). *Practical Optimization Methods*. N. York: John Wiley & Sons.

Fling, R. S. (1970). Design of steel bearing plates. *AISC Engineering Journal* 7: 37-40.

Hyer, M. W., and Klang, E. C. (1985). Contact stresses in pin-loaded orthotropic plates. *International Journal of Solids Structures* 21:957-975.

Jaspart, J.-P. (1994): Numerical simulation of a T-stub. Experimental Data. *COST C1 Numerical simulation working group* C1WD6/94/09:1-9.

Jaspart, J.-P. and Bursi, O. (1997). Calibration of a finite element model for bolted endplate steel connections. *Journal of Constructional Steel Research* 44:225-262.

Kato, B. and McGuire, W. (1973). Analysis of T-stub flange to column connection. *ASCE Journal of Structural Division* 99:865-888.

Kontoleon, M., Mistakidis, E. S., Baniotopoulos, C. C., and Panagiotopoulos, P. D. (1999). Parametric analysis of the structural response of steel base-plate connections. *Computers and Structures* 71:87-103.

Korol, R., and Packer, J. (1992). *Behaviour of RHS Gap K-joints at Service Loads*. University of Toronto, Toronto.

Koskimaki, M., and Niemi, E. (1990). *Finite Element Studies on the Behavior of Rectangular Hollow Section K-joints*. Lapeenranta University of Technology, Rovaniemi.

Krishnamurthy, N., and Graddy, D. (1976). Correlation between 2-and 3-dimensional finite element analysis of steel bolted end plate connections. *Computers and Structures* 6:381-389.

Krishnamurthy, N. (1978), A fresh look at bolted steel end-plate behavior and design. *AISC Engineering Journal* 15: 39-49.

Künzi, H., and Krelle, W. (1962). *Nichtlineare Programmierung*. Berlin: Springer.

Kukreti, A. R., Murray, T. M., and Abolmaali, A. (1987). End plate connection moment-rotation relationship. *Journal of Constructional Steel Research* 8:137-157.

Maier, G. (1968). A quadratic programming approach for certain classes of non-linear structural problems. *Meccanica*, 2:121-130.

Mistakidis, E. S., Baniotopoulos, C. C., and Panagiotopoulos, P.D. (1996a): A numerical method for the analysis of semi-rigid base plate connections. In *Proceedings of ECCOMAS 96*, Paris: J. Wiley & Sons Ltd, 842-848.

Mistakidis, E. S., Baniotopoulos, C. C., Bisbos, C. D., and Panagiotopoulos, P. D. (1996b). A 2-D numerical method for the analysis of steel T-stub connections. In *Proceedings of the 2nd Greek Conference on Computational Mechanics*, Chania: GRACM, 777-748.

Mistakidis, E. S., Baniotopoulos, C. C. and Panagiotopoulos, P. D. (1998). An effective two-dimensional numerical method for the analysis of a class of steel connections. *Computational Mechanics* 21:363-371.

Mistakidis, E. S., Baniotopoulos, C. C., Bisbos, C. D., and Panagiotopoulos, P.D. (1997). Steel T-stub Connections Under Static Loading: an Effective 2-D Numerical Model. *Journal of Constructional Steel Research* 44: 51-67.

Moreau, J. J., Panagiotopoulos, P. D., and Strang, G. (eds) (1988). *Topics in Nonsmooth Mechanics*. Basel, Boston: Birkhäuser.

Moreau, J. J., and Panagiotopoulos, P.D. (eds) (1988). *Nonsmooth Mechanics and Applications*. CISM Lecture Notes 302. Wien, New York: Springer.

Murray, T.M. (1983), Design of lightly loaded steel column base plates. *AISC Engineering Journal* 20: 143-152.

Necas, J., Jarusek, J., and Haslinger, J. (1980). On the solution of the variational inequality to the Signorini problem with small friction. *Bulletino U.M.I.* 17B:796-811.

Paker, S. A., and Morris, L.J. (1977). A limit state design method for tension of bolted column connections. *Structural Engineer* 55:876-889.

Paker, J. A., Wardenier, J., Kurobane, Y., Dutta, D. and Yeomans, N. (1992). *Design Guide for Rectangular Section (RHS) Joints under Predominantly Static Loading*. CIDECT, Verlag TUV Reinland, Kologne.

Panagiotopoulos, P. D. (1975). A nonlinear programming approach to the unilateral contact and friction boundary value problem in the theory of Elasticity. *Ingenieur Archiv* 44:421-432.

Panagiotopoulos, P. D. (1976). Convex analysis and unilateral static problems. *Ingenieur Archiv* 45:55-68.

Panagiotopoulos, P. D. (1985). *Inequality Problems in Mechanics and Applications. Convex and Nonconvex Energy Functions*. Basel, Boston: Birkhäuser ((1989), Russian translation, Moscow: Mir).

Panagiotopoulos, P. D. (1988). *Hemivariational Inequalities. Applications in Mechanics and Engineering*. Boston-Basel: Birkhäuser.

Panagiotopoulos, P. D., and Talaslidis, D. (1980). A linear analysis approach to the solution of certain classes of variational inequality problems in Structural Analysis. *International Journal of Solids and Structures* 16:991-1005.

Panagiotopoulos, P. D., Baniotopoulos, C. C. and Avdelas, A. V. (1984). Certain propositions on the activation of yield modes in Elastoplasticity and their applications to deterministic and stochastic problems. *Zeitschrift fuer Angewandte Mathematik und Mechanik* 64:491-501.

Panagiotopoulos, P. D., and Strang, G. (1988). *Topics in Nonsmooth Mechanics*. Boston-Basel: Birkhäuser.

Raffa, F., and Strona, P. (1984). Boundary Element Method application to bolted joint analysis. *Engineering Analysis* 1:78-89.

Sakellariadou, H. I., Bisbos, C. D., Thomopoulos, K., and Panagiotopoulos, P. D. (1996). A 3-D numerical study of the T-stub problem taking into account the interfacial unilateral contact effects. In *Proceedings of the Conference in Advances in Computational Methods for Simulation*. Civil Comp Press, Edinburgh: 115-124.

Soh, A., and Soh, C. (1994). *On the Compatibility Conditions for Modeling Tubular Joints Using Different Types of Elements*. Nanyang University of Singapore, Singapore.

Stockwell, Jr., F. W. (1975). Preliminary base plate selection. *AISC Engineering Journal* 12: 92-99.

Mistakidis, E. S., Baniotopoulos, C. C., Bisbos, C. D., and Panagiotopoulos, P.D. (1997). Steel T-stub Connections Under Static Loading: an Effective 2-D Numerical Model. *Journal of Constructional Steel Research* 44: 51-67.

Moreau, J. J., Panagiotopoulos, P. D., and Strang, G. (eds) (1988). *Topics in Nonsmooth Mechanics*. Basel, Boston: Birkhäuser.

Moreau, J. J., and Panagiotopoulos, P.D. (eds) (1988). *Nonsmooth Mechanics and Applications*. CISM Lecture Notes 302. Wien, New York: Springer.

Murray, T.M. (1983), Design of lightly loaded steel column base plates. *AISC Engineering Journal* 20: 143-152.

Necas, J., Jarusek, J., and Haslinger, J. (1980). On the solution of the variational inequality to the Signorini problem with small friction. *Bulletino U.M.I.* 17B:796-811.

Paker, S. A., and Morris, L.J. (1977). A limit state design method for tension of bolted column connections. *Structural Engineer* 55:876-889.

Paker, J. A., Wardenier, J., Kurobane, Y., Dutta, D. and Yeomans, N. (1992). *Design Guide for Rectangular Section (RHS) Joints under Predominantly Static Loading*. CIDECT, Verlag TUV Reinland, Kologne.

Panagiotopoulos, P. D. (1975). A nonlinear programming approach to the unilateral contact and friction boundary value problem in the theory of Elasticity. *Ingenieur Archiv* 44:421-432.

Panagiotopoulos, P. D. (1976). Convex analysis and unilateral static problems. *Ingenieur Archiv* 45:55-68.

Panagiotopoulos, P. D. (1985). *Inequality Problems in Mechanics and Applications. Convex and Nonconvex Energy Functions*. Basel, Boston: Birkhäuser ((1989), Russian translation, Moscow: Mir).

Panagiotopoulos, P. D. (1988). *Hemivariational Inequalities. Applications in Mechanics and Engineering*. Boston-Basel: Birkhäuser.

Panagiotopoulos, P. D., and Talaslidis, D. (1980). A linear analysis approach to the solution of certain classes of variational inequality problems in Structural Analysis. *International Journal of Solids and Structures* 16:991-1005.

Panagiotopoulos, P. D., Baniotopoulos, C. C. and Avdelas, A. V. (1984). Certain propositions on the activation of yield modes in Elastoplasticity and their applications to deterministic and stochastic problems. *Zeitschrift fuer Angewandte Mathematik und Mechanik* 64:491-501.

Panagiotopoulos, P. D., and Strang, G. (1988). *Topics in Nonsmooth Mechanics*. Boston-Basel: Birkhäuser.

Raffa, F., and Strona, P. (1984). Boundary Element Method application to bolted joint analysis. *Engineering Analysis* 1:78-89.

Sakellariadou, H. I., Bisbos, C. D., Thomopoulos, K., and Panagiotopoulos, P. D. (1996). A 3-D numerical study of the T-stub problem taking into account the interfacial unilateral contact effects. In *Proceedings of the Conference in Advances in Computational Methods for Simulation*. Civil Comp Press, Edinburgh: 115-124.

Soh, A., and Soh, C. (1994). *On the Compatibility Conditions for Modeling Tubular Joints Using Different Types of Elements*. Nanyang University of Singapore, Singapore.

Stockwell, Jr., F. W. (1975). Preliminary base plate selection. *AISC Engineering Journal* 12: 92-99.

Talaslidis, D., and Panagiotopoulos, P. D. (1982). A linear finite element approach to the solution of variational inequalities arising in contact problems of Structural Dynamics. *Inernational Journal of Numerical Methods in Engineering* 18:1505-1520.

Thambiratnam, D. P., and Paramasivam, P. (1986). Base plates under axial loads and moments. *ASCE Journal of Structural Division* 112:1166-1181.

Thambiratnam, D. P., and Krishnamurthy, N. (1989). Computer analysis of column base plates. *Computers and Structures* 33:839-850.

Thomopoulos, K. (1985). Improvement of the design method for steel column base plates via an inequality approach. *Civil Engineering for Practicing and Design Engineers* 4:923-933.

Thomopoulos, K. (1986). A new method for rectangular hollow section splices via contact analysis. *Civil Engineering for Practicing and Design Engineers* 5:443-452.

Vogt, F. (1947). *The Load Distribution in Bolted or Riveted Joints in Light-Alloy Structures*. Technical Report TM 1135. Washington DC: NACA.

Wald, F. (1993). *Column Base Connections. A Comprehensive State of the Art*. Prague: CVUT.

Wald, F. (1995). *Patky Sloupu, Column Bases*. Prague : CVUT.

Wardenier, J. (1984). *Hollow Section Joints.* Delft: Delft University Press.

MIX
Papier aus verantwortungsvollen Quellen
Paper from responsible sources
FSC® C105338

If you have any concerns about our products,
you can contact us on
**ProductSafety@springernature.com**

In case Publisher is established outside the EU,
the EU authorized representative is:
**Springer Nature Customer Service Center GmbH**
**Europaplatz 3, 69115 Heidelberg, Germany**

Printed by Libri Plureos GmbH
in Hamburg, Germany